Liquid-Crystalline Polymer Systems

ACS SYMPOSIUM SERIES **632**

Liquid-Crystalline Polymer Systems

Technological Advances

Avraam I. Isayev, EDITOR
University of Akron

Thein Kyu, EDITOR
University of Akron

Stephen Z. D. Cheng, EDITOR
University of Akron

Developed from a symposium sponsored
by the Division of Polymeric Materials:
Science and Engineering, Inc.,
at the 209th National Meeting
of the American Chemical Society,
Anaheim, California,
April 2–7, 1995

American Chemical Society, Washington, DC

Library of Congress Cataloging-in-Publication Data

Liquid-crystalline polymer systems: technological advances / Avraam I. Isayev, editor, Thein Kyu, editor, Stephen Z.D. Cheng, editor.

p. cm.—(ACS symposium series, ISSN 0097–6156; 632)

"Developed from a symposium sponsored by the Division of Polymeric Materials: Science and Engineering, Inc., at the 209th National Meeting of the American Chemical Society, Anaheim, California, April 2–6, 1995."

Includes bibliographical references and indexes.

ISBN 0–8412–3408–6

1. Polymer liquid crystals—Congresses.

I. Isayev, Avraam I., 1942– . II. Kyu, Thien, 1948– . III. Cheng, Stephen Z. D., 1949– . IV. American Chemical Society. Division of Polymeric Materials: Science and Engineering. V. American Chemical Society. Meeting (209th: 1995: Anaheim, Calif.) VI. Series.

QD923.L5327 1996
668.9—dc20 96–20169
CIP

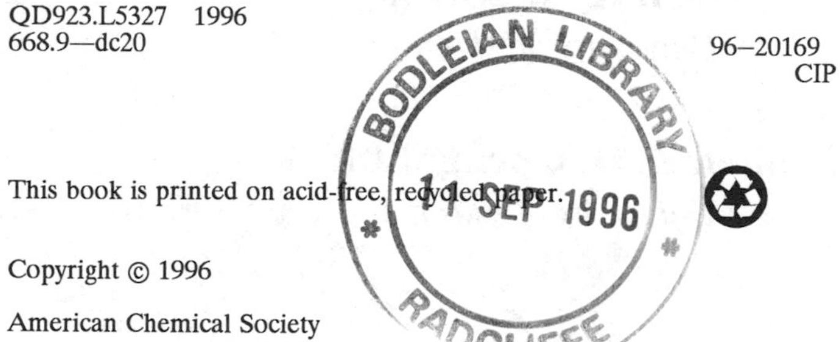

This book is printed on acid-free, recycled paper.

PRINTED IN THE UNITED STATES OF AMERICA

Foreword

THE ACS SYMPOSIUM SERIES was first published in 1974 to provide a mechanism for publishing symposia quickly in book form. The purpose of this series is to publish comprehensive books developed from symposia, which are usually "snapshots in time" of the current research being done on a topic, plus some review material on the topic. For this reason, it is necessary that the papers be published as quickly as possible.

Before a symposium-based book is put under contract, the proposed table of contents is reviewed for appropriateness to the topic and for comprehensiveness of the collection. Some papers are excluded at this point, and others are added to round out the scope of the volume. In addition, a draft of each paper is peer-reviewed prior to final acceptance or rejection. This anonymous review process is supervised by the organizer(s) of the symposium, who become the editor(s) of the book. The authors then revise their papers according to the recommendations of both the reviewers and the editors, prepare camera-ready copy, and submit the final papers to the editors, who check that all necessary revisions have been made.

As a rule, only original research papers and original review papers are included in the volumes. Verbatim reproductions of previously published papers are not accepted.

ACS BOOKS DEPARTMENT

Contents

SYNTHESIS AND PROPERTIES

INDEXES

Preface

UNDERSTANDING THE PROCESSING, structure, and properties of liquid-crystalline polymers (LCPs) has advanced significantly during the past five years. New and exciting investigations have been made worldwide by various scientific groups in several key areas of polymer science and engineering of low-molecular-mass liquid crystals, LCPs, and their blends with various thermoplastic polymers.

This book provides up-to-date-information concerning the synthesis, structure, rheology, processing, performance, and applications of LCPs. This information will be useful to those industries using LCPs for electronics; telecommunications and fiber optics; appliances; consumer goods; and automotive, marine, and aerospace applications.

The first section of the book describes recent advances in the processing and properties of molecular composites made of rigid–flexible block copolymers and of flexible polymers and rigid-rod LCPs through solution blending.

The second section is devoted to self-reinforced composites prepared through the melt processing of thermoplastics and thermotropic LCPs. The individual chapters address various aspects of rheology, processing compatibilization, and performance characteristics of in situ composites. Attention is directed to preparation of fibers and molded products, with the special objective of optimal performance properties through processing.

The third section of the book addresses recent research efforts in making polymer-dispersed liquid crystals consisting of nematic or cholesteric low- molecular-mass liquid crystals in flexible-chain polymers or LCPs. These systems are prepared by polymerization of reactive monomers in the presence of liquid crystals stabilized by flexible-chain polymer or LCP. Special attention has been paid to the use of these materials in display and electro-optical devices.

The next section is concerned with recent progress in characterization of the structure and phase behavior of mesophases, including their melting behavior, conformational order, crystallographic modifications, and solid-state proton NMR relaxation behavior and the formation of supramolecular entities arising from interaction in amphiphilic complexes.

The final section describes the synthesis and properties of new main-chain and side-chain liquid-crystalline block copolymers, mesogen-jacketed LCPs, and liquid-crystalline thermosets, which offer a unique class of materials for the fabrication of advanced composites and optical applications.

This volume is a comprehensive description of the recent advances in the science and technology of LCPs. At present, consumption of these polymers is rapidly expanding and is expected to increase in the coming years. This book will help newcomers obtain up-to-date developments in this field. It will also be useful for specialists engaged in various areas of research on liquid crystals.

Acknowledgments

We express our appreciation to the International Science Foundation, the Petroleum Research Fund, the ACS Division of Polymeric Materials: Science and Engineering, Inc., the Hoechst Celanese Research Company, and the Edison Polymer Innovation Corporation for providing financial support for travel for a number of foreign scientists.

AVRAAM I. ISAYEV
Institute of Polymer Engineering
University of Akron
Akron, OH 44325–0301

THEIN KYU
Institute of Polymer Engineering
University of Akron
Akron, OH 44325–0301

STEPHEN Z. D. CHENG
Maurice Morton Institute of Polymer Science
University of Akron
Akron, OH 44325–3909

February 20, 1996

Chapter 1

Self-Reinforced Composites Involving Liquid-Crystalline Polymers

Overview of Development and Applications

Avraam I. Isayev

Institute of Polymer Engineering, University of Akron, Akron, OH 44325–0301

A brief overview of the novel technology of self-reinforced composites based on liquid crystalline polymer (LCP)/thermoplastic and LCP/LCP blends is presented. Various aspects of the composite technology are considered including mixing, transition temperatures, factors governing LCP fibrillations, enhancement of crystallization, use of LCP as a processing aid, compatibilization, and the possibility of utilizing of ternary blends. Novel approaches to reduce anisotropy of products made of self-reinforced composites are discussed. Various possibilities and directions for future research and industrial applications of self-reinforced composites are also described.

The growth of the industrial activities around the world has prompted the development of newer and better materials. Newer technologies have brought in new consumer and industrial commodities. Naturally, newer materials are expected to be more durable, longer lasting and easily proccessable and recyclable. Many polymeric materials, evolved into excellent alternative to wood, glass and metals, are obtained by combinations of different types of polymers. Polymer alloys and blends have proven to be useful in ameliorating some of the problems of utility by taking advantage of the good properties of the polymers combined. Thus, polymer blends and composites have evolved into a major scientific and industrial field. Among the various classes of polymer composites, composites involving LCP have been developed. These materials exhibit high mechanical strength, excellent thermal and chemical resistance, along with ease of processing and are making rapid strides, particularly in the last few years. Here, LCP, which is blended with a thermoplastic, has the ability to form microfibrils or fibers in the matrix of thermoplastic. This occurs during the processing stage itself. The process provides composites that resemble glass reinforced polymers, though the effort and time involved in their

0097–6156/96/0632–0001$15.00/0

fabrication is much less than in the latter case. Since they are the self-reinforcing, research and development of LCP based composites is gaining intensive interest in the industrial and academic communities.

In recent years a number of books (1-5) and reviews (6-13) have appeared in scientific literature related to thermotropic LCP's. This literature provides information on various aspects of LCP's such as their synthesis, classification and properties. Also, a recent review of the effect of processing conditions on properties of LCP's is given in Ref. 14. Moreover, a brief overview of the science and technology of high temperature polymer blends including blends involving LCP's is presented in Ref. 15, which also gives some important technological trends in this emerging field. Thermotropic LCP's were first synthesized by the du Pont Chemical Company in 1972 (16,17) and were later commercialized as Ekkcel I-2000. This was a high melting point aromatic copolyester consisting of p-hydroxybenzoic acid (HBA), terephthalic acid (TPA) and 4,4' dioxydiphenol (DODP). This copolyester was injection moldable at temperatures above 400°C. Evidently, this copolyester later became the originator of a new family of LCP's presently known under the trade name Xydar. It was first commercialized by Dartco Manufacturing Company in 1985 and is now manufactured by Amoco Chemical Company as a family of LCP's with various melting points above 300°C. In 1973 another copolyester was synthesized based on HBA and polyethylene terephthalate (PET) by Tennesee Eastman Chemical Company (18,19). This polymer was well characterized and marketed under the code name X7G. The modified versions of this LCP are presently commercially available from the Unitika Chemical Company under trade name Rodrun. Depending on HBA and PET ratios in macromolecular chains, these materials exhibit a melting point between 230°C to 300°C. Development of the thermotropic LCP's by du Pont and Eastman Co. has led the way for further progress in the synthesis of various LCP's which are important from the point of view of industrial applications. The major milestone in this direction was a synthesis of a family of thermotropic LCP's based on HBA and hydroxy naphthoic acid (HNA) by Celanese Research Company (20-22). These LCP's were first commercialized in 1985 under a trade name Vectra. An introduction of Vectra and Xydar allowed further expansion of the LCP technology. Information on various LCP's started to appear in a number of trade journals and conferences (23-27). Since 1985 a number of other companies introduced LCP's including Rodrun by Unitika in 1986, KV by Baer AG, Ultrax by BASF, Victrex by ICI in 1987, Granlar by Granmont Inc. in 1988, Zenite by du Pont in 1995, Novacor by Mitsubishi, Econol by Sumitomo and Rhodester by Rhone-Poulenc. At first, the price of LCP was prohibitively high at about $20-30 per pound. Nowadays, their price has substantially dropped to as low as $8 per pound. Information on the exact molecular characteristics of commercialized LCP's is considered as proprietory by the manufacturers. However, it can be approximately deduced from the patent literature or obtained by the direct investigation of their molecular architecture. Although the price of these materials has dropped in recent years, a number of companies have already announced that they are out of the LCP business. These include BASF, ICI and Bayer. Their dropout from the market is explained by a limited demand. However, available statistics indicates a significant growth of LCP consumption . In particular, the LCP consumption in the USA, Europe, and Japan in the last few years has increased at a double digit rate (28). One possible area for further expansion of LCP consumption includes use of LCP in blends with engineering and conventional thermoplastics. Therefore, this subject is the main aim of the present review.

Mixing Techniques

Mixing is a major step in making new products from polymer blends. The research on polymer blends and alloys initiated in early 1960's and had led to the discovery of many new materials with unique properties (29-31). Since thermotropic LCP's are latecomers to the market place, the research on their blending with thermoplastics is a quite recent issue. Since the early 1980's, patents on thermotropic LCP/thermoplastic blends have started to appear (32,33). At the present time, mixing of thermotropic LCP with thermoplastics can be considered as the art of mixing, since there is no scientific foundation for the choice of the components, mixing machines and conditions of mixing. As in the case of conventional blends, most LCP/thermoplastic pairs are immiscible. Thus, the problem of interface enhancement becomes extremely important in these blends. A major expansion of research in blending of LCP with thermoplastics started after the first discovery (34) and publications (35-38) indicating that one can obtain LCP/thermoplastic blends with good or synergistic mechanical properties even in the absence of the miscibility of components. This synergism is achieved due to a so-called self-reinforcing effect of LCP. In particular, during the process of mixing and shaping, LCP creates in-situ high strength and modulus fibers within thermoplastic matrix. Thus, the concept of the self-reinforced or in-situ composites has appeared. Evidently, in self-reinforced composites, fibers can easily transfer mechanical load, even in the absence of good adhesion at the interface between the LCP fibers and the matrix. It is also obvious that properties of these self-reinforced composites can be further enhanced by an improvement of the interfacial interaction. However, it should be kept in mind that too much improvement of the interfacial interaction between components may reduce the efficiacy of the fibrillation process.

Let us now discuss the use of various mixing techniques. The available literature does not give a comparison of the effectiveness of various mixing machines concerning the capability of creating in-situ LCP fibers in blends. In this respect, the available information is very sketchy. Usually, each investigator utilizes a mixing machine which is available in their own laboratory. Accordingly, a comparison of the efficiency of various mixing machines to promote fibrillation is impossible to make. The mixing machines utilized in blending are internal mixers, conventional single or twin screw extruders, single screw extruder with static mixer attachment. Moreover, mixing by multiple passes of physically mixed LCP and thermoplastic pellets through a capillary rheometer is utilized. LCP/thermoplastic pellets are also dry blended and shaped into articles by an injection molding machine. However, these mixing techniques do not provide any evidence concerning an influence of processing conditions during mixing on the efficiacy of fibrilation. The effect of mixing, the static mixer vs. the internal mixer in the case of polycarbonate (PC)/LCP blends, indicated the better tensile properties can be obtained in the static mixer blends (35). Furthermore, two studies were reported on mixing polyetherimide(PEI)/LCP blends (39, 40) indicating a change in the quality of mixing. However, no change in the mechanical properties was found. Recently, some comparison was made between various mixing methods (41). In this case, the PC/LCP blends were prepared using an internal mixer, a single screw extruder with a static mixer attachment and a corotating twin screw extruder. Based on rheological, morphological and mechanical properties of these blends, it was concluded that the best combination of mechanical properties can be obtained in the case of mixing by the twin screw extruder. The internal mixer blends show inferior properties. This finding was very well correlated with the amount and distribution of the LCP fibers in the blends. The blends from the twin screw extruder showed the most uniform distribution of the fibers. Some direct comparisons between the efficiacy of mixing using a corotating twin screw extruder and single screw extruder with a static mixer attachment were also made earlier (42) for the case of PEI/LCP blends. This earlier study also indicated better mechanical properties in the blends prepared by the twin screw extruder. Morphology of polypropylene (PP)/LCP blends was also found to be affected by the use of different mixing equipment (43). In particular, the blends prepared by the Buss-Co-Kneader extruder exhibited the finest dispersion of LCP phase. This is in contrast to the results reported in (37) which indicated no effect on morphology and properties, using different mixing techniques. Studies on the effect of two different combinations of screw elements in intermeshing corotating twin screw extruder on mixing of PC with LCP were also reported in (44). However, in this case insignificant differences in morphology of resulting pellets were also obtained. In addition, differences in properties of injection molded samples of these blends and of blends premixed in the pellet form were also insignificant.

The majority of thermotropic LCP's exhibits the melting point which is high in comparison with the processing temperature of thermoplastics. Thus, blending of LCP's with thermoplastics by means of one extruder is difficult to carry out due to the possibility of degradation of the low melting point thermoplastics at elevated temperatures. In order to overcome this problem a novel mixing technique is developed which is called the dual extruder method (45,46). In this method, the thermoplastic and LCP are extruded separately by using two conventional single screw extruders connected to a static mixer with a die attachment and a take up device. The barrel temperatures of each extruder are kept different to accommodate the differences in the melting point of the LCP and thermoplastic. The temperature of the mixer is kept in between these two temperatures. The effect of varying the number of mixing elements in the Kenics static mixer was studied. In particular, based on morphological observations it was concluded that the presence of three and nine mixing elements was insufficient for obtaining good quality of mixing. Thus, the mixer with eighteen elements was utilized. In addition to dual extrusion method, the blends were prepared by using dry blended pellets and a single screw extruder at elevated temperature with similar die attachment and take up device. Three different LCP's were mixed with PET and polypropylene. It was concluded that the dual-extruder mixing method has significant advantage over the blending in a single screw extruder especially with respect to formation of the in-situ fibers and accordingly with respect to mechanical properties. In particular, single screw extruder blends showed a skin-core fibril-droplet structure, while the dual extruder blends indicated continuous LCP fibers.

As already mentioned the interfacial interaction between components is an important factor if one wishes to obtain a product with good mechanical properties. This can be realized by the proper choice of the components for blending with LCP. Since the majority of LCP's are wholly aromatic polyesters containing phenyl or ester groups, it is important to have similar contrasting functionality such as electron donor versus acceptor groups in the thermoplastic component. Also, a possibility of utilizing formation of hydrogen interaction between LCP and thermoplastic should be considered as a step towards an improvement of the interfacial interaction. In addition, use of block copolymers and polymers with functional groups well known in blending technology of flexible chain polymers (29-31) is also a possible direction for obtaining LCP/thermoplastic blends with improved properties. In this regard, some information already exists concerning the importance of the solid state reaction (47), transesterification (48), solid state polymerization (49), melt polymerization (50) and reaction in the melt (51), LCP and charged flexible polymers (52), hydrogen bonding (53,54) among others.

Transition Temperatures

When a new polymer blend is prepared, it is customary to carry out a set of characterization tests to determine the transition temperatures including melting point, T_m, glass transition temperature, T_g, and secondary transition temperatures. Useful techniques are differential scanning calorimetry (DSC), dynamic mechanical thermal analyser (DMTA), and dielectric thermal analyser (DTA). These techniques usually supplement each other. For example, in the LCP, it is often difficult to resolve their T_g and secondary transitions by the DSC method. However, using DSC it is easy to measure the melting points, T_m, and heat of fusion. Thus, T_g and secondary transitions of these polymers can be measured by the DMTA or DTA. In addition, the DMTA can measure the storage and loss moduli and energy absorption characteristic. In the case of LCP/thermoplastic blends, the data obtained by these two techniques give indirect information concerning the miscibility of components and direct information concerning the transition temperatures. The majority of studies on LCP/thermoplastic blends indicate the immiscibility between the components. This includes blends of various LCP with PC (35,37,55-67), PEI (37,42,68-70), PET (65,71-81), polyetheretherketone, PEEK (37,82-85), polyphenylene-oxide, PPO (86-92), polyphenylene sulfide, PPS (93-100), polystyrene, PS/PPO alloy (88,101), PP (43,97, 102-114), Nylon (36,37,53,115-120), PS (38,61,85,88,91,92,121-124), polyethylene, PE (125,126), polyether sulfone, PES (37,67,127-136), polyamideimide, PAI (137-139), thermoplastic polyimide (140,141), polybutylene terephthalate, PBT (37,117,142-146), polyarylate, PAR (37,100,147-151), polysulfone, PSU (152-154), polyvinylchloride,

PVC (155,156), chlorinated PVC, CPVC (157), polymethylmethecrylate, PMMA (158), SEBS (159,160) and Santoprene (161,162) thermoplastics elastomer, fluoropolymer (163,164).

Factors Governing Fibrillation

Rheological properties, shear rate, temperature, concentration and chemical nature and compatibility of components play important roles in LCP fibrillation during processing of LCP/thermoplastic blends (7,9-11,35-38,46,64,66,165-168). The most important parameter governing the fibrillation is the viscosity ratio of thermoplastic to LCP and strain rate. In the case of the flexible chain polymer blends, deformation and breakup of the dispersed phase is governed by the viscosity ratio and the Weber or capillary number (169-175). The Weber number is a ratio of the viscous forces in the fluid to the interfacial forces. In the case of LCP/thermoplastic blends, the majority of the available information indicates that the effect of fibrillation is enhanced as the viscosity ratio of LCP to thermoplastic becomes less than unity (35,36,42,82,88,97-99,101,117,176,177), although there are some instances where fibrillation is not evident at the viscosity ratio lower than unity (178). The latter was explained as being due to miscibility or reaction between the components. At the viscosity ratio greater than unity fibrillation is not observed (84,93,99). Available information is very sketchy concerning the effect of the Weber number on the fibrillation, since the interfacial tension values for LCP/thermoplastic melts are not readily available. It is suggested in Ref. 66, that the high interfacial tension stabilizes the LCP fibrils. In addition to the above mentioned factors, the fibrillation is also affected by the type of the deformation, shear or elongation. Although the deformation mechanism of LCP phase in thermoplastic surrounding is not well understood, the prevailing opinion is that the elongational flow causes more extensive fibrillation than the shear flow (10-13, 35, 36, 60, 98, 121, 125, 163, 164, 179, 180). With increasing extension ratio, the LCP phase is transformed from spherical to globular and then to a fibrillar form. Accordingly, fibers form in elongational flow at any viscosity ratio. Experiments also show that more LCP fibrils are created in dies having a low L/D ratio, in converging dies and in processes involving drawing. Also, in the skin layer of injection molded products a stronger fibrillation is observed due to the fountain flow effect governed by elongational flow. Furthermore, the chemical nature of the LCP molecular chain may also affect fibrillation. In particular, it is expected that the higher rigidity of the molecular chain can lead to a more effective fibrillation. However, this hypothesis has not been verified. Similarly, the effect of the normal stresses developed in LCP and thermoplastic melts on fibrillation has not been studied. Concentration of LCP also plays a significant role in the morphology developed blends. There are many cases where LCP fibrils are found in blends at low concentrations (34-38,53,64,87,88,103, 112, 181-183) and spherical domains or particles at high concentrations (34,35) and vice versa (36). Apparently, the effect of the concentration on fibrillation in blends depends on the nature of components. However, in a majority of the cases reported in the literature a strong fibrillation was observed at high concentrations of LCP.

The concentration corresponding to phase inversion is different for various pairs of polymers. Finally, it is noted that the optimal conditions for LCP fibrillation for various LCP/thermoplastic pairs are not established. However, it was claimed for LCP/PP blend that an optimal fiber formation was achieved in the range of the viscosity ratios from about 0.5 to 1.0 (43). In some cases, fibrillation becomes so intensive that extrusion and subsequent stretching of LCP/thermoplastic sheets leads to a formation of a continous open work web comprising polymer fibers with interstices between the fibers (184).

Enhancement of Crystallization

An addition of a small amount of LCP into some semi-crystalline thermoplastics leads to the enhancement of crystallization processes of thermoplastic phase (74,75,82,84,102,124,140,185-187). Apparently, crystals nucleates on the surface of the LCP domains, i.e., LCP acts as a nucleating agent. This effect is verified by an increase in the amount of exotherm during cold crystallization and a higher heat of fusion during melting. Similarly, the enhancement of crystallization is indicated during melt crystallization by a larger value of crystallization exotherm in cooling of the thermoplastic melt in the presence of LCP.

LCP as Processing Aid

Within the last decades during which LCP research gained momentum, it became clear that an addition of LCP to thermoplastics has a strong effect on viscosity. In particular, it was discovered that an addition of small amount of LCP reduces the viscosity of the thermoplastic matrix. Therefore, it was proposed to use LCP as a processing aid (32). However, later it was shown that depending on concentration the viscosity of LCP/thermoplastic blends can be between (7-13, 35-42,49-53,61-63,82-87,101,117,120,124,133,135,138,139,180,188,189), below (abnormal behavior) (35,67,83,92,101,102,116,119,135) or above (35,93,135) those of components. When viscosity of LCP is above that of thermoplastic the blend viscosity becomes higher than the viscosity of thermoplastics (82,136,189). These effects of the reduction or increase of the viscosity of thermoplastic on addition of LCP depends on the nature of the components, concentration and shear rate value. Concerning the viscosity reduction of thermoplastic on an addition of LCP, there are several opinions explaining this effect (101): (i) low viscosity LCP melt during flow migrates toward the wall; (ii) the oriented fibers are presumed to slide past one another; (iii) the interfacial slippage occurs between the two melts. In fact, measurements of LCP content in the gapwise direction in LCP/PP injection moldings indicate that the skin layer is LCP rich (94). A migration of the LCP particles towards the capillary wall and towards axis was observed by some authors (38,119). In all these cases, LCP acts as a lubricant leading to a reduction of flow resistance. Furthermore, an attempt has been made to approximate the abnormal viscosity behavior by some theoretical consideration (67). However, whatever underlying reason for the reduction of the viscosity, the effect is beneficial in processing since it reduces the energy consumption. This effect can be easily applied to industrial practice in processing thermoplastics which are otherwise difficult to shape.

LCP-LCP Blends

In recent years, some attention was also paid to blending of two LCP's. The major objective of these studies was to obtain new LCP materials with improved mechanical properties. Number of investigations was carried out to look at the phase behavior, rheological properties, morphology and mechanical properties. Various pairs of LCP/LCP polymers were prepared. This includes Vectra A900 with PET/60HBA (190), LCP KU-9211 (also called K-161) with PET/60PHB (191), Vectra A950 with copolymer HNA/TA/HQ (192) and with HBA/HNA/TA/HQ (193) and with LCP of different mole fraction ratios of HBA/HNA (194), blends of PET/60HBA with PET/80HBA (195), Vectra A950 with Vectra (186), and with Ultrax KR4002 (197-199) and with Ultrax KR 4003 (200), blends of main chain LCP with side chain LCP (201). In particular, it was found that Vectra A950 with PET/60HBA phase separates in the entire range of composition (190). On the other hand, blends of main chain and side chain LCP showed a broadening of melting endotherm (201). Blends of Vectra A950 with HBA/HNA/TA/HQ (193) and with LCP of various mole fraction ratios of HBA/HNA (194) indicated a compatibility in the solid state. At the same time, blends of Vectra A950 with HNA/TA/HQ showed a miscibility at low concentration and immiscibility at high concentration of Vectra A950 (192). Blends of PET/60PHB with PET/80PHB indicated miscibility at all temperatures and concentrations. The existence of the molecular exchange reaction in blends of LCP K-161 and PET/60PHB was observed (191). Blends of Vectra A950 with Ultrax KR-4002 and with Ultrax KR-4003 were found to be immiscible (197-200). The miscibility of two nematic LCP's in the mesomophic state along with the phase separation in the solid state was found (202). Concerning rheology of LCP/LCP blends, some authors (190,193,194,196-200) reported results indicating a possibility of a modification of the viscosity of LCP by an addition of another LCP. This is important for processibility of LCP blends.

Morphological studies revealed, in many cases, the existence of two phases of heterogeneous structure in blends (191,192,197-201). Injection molded samples in some cases indicated some unusual morphology with a highly oriented skin layer encasing the core region (199). An enhancement of fibrillation and orientation of one LCP in the presence of other LCP is also postulated (196-200). In these cases, a significant synergism of mechanical properties of moldings were found. LCP/LCP blends were also spun into fibers. It was found that the major factor affecting fiber modulus and the order parameter was the draw ratio (203). Further studies on LCP/LCP blends will undoubtedly lead to creation of new materials with performance characteristics superior to those of high performance engineering thermoplastics.

Compatibilization and Ternary Blends

A number of processing technologies can be developed based on the a concept of self-reinforcing or in-situ composites involving LCP. This includes fiber spinning, extrusion, film casting, thermoforming, injection, compression and blow molding. In particular, injection molding and extrusion are the two predominant processing operations. However, the main problem here is to find suitable molding conditions to realize the full potential for self-reinforcement. This can be attained through findings the optimal LCP fibrillation conditions, and through improvement of the compatibility of LCP-thermoplastic pairs or LCP-LCP pairs. In the case of LCP/thermoplastic blends, the major direction of research is toward finding a compatibilizer or a third component to improve the interaction at the interface between the components (91,92,128,204-210). Some positive results have already been obtained. In particular, an addition of ethylene-based terpolymer is found to improve the impact strength of PP/LCP blends (211). The second LCP is also added to improve the adhesion between components and the dispersion of LCP in LCP/thermoplastic blends (149,212-214). An addition of PES oligomer with the reactive functions and groups to PES/LCP blend resulted in improved fibrillation (128). Some studies to improve mechanical properties of LCP/thermoplastic blends are also performed. These include adding thermotropic block graft copolymer (91,92), functionalized polymer (209), block copolymer (215,216), maleanated PP (104,221). A novel technology is also proposed in which in-situ polymerization of monomer in the presence of LCP is carried out to obtain a more homogeneous blend (216). The possibility of modifying the molecular structure of LCP to improve the interfacial adhesion with PC is also explored (217). An addition of LCP to improve blends of engineering thermoplastics such as PEI/PEEK was also carried out (182,208). In fact, all the above studies on blends involving LCP follow the footsteps similar to those taken for several decades by various researchers working on blends and alloys of flexible chain polymers. Future systematic research related to compatibilization and ternary blends involving LCP is expected to produce new products with the high payback potential.

Novel Technologies

The main disadvantage of using LCP as a reinforcing agent in a thermoplastic or another LCP is the anisotropy of the mechanical properties of the product made from the blends. Similar to pure LCP's, the molded products made from the self-reinforced or in-situ composites exhibit high mechanical properties in the flow direction, while the properties in the direction transverse to the flow are poor. A number of novel approaches are proposed to control or reduce anisotropy of mechanical properties. All the approaches which are suitable to both LCP/thermoplastic and LCP/LCP blends, utilize the difference in melting points of components (87,105,106,121,163,164, 200,219-221). These technologies include manufacturing fibers, strands, films, or prepregs (sheets) of LCP/thermoplastic or LCP/LCP blends at a temperature range at which both components are melt processable. In one approach, the prepared fibers or strands are placed into a mold at the desired orientation of fibers or at the random orientation and then they are compression molded into a product at temperatures below the melting point of LCP but above the melting point of thermoplastic in the case of LCP/thermoplastic fibers or below the melting point of one LCP but above the melting point of the other LCP in the case of LCP/LCP fibers. The processing at these temperatures allows to fuse fibers together and does not lead to a disintegration or melting of LCP fibrous structure achieved in the first processing step. Therefore, the second processing step allows one to achieve the controlled anisotropy of mechanical properties of the product. Properties of the obtained products are found to depend on the extension ratio in fiber spinning process, concentration and chemical nature of components. In the second approach, which is applicabe to drawn films or prepregs, multilayers of films are stacked in preferred orientation of each layer required for the product (106,163,164,200,219-221). Then, lamination or consolidation is carried out at temperatures above the melting point of thermoplastic in the case of LCP/thermoplastic films or matrix LCP in the case of LCP/LCP films but below the melting point of the other polymer. The properties of laminates are found to depend on the compression or reduction ratio during consolidation step and orientation of layers in addition to other parameters mentioned above in the case of compression molding of fibers. Products with good isotropic properties are obtained. It is also noted that these laminates can be utilized for further shaping such as thermoforming. These novel technologies for making laminates were also described in the patent literature (222-225).

In addition to these two approaches, the LCP/thermoplastic or LCP/LCP fibers or prepregs manufactured in the first step are injection molded or extruded at temperatures corresponding to the processing temperature range of matrix component in the fibers or prepregs, but below melting point of reinforcing LCP components (105,221). Products made using systems of pregenerated fibers show much better mechanical properties than those made by the direct processing of blends. However, these products made using the technology of pregenerated fibers would probably exhibit some deficiency concerning anisotropy of mechanical properties similar to that of in-situ composites. Further studies are required to utilize full potentials of this technology.

The third novel technology which allows one to reduce anisotropy, especially in the case of manufacturing LCP/thermoplastic films and tubes, is to utilize extrusion using a counter-rotating annular die having a rotating mandrel. This process induces biaxial orientation into extruded films or tubes by controlling the rate of extrusion, takeup velocity and the speed of mandrel rotation (140,226-230). The other approach which is suitable to reduce anisotropy of injection molded articles made from LCP/thermoplastic blends is to utilize push-pull injection molding technology (231,232). This technology has already shown a promise concerning an improvement of the weld-line strength in injection molding of pure LCP's.

Applications

Self-reinforced or in-situ composites involving LCP exhibit high strength, modulus, improved temperature and chemical resistance, low coefficient of thermal expansion and improved barrier properties. These composites can be utilized for making products which are currently manufactured using pure LCP. In particular, taylor-made LCP/thermoplastic or LCP/LCP alloys for specific applications can be developed using technology of self-reinforced composites. This technology can be utilized in two possible ways. One way is to carry out a modification of flexible chain thermoplastics, including both conventional and high performance thermoplastics, by addition of LCP (34). The other way is to perform modification of LCP by addition of thermoplastic or other LCP (197,233-236). This last modification is especially important in the case of high priced engineering thermoplastics, such as PEEK or polyimides. Numerous possibilities for applications of blends involving LCP are also described in recent patent literature (237-283). In particular, several areas of applications are already identified including a manufacture of high barrier containers and films, electronic and electrical components, tubing and light-weight structures, health care and industrial and consumer goods. In area of high barrier containers and films, use of self-reinforced composites is targeted for food and beverage packaging, fuel and chemical storage tanks, pumps, metering devices, valve liners, films for vacuum insulation panels. In electronic circuits and connectors, polymeric materials with a low coefficient of thermal expansion and high temperature stability are required. Self-reinforced composites satisfy these requirements quite well. They also fulfill specifications for the IR and vapor soldering. Products having tight tolerances and thin walls can be obtained. Miniaturization capability is another advantage of these products. This is in addition to excellent moldability and long flow length of self-reinforced composites. Other electrical and electronic applications include a possibility to manufacture color-coded connectors for easy identification, sockets, bobbins, switches, connectors and cables for fiber optics, chip carriers, printed circuit boards, surface mount technology parts, transmit/receive modules, circuits for smart credit cards, dielectric layers in high temperature capacitors. Self-reinforced composites are also suitable for manufacturing health care products, such as sterialization trays, dental tools, laparoscopic and other surgical instruments, dental bar caddies and pharmaceutical packaging. In industrial and consumer goods area, self-reinforced composites can be utilized as materials for printers, copies, fax machines, watch components and business machine housings. Tube and light-weight structural applications includes surgical instruments and medical tubing such as endoscopic instruments and catheters, small bore tubing in fiber optic devices and honeycomb cores satisfying aerospace requirements. Finally, these materials are suitable for making structural component in cars, satisfying requirements for multiple recyclability. Self-reinforced composites offer a variety of possible choices for today and future applications and the plastic industry is getting ready to embark on these new challenges.

References

1. Weiss, R. A., and C. K. Ober eds., "Liquid Crystalline Polymers," ACS Symposium Series 435, ACS, Washington D.C., 1990.
2. Donald, A. M. and A. H. Windle, "Liquid Crystalline Polymers," Cambridge University Press, Cambridge, 1992.
3. Plate, N. A., ed., "Liquid-Crystal Polymers," Plenum, New York, 1993.
4. La Mantia, L. P., ed., "Thermotropic Liquid Crystal Polymer Blends", Technomic, Lancaster, 1993.
5. Carfagna, C., ed., "Liquid Crystalline Polymers," Pergamon, Oxford, 1994.
6. Brostow, W., Kunststoffe, 78, 411 (1988).
7. Isayev, A. I., and T. Limtasiri, in "International Encyclopedia of Composites," S. M. Lee, ed., VCH, New York, v.3, p.55 (1990).
8. LaMantia, F.P. and A. Valenza, Macromol. Chem., Macromol. Symp., 56, 151 (1992).
9. Kulichikhin, V.G. and N. A. Plate, Polymer Sci., USSR, 33A, 3 (1991).
10. Dutta, D., H. Fruitwala, A. Kohli and R. A. Weiss, Polym. Eng. Sci., 30, 1005 (1990).
11. Handlos, A. A. and D. G. Baird, Marcromol. Reviews, Chem. Phys., C35, 183 (1995).
12. Brown, C. S. and P. T. Alder, in "Polymer Blends and Alloys," Chapman and Hall, London, 1993, Chapter 8, p. 195.
13. Pawlikowski, G.T., D. Dutta and R.A. Weiss, Annual Rev. Mater. Sci., 21, 159 (1991).
14. Acierno, D. and M. R. Nobile, in "Thermotropic Liquid Crystal Polymer Blends," F. P. LaMantia, ed., Technomic, Lancaster, 1993.
15. Jaffe, M., P. Choe, T.S. Chung and S. Makhija, Adv. Polym. Sci., 117, 297 (1994).
16. Gotsis, S. G., J. Economy, and B. E. Novak, US Patent 3,637, 795 (1972).
17. Gotis, S. G., J. Economy, and L. C. Wohrer, US Patent 3,975,487 (1976).
18. Jackson, W.J., Jr. and H.F. Kuhfuss, US Patent 3,778,410 (1973).
19. Jackson, W.J., Jr. and H.F. Kuhfuss, J. Poy. Sci., Polym. Chem. ed., 14. 2043 (1976).
20. Calundann, G.W., US Patent 4,067,852 (1978).
21. East, A.J. and G.W. Calundann, US Patent 4,318,841 (1982); 4,318,842 (1982).
22. Calundann, G.W. and M. Jaffe, in Proceedings of Robert A. Welch Conference on Chemical Research, Synthesis of Polymers, 36, 247 (1982).
23. Zanetti, R., S. Ushio and S. McQueen, Chem. Eng. 91, No. 15, 14 (1984).
24. Zanetti, R., S. Ushio and S. McQueen, Brit. Plast. Pub., No. 7/8, 25 (1985).
25. Zanetti, R., S. Ushio and S. McQueen, Plastic World, No. 12, 8 (1984).
26. Lenz, R. W. and J. L. Jin, Polymer News, 11, No. 7, 200 (1986).
27. Fein, M.M., Advances Technol. Mater. Process., 30th National SAMPE Symp., USA, p. 556 (1985).
28. Romer, M., Kunststoffe, 83, 785 (1993).

29. Paul, D. and S. Newman, eds., "Polymer Blends," Academic Press, New York, 1978.
30. Utracki, L., "Polymer Alloys and Blends: Thermodynamic and Rheology," Hanser, Munich, 1989.
31. Folkes, M.J. and P.S. Hope, eds., "Polymer Blends and Alloys," Chapman and Hall, London, 1993.
32. Cogswell, F.N., B.P. Griffin and J.B. Rose, US Patent 4,386,174 (1981); 4,438,236 (1984).
33. Froix, M., US Patent 4,460,735 (1984).
34. Isayev, A.I. and M. Modic, US Patent 4,728,698 (1988).
35. Isayev, A.I. and M. Modic, Polym. Compos., 8, 158 (1987); SPE ANTEC, 32, 573 (1986).
36. Blizard, K.G. and D.G. Baird, Polym. Eng. Sci., 27, 653 (1987); SPE ANTEC, 32, 311 (1986).
37. Kiss, G., Polym. Eng. Sci., 27, 410 (1987).
38. Weiss, R.A., W. Huh, and L. Nicolais, Polym. Eng. Sci., 27, 684 (1987); SPE ANTEC, 32, 307 (1986).
39. Bafna, S.S., J. P. Desouza, T. Sun and D. G. Baird, Polym. Eng. Sci., 33, 808 (1993).
40. Bafna, S.S., T. Sun and D. G. Baird, Polymer, 34, 708 (1993).
41. Viswanathan, R., MS Thesis, The University of Akron (1993).
42. Isayev, A. I. and S. Swaminathan, in Advanced Composites. III Expanding Technology, ASM, p.259 (1987).
43. Heino, M.T., P.T. Hietaoja, T.P. Vainio and J. V. Seppala, J. Appl. Polym. Sci., 51, 259 (1994).
44. Michaeli, W., T. Brinkmann, P. Heidemeyer, P. Hock, and S. Witte, Kunststoffe, 82, 1136 (1992).
45. Baird, D.G. and A.M. Sukhadia, US Patent 5,225,488 (1993).
46. Sukhadia, A.M., A. Datta, and D.G. Baird, Intern. Polym. Process., 7, 218 (1992).
47. George, E.R., R.S. Porter and A.C. Griffin, Mol. Cryst. Liquid Cryst. 110, 27 (1984).
48. Paci, M., S. Carozinno, M. Liu, and P.L. Magagnini, J. Calorim. Anal. Therm. Thermodyn. Chim 17, 533 (1986).
49. Ginnings, P.R., Europ. Patent 281,496 A2 (1988).
50. Shin, B.Y. and I.J. Chung, Polym. Eng. Sci., 30, 13 (1990).
51. Croteau, J.F. and G.V. Laivins, J. Appl. Polym. Sci., 39, 2377 (1990).
52. Nyrkova, A.I. and A.R. Khokhlov, Polym. Sci. USSR, B, 31, 375 (1989).
53. Siegmann, A., A. Dagan, and S. Kenig, Polymer, 26, 1325 (1985).
54. Paci, M., D. Lupinacci, and B. Bresci, Thermochimica Acta, 122, 181 (1987).
55. He, J., W. Bu, H. Zhang, P. Xie and X. Xu, Intern. Polym. Process., 8, 129 (1993).
56. Amendola, E., C. Carfagna, P.Netti, L. Nicolais and S. Saiello, J. Appl. Polym Sci., 50, 83 (1993).

57. Chapleau, N., P.J. Carreau, C. Peleteiro, P. A. Lavoie, and T.M. Malik, Polym. Eng. Sci., 32, 1876 (1992).
58. Lin Q., J. Jho, and A.F. Yee, Polym. Eng. Sci..33, 789 (1993).
59. Malik, T.M., P. J. Carreau and N. Chapleau, Polym. Eng. Sci., 29, 600 (1989).
60. Kenig, S., Polym. Adv. Technol., 2, 20 (1991).
61. Zhuang, P., T. Kyu, and J.L. White, Polym. Eng. Sci 28, 1095 (1988).
62. Nobile, M.R., A. Amendola, L. Nicolais, D. Acierno, and C. Carfagna, Polym. Eng. Sci, 29, 573 (1989).
63. Kohli, A., N. Chung and R.A. Weiss, Polym. Eng. Sci., 29, 573 (1989).
64. Blizard, K.G., C. Federici, O. Federico, and L.L. Chapoy, Polym. Eng. Sci., 30, 1442 (1990).
65. Joslin, S., W. Jackson, and R. Farris, J. Appl. Polym. Sci., 54, 289 (1994).
66. Beery, D., S. Kenig, A. Seigmann and M. Narkis, Polym. Eng. Sci., 33, 1548 (1993).
67. Shi, F. and X.S. Yi, Intern. J. Polym. Mater., 25, 243 (1994).
68. Bafna, S.S., T. Sun, J. P. deSouza and D.G. Baird, Polymer, 36, 259 (1995).
69. Ryu, C., Y. Seo, S. S. Hwang, S. M. Hong, T.S. Park, and K.U. Kim, Intern. Polym. Process., 9, 266 (1994).
70. Lee, S., S.M. Hong, Y. Seo, T.S. Park, S.S. Hwang, K.U. Kim, and J.W. Lee, Polymer, 36, 519 (1994).
71. Joslin, S., W. Jackson, and R. Farris, J. Appl. Polym. Sci., 54, 439 (1994).
72. Kyu, T. and P. Zhuang, Polymer Commun., 29, 99 (1988).
73. Mehta, S. and B.L. Deopura, Polym. Eng. Sci., 33, 931 (1993).
74. Paci, M., C. Barone, and P. Magagnimi, J. Polym. Sci., Polym. Phys. Ed., 25, 1595 (1987).
75. Joseph, E.G., G.L. Wilkes, and D.G. Baird, in "Polymer Liquid Crystals," A. Blumstein, Ed., Planner Press, New York, 1985.
76. Kyotani, M., A. Kaito and K. Nakayama, Polymer, 33, 4756 (1992).
77. Mithal, A.K. and A. Tayebi, Polym. Eng. Sci., 31,. 1533 (1991).
78. Schleeh, T., L. Salamon, G. Hinrichsen, and G. Koβmehl, Macromol. Chem., 194, 2771 (1993).
79. Melot, D. and W. J. MacKnight, Polym. Adv. Technol., 3, 383 (1992).
80. Brostow, W., T.S. Dziemianowcz, J. Romanski, and W. Weber, Polym. Eng. Sci., 28, 785 (1988).
81. Lee, W.C. and A.T. DiBenedetto, Polymer, 34, 684 (1993).
82. Mehta, A. and A. I. Isayev, Polym. Eng. Sci., 31, 963 (1991).
83. Mehta, A. and A. I. Isayev, Polym. Eng. Sci., 31, 971 (1991).
84. Isayev, A.I. and P.R. Subramanian, Polym. Eng. Sci., 32, 85 (1992).
85. James, S.G., A.M. Donald and W.A. Macdonald, Mol. Cryst. Liq. Cryst., 153, 491 (1987).
86. Limtasiri, T. and A. I. Isayev, J. Appl. Polym. Sci., 42, 2923 (1991).
87. Crevecoeur, G. and G. Groeninckx, Polym. Eng. Sci., 33, 937 (1993).
88. Crevecoeur, G. and G. Groeninckx, Polym. Eng. Sci., 30, 532 (1990).
89. Crevecoeur, G. and G. Groeninckx, Polym. Compos., 13, 244 (1993).
90. Crevecoeur, G. and G. Groeninckx, J. Appl. Polym. Sci. 49, 839 (1993).

91. Kobayashi, T., M. Sato, N. Takeno and K. Mukaido, Macromol. Chem. Phys., 195, 2771 (1994).
92. Kobayashi, T., M. Sato, N. Takeno and K. Mukaido, Europ. Polym. J., 29, 1625 (1993).
93. Subramanian, P.R. and A. I. Isayev, Polymer, 32, 1961 (1991).
94. Shonaike, G.O., H. Hamada, S. Yamaguchi, N. Nakamichi, and Z. Maekawa, J. Appl. Polym. Sci., 54, 881 (1994).
95. Shonaike, G.O., S. Yamaguchi, M. Ohta, H. Hamada, Z. Maekawa, M. Nakamichi, and W. Kosaka, Europ. Polym. J., 30, 413 (1994).
96. Baird, D.G. and T. Sun, in "Liquid Crystalline Polymers," R.A. Weiss, and C.K. Ober, eds., ACS Symp. Ser. 435, ACS, Washington, DC., 1990.
97. Seppala, J.V., T.M. Heino, and C. Kapanen, J. Appl. Polym. Sci., 44, 1051 (1992).
98. `Heino, M.T. and J.V. Seppala, J. Appl. Polym. Sci., 44. 2185 (1992).
99. Valenza, A., F.P. LaMantia, L. I. Minkova, S. DePedris, M. Paci, and P.L. Magagnini, J. Appl. Polym. Sci., 52, 1653 (1994).
100. Kim, B.C. and S. M. Hong, Mol. Cryst. Liq. Cryst., 254, 251 (1994).
101. Viswanathan, R. and A. I. Isayev, J. Appl. Polym. Sci., 55, 1117 (1995).
102. Tjong, S.C., S. Liu, R.K.Y. Li, J. Mater Sci., 30, 353 (1995).
103. Datta, A., H.H. Chen, and D.G. Baird, Polymer, 34, 759 (1993).
104. Datta, A. and D.G. Baird, Polymer, 36, 505 (1995).
105. Isayev, A.I., US Patent, 5,260,380 (1993).
106. Isayev, A.I., Y. Holdengreber, R. Viswanathan and S. Akhtar, Polym. Compos., 15, 254 (1994).
107. Yazaki, F., Y. Tsubouchi, and R. Yosomiya, Polym. Polym. Compos.,1, 183 (1993).
108. Quin, Y., D.L. Brydon, R.R. Mather and R.H. Wardman, Polymer, 34, 1996 (1993).
109. Qin, Y. D.L. Brydon, R.R. Mather, and R.R. Wardman, Polymer, 34, 3597 (1993).
110. Heino, M.T. and J.V. Seppala, J. Appl. Polym. Sci., 48, 1677 (1993).
111. Heino, M.T. and J.V. Seppala, J. Appl. Polym. Sci., 44, 1677 (1993).
112. Datta, A., H.H. Chen and D. G. Baird, Polymer, 34, 759 (1993).
113. LaMantia, F.P., Y. Yongcheng, A. Valenza, V. Citta, U. Pedretti, and A. Roggero, Eur. Polym. J., 27, 723 (1991).
114. Qin, Y., J. Appl. Polym. Sci., 54, 873 (1994).
115. LaMantia, F.P., M. Saiu, A. Valenza, M. Paci, and P.L. Magagnini, Eur. Polym., 26, 323 (1990).
116. Jang, S.H. and S.C. Kim, Polym. Eng. Sci., 34, 847 (1994).
117. Beery, D., S. Kenig, and A. Siegmann, Polym. Eng. Sci., 31, 459 (1991).
118. Shin, B.Y. and I. J. Chung, Polym. Eng. Sci., 30, 22 (1990).
119. Chung, T.S., Plast. Eng., 43, 39, Oct. 1987.
120. LaMantia, F.P., A. Valenza, and F. Scargiale, Polym. Eng. Sci., 34, 799 (1994).
121. Bassett, B.R. and A.F. Yee, Polym. Compos, 11, 10 (1990).

122. Pant, B.G., S.S. Kulkarni, D.G. Panse, and S.G. Joshi, Polym. Eng. Sci., 35, 2549 (1994).
123. Ogata, N., T. Tanaka, T. Ogihara, K. Yoshida, Y. Kondou, K. Hayashi, and N. Yoshida, J. Appl. Polym. Sci., 48, 383 (1991).
124. Incarnato, L., M.R. Nobile, M. Frigione, O. Motta, and D. Acierno, Intern. Polym. Procc., 8, 191 (1993).
125. Hsu, T.C., A.M. Lichkus, and I.R. Harrison, Polym. Eng. Sci., 33, 860 (1993).
126. Huang, X., MS Thesis, The University of Akron (1995).
127. Yang, Y., J. Yin, B. Li, G. Zhuang, and G. Li, J. Appl. Polym. Sci., 52, 1365 (1994).
128. Yazaki, F., A. Kohara, R. Yosomiya, Polym. Eng. Sci.,34 , 1129 (1994).
129. He, J. and W. Bu. Polymer, 35, 5061 (1994).
130. Shi, F., Intern. J. Polym. Mater., 23, 207 (1994).
131. Shi, F., Polym. Plast. Technol. Eng., 33, 445 (1994).
132. Engberg, K., O. Stromberg, J. Martinsson, and U.W. Gedde, Polym. Eng. Sci., 34, 1336 (1994).
133. Li, G., J. Yin, B. Li, G. Zhang, and Y. Yang, Polym. Eng. Sci., 35, 658 (1995).
134. Zheng, J.Q. and T. Kyu, Polym. Eng. Sci., 32, 1004 (1992).
135. Shumsky, V.F., Yu. S. Lipatov, V.G. Kulichikhin and I.P. Getmanchuk, Rheol. Acta, 32, 352 (1993).
136. LaMantia, F.P., A. Valenza, M. Paci and P.L. Magagnini, Rheol. Acta, 30, 7 (1990).
137. Lai, X.Y., D.F. Zhao, and F. Lai, SPE ANTEC, 39, 2676 (1993).
138. Lai, X.Y., D.F. Zhao, and F. Lai, SPE ANTEC, 38, 382 (1992).
139. Varma, T., MS Thesis, The University of Akron (1993).
140. Blizard, K. G. and R.R. Haghighat, Polym. Eng. Sci., 33, 799 (1993).
141. Aihara, Y., and P. Cebe, Polym. Eng. Sci., 34, 1275 (1994).
142. Engberg, K., M. Ekblad, P.E. Werner, and U.W. Cedde, Polym. Eng. Sci., 34, 1346 (1994).
143. Darragas, K., G. Groeninckx, H. Reynaers, and C. Samyn, Europ. Polym. J., 30, 1165 (1994).
144. Heino, M.T. and J.V. Seppala, Polym. Bull., 30, 353 (1993).
145. Ajji, A., J. Brisson, and Y. Qu, J. Appl. Polym. Sci., 30 505 (1992).
146. Wang, L.H. and R.S. Porter, J. Polym. Sci., 31, 1067 (1993).
147. Hong, S.M., B.C. Kim, S.S. Hwang, and K.U. Kim, Polym. Eng. Sci., 33, 630 (1993).
148. Zaldua, A., M. E. Munoz, J.J. Pena, and A. Santamaria, Polym. Eng. Sci., 32, 43 (1991).
149. Sun, L.M., T. Sakoda, S. Ueta, K. Koga, and M. Takayanagi, Polym. J., 26, 961 (1994).
150. Hong, S.M. and B.C. Kim, Polym. Eng. Sci., 34, 1605 (1994).
151. Zaldua, A., M.E. Munoz, J.J. Pena, A. Santamaria, Polymer, 33, 2007 (1992).
152. Kulichikhin, V.G., O.V. Vasil'eva, I.A. Litvinov, E.M. Antipov, I.L. Parsamyan, and N.A. Plate, J. Appl. Polym. Sci., 42, 363 (1991).
153. Golovoy, A., M. Kozlowski, and M. Narkis, Polym. Eng. Sci., 32, 854 (1992).

154. Skovby, M.H.B., J. Kops, and R.A. Weiss, Polym. Eng. Sci., 31, 954 (1991).
155. Lei, H. and Y. Zhao, Polymer, 35, 104 (1994).
156. Lee, B.L., Polym. Eng. Sci., 32, 1028 (1988).
157. Lee, B.L., Polym. Eng. Sci., 28, 1108 (1988).
158. Schild, H.G., E.S. Kolb, R.A. Gaudiana, Y. Chang, and W.C. Schwarzel, J. Appl. Polym. Sci., 46, 959 (1992).
159. Verhoogt, H., C.R.J. Willems, J. VanDam, and A. Posthuma DeBoer, Polym. Eng. Sci., 34, 453 (1994).
160. Verhoogt, H., H.C. Langelaan, J. VanDam, and L. Posthuma DeBoer, Polym. Eng. Sci., 33, 754 (1993).
161. Qian, X. and A.I. Isayev, SPE ANTEC, 35, 1744 (1989).
162 Isayev, A.I., US Patent 5,021,475 (1991).
163. Dutta, D., R.A. Weiss and K. Kristal, Polym. Eng. Sci., 33, 838 (1993).
164. Dutta, D., R.A. Weiss and K. Kristal, Polym. Compos., 13, 394 (1992).
165. Carfagna, C., E. Amendola and M.R. Nobile, in "International Encyclopedia of Composites", S.M. Lee ed., VCH, New York, V.2, 350 (1990).
166. Silverstein, M.S., A. Hiltner and E. Baer, J. Appl. Polym. Sci., 43, 157 (1991)
167. Sukhadia, A.M., D. Done, and D.G. Baird, Polym. Eng. Sci., 30, 519 (1990).
168. Baird, D.G. and R. Ramanathan, "Multiphase Macromolecular Systems, B.M. Culberston, ed., Plenum, New York, 1989.
169. Vinogradov, G.V. and A. Ya. Malkin, "Rheology of Polymers," Mir Publishers, Moscow, 1980.
170. Min, K., J.L. White and J.F. Fellers, Polym. Eng. Sci., 24, 1327 (1984).
171 Grace, H.P., Chem. Eng. Commun., 14, 225 (1982).
172. Alle, N. and J. Lyngaae-Jorgensen, Rheol. Acta, 19, 104 (1980).
173. Karam, H. and J. L. Bellinger, Ind. Eng. Chem. Fund., 7, 576 (1968).
174. Han, C.D., "Multiphase Flow," Academic Press, New York, 1981.
175. Vinogradov, G.V., B.V. Yarlykov, M.V. Tsebrenko, A.V. Yudin and T.I. Ablazova, Polymer, 16, 609 (1975).
176. Ramanathan, R., K.G. Blizzard and D.G. Baird, SPE ANTEC, 36, 1399 (1987).
177 Acierno, D., E. Amendola, L. Nicholais and R. Nobile, Mol. Cryst. Liq. Cryst., 153, 533 (1987).
178 Ajji, A. and P.A. Gignac Polym. Eng. Sci.,32, 903 (1992).
179 Lin, Q. and A.F. Yee, Polymer, 35, 3463 (1994).
180. Nobile, M.R., E. Amendola, L. Nicolais, D. Acierno and C. Cafagna, Polym. Eng. Sci., 29, 244 (1989).
181. Shin, B.Y. and I. J. Chung, Polymer J., 21, 851 (1989).
182. Bretas, R.E.S. and D. G. Baird, Polymer, 33, 5233 (1992).
183. Brostow, W., Polymer, 31, 979 (1990).
184. Isayev, A.I. and P.R. Subramanian, US Patent 5,032,433 (1991).
185. Sharma, S.K., A. Tendolkar and A. Misra, Molec. Cryst. Liq. Cryst. 157, 597 (1988).
186. Bhattacharya, S., K.A. Tendolkar and A. Misra, Molec. Cryst. Liq. Cryst. 153, 501 (1987).

187. Pracella, M., E. Chiellini and D. Dainelli, Macromol. Chem., 190, 175 (1989).
188. Nobile, M.R., D. Acierno, L. Incarnato, E. Amendola and L. Nicolais, J. Appl. Polym. Sci, 41, 2723 (1990).
189. Engberg, K., M. Ekblad, P.E. Webner and U. W. Gedde, Polym. Eng. Sci., 34, 1346 (1994).
190 Lin, Y.G. and H.H. Winter, Polym. Eng. Sci., 32, 773 (1992).
191. Lee, W.C. and A.T. DiBenedetto, Polym. Eng. Sci., 32, 400 (1992).
192. DeMeuse, M.T. and M. Jaffe, Polym. Adv. Technol., 1, 81 (1991).
193. DeMeuse, M.T. and M. Jaffe, SPE ANTEC, 36, 905 (1991).
194. DeMeuse, M.T. and M. Jaffe, Mol. Crys. Liq. Cryst., 157, 535 (1988).
195. Ramanathan, R., D.S. Done, and D.G. Baird, SPE ANTEC, 34, 1716 (1989).
196. Kenig, S., M.T. DeMeuse, and J. Jaffe, Polym. Adv. Technol, 2, 25 (1991).
197. Isayev, A.I. and P.R. Subramanian, US, Patent 5,070,157 (1991).
198. Isayev, A.I., SPE ANTEC, 36, 908 (1991).
199. Akhtar, S. and A.I. Isayev, Polym. Eng. Sci., 33, 32 (1993).
200. Ding, R. and A.I. Isayev, J. Thermoplast. Compos. Mater, 8, 208 (1995).
201. Cheng, W.M., H.A.A. Rasoul, and R.W. Atackman, Polym. Reprints, 32, 50 (1991).
202. Jin, J.I., E.J. Choi, and K.Y. Lee, Polym.J., 18, 99 (1986).
203. Lee, W.C., A.T.DiBenedetto, J.M. Gromek, M.R. Nobile, and D. Acierno, Polym. Eng. Sci., 33, 156 (1993).
204. Amendola, E., C. Carfagna, P. Netti, L. Nicolais, and S. Saiello, J. Appl. Polym. Sci., 50, 83 (1993).
205. Cifferi, A., Polym. Eng. Sci., 34, 377 (1994).
206. Dutta, D., H.H. Chen, and D.G. Baird, Polymer, 34, 799 (1993).
207. Heino, M., J.V. Seppala, M. Westman, Patent Application, WO 92-17545 (1992).
208. Kim, B.C. and S.M. Hong, Mol. Cryst. Liq. Cryst., 254, 251 (1994).
209. Miller, M.M., D.L. Brydon, J.M.G. Gowie, and R.R. Mather, Macr. Rapid Comm., 15 857 (1994).
210. Takashi, T., O. Masakatsu, and A. Daisaburou, Europ. Patent 566,149 A2 (1993).
211. Heino, M.T., and J.V. Seppala,J. Appl. Polym Sci., 48, 1677 (1993)
212. Lee, W.C. and A.T. DiBenedetto, Polymer, 34, 684 (1993).
213. Sun, L.-M., T. Sakamoto, S. Ueta, K. Koga, and M. Takayanagi, Polym. J., 26, 939 (1994).
214. Sun, L.-M., T. Sakamoto, S. Ueta, K. Koga, and M. Takayanagi, Polym. J., 26, 953 (1994).
215. Hong, C.M. and B.C. Kim, Polym. Eng. Sci., 34, 1605 (1994).
216. Gupta, B., G. Calundann, L.F. Charbonneau, H.C. Linstic, J.P. Shepherd and L.C. Sawyer, J. Appl. Polym. Sci., 53, 575 (1994).
217. Kwon, S.K. and I. J. Chung, Polym. Compos., 16, 297 (1995).
218. Bretas, R.E.S., D. Collias and D. G. Baird, Polym. Eng. Sci., 34, 1492 (1994).
219. Isayev, A. I. and R. Viswanathan, Polymer, 36, 1585 (1995).
220. Kyotani, M., A. Kaito and K. Nakayama, Polymer, 35, 5138 (1994).

221. Sabol, E.A., A.A. Handlos and D.G. Baird, Polym. Compos., 16, 330 (1995).
222. Isayev, A.I., Eur. Patent Appl. WO 91/01879 (1991).
223. Isayev, A.I., US Patent 5,275,877 (1994).
224. Isayev, A.I., US Patent 5,238,638 (1993).
225. Isayev, A.I., U.S. Patent 5,268,225 (1993).
226. Miller, B., Plastic World, April 1994, p.37.
227. Harvey, A.C., R.W. Lusignea, F.L. Racich, D.M. Baars, D.D. Bretches and R.B. Davis Lusignea, R.W., et. al., US Patent 4,963,428 (1990).
228. Lusignea, R.W.,T.L. Racich, A.C. Harvey and R.R. Ruby et. al., US Patent 4,975,312 (1990).
229. Harvey, A.C., R.W. Lusignea and F.L. Racich, US Patent 4,966,807 (1990).
230. Lusignea, R.W., et. al., US Patent 5,202,165 (1993).
231. Allan, P.S. and M.J. Bevis, British Patent 2170140B.
232. Wang, L., P.S. Allan and M. J. Bevis, Plast. Rub. Compos. Process. Appl., 23, 139 (1995).
233. Isayev, A.I. and S. Swaminathan, US Patent 4,835,047 (1989).
234. Isayev, A.I., US Patent 5,006,402 (1991).
235. Isayev, A.I., US Patent 5,006,403 (1991).
236. Isayev, A.I., US Patent 5,283,114 (1994).
237 Rosenau, B., B. Hisgen, G. Heinz, H.G. Braun, D. Lausberg and H. Zeiner, US Patent 5,011,884 (1991).
238. Borowczak, M., D.J. Burlett, R.G. Bauer, and J.W. Miller, US Patent 5,222,457 (1993).
239. Tung, W.C.T., D.A. Tung, M.M. Kelley, D.D. Callander and R.G. Bauer, US Patent 5,120,965 (1992)
240. Tagima, Y. and K. Miyawaki, US Patent 5,314,946 (1994).
241. Sasaki, K., M. Hara, T. Tomita and M. Shinomori, Eur. Patent 524655 (1993).
242 Muelhaupt, R., J. Roesch, S. Hopperdietzel, E. Weinberg and H. Klein, Eur. Patent 535526 (1993).
243 Yesair, D.W., Eur. Patent 543943 (1994)
244. Itoyama, K., N. Horita, S. Seki and S. Minami, US Patent 4,798,875 (1989).
245. Rock, J.A., US Patent 4,871,817 (1989).
246. Minamisawa, T., K. Endo, E. Ikegami and S. Nezu, US Patent 4,874,800 (1989).
247. Baba, F., U.S. Patent 4,880,591 (1989).
248. Capp, F.W., D.J. Gerhardt and R.A. Little, U.S. Patent 4,894,193 (1989).
249. Shikae, T. and K. Hijikata, US Patent 4,904,752 (1989).
250. Okey, D.W., H. Saatchi and J.F. Scanlon, US Patent 4,933,131 (1990).
251. Gerbhardt, D.J., R.A. Little and W. Capp, US Patent 4,933,658 (1990).
252. Schilo, D. and W. Birkenfeld, US Patent 4,943,481 (1990).
253. Schultze, H.J. and H.J. Liedloff, US Patent 4,992,514 (1991).
254. Tsuruta, A., H. Kawaguchi, T. Ishikawa and Y. Kondo, US Patent 5,043,400 (1991).
255. Scanlon, J.F., H. Saatchi, D.W. Okey, J.S. Church and G.A. Wigell, US Patent 5,049,342 (1991)

256. Brooks, G.T., US Patent 5,098,940 (1992).
257. Bartlet, P. and P. Moissinnie, US Patent 5,100,605 (1992).
258. Ueno, T., H. Tanabe, R. Takagawa, Y. Eguchi, K. Tsutsui, N. Yabuuchi, M. Maruta and A. Kashihara, US Patent 5,112,689 (1992).
259. Baba, F., US. Patent 5,149,486 (1992).
260. Zdrahala, R.J., US Patent 5,156,785 (1992).
261. P.K. Handa, C.M. Lansinger, V.R. Parameswaran, G.R. Schorr, US Patent 5,158,725 (1992).
262. Farris, R.J., S.L. Joslin and R. Giesa, US Patent 5,232,778 (1993).
263. Carter, J.D. and R.R. Smith, US Patent 5,232,977 (1993).
264. Zdrahala, R.J., US Patent 5,248,305 (1993).
265. Poll, G., J. Finke, H. Beyer and H. Modler, US Patent 5,258,470 (1993).
266. Cottis, S., H. Chin, W-H. Shiau and D. Shopland, US Patent 5,262,473 (1993).
267. Dashevsky, S., K. Kim and S.W. Palmaka, US Patent 5,266,658 (1993).
268. Tsutsumi, T., T. Nakakura, S. Morikawa, K. Shimomura, T. Takahashi, A. Morito, N. Noga, A. Yamaguchi and M. Ohta, US Patent 5,312,866 (1994).
269. Suenaga, J, US Patent 5,324,795 (1994).
270. Wong, C.P., US Patent 5,330,697 (1994).
271. Kaku, M. and R.R. Luise, US Patent 5,346,969 (1994).
272. Dashevsky, S., K. Kim, S.W. Palmaka, R.L. Johnston, L.A.G. Busscher and J.A. Juijn, US Patent 5,346,970 (1994).
273. Baird, D.G. and A. Datta, Eur. Patent Appl. WO 92/18568 (1992).
274. Allington, I.N.H., Eur. Patent Appl. WO 93/11024 (1993).
275. Tendolkar, A., S. Narayan, S.W. Kantor, R.W. Lenz, R.J. Ferris, S.L. Joslin and R. Giesa, Eur. Patent Appl. WO 94/09194 (1994).
276. Kim, K.U., B.C. Kim and S.M. Hong, US Patent 5,276,107 (1994).
277. Heino, M., J. Seppala and M. Westman, Eur. Patent Appl. WO 93/24574 (1993).
278. Postema, A.R., G.P. Schipper and J.A.N. Scott, Eur. Patent Appl. 636,644 A1 (1995).
279. Haghighat, R.R., R.W. Lusignea and L. Elandjian, Eur. Patent Appl. WO 92/17545 (1992).
280. Citta, V., U. Pedretti and A. Roggero, Eur. Patent Appl 612,802 A1 (1994).
281. Moria, A., Eur. Patent Appl. 612 610 A1 (1994).
282. Toy, L.T., A.N.K. Lau and C.W. Leong, Eur. Patent Appl. WO 94/14890 (1994).
283. Tomita, T., M. Ohsugi and D. Adachi, Eur. Patent Appl. 566 149 A2 (1993).

MOLECULAR COMPOSITES

Chapter 2

Molecular Composites of Liquid-Crystalline Polyamides and Amorphous Polyimides

Synthesis, Rheological Properties, and Processing

Ken-Yuan Chang[1] and Yu-Der Lee[2,3]

[1]Union Chemical Laboratories, Industrial Technology Research Institute, 321 Kuang Fu Road, Section 2, Hsinchu 300, Taiwan
[2]Department of Chemical Engineering, National Tsing Hua University, Hsinchu 30043, Taiwan

Novel block copolymers of a lyotropic polyamide (PBTA) and two amorphous polyimides (PI) were prepared, and were characterized to exhibit liquid-crystalline behavior. The rheology of PBTA and its copolymers were studied by a cone-and-plate rheometer. Additionally, the experimental relationship between shear viscosity and shear rate could be fitted by power-law model for isotropic solutions and Carreau model for anisotropic solutions. Moreover, solution processing of PBTA homopolymer and PBTA/PI block copolymers was carried out by a wet spinning apparatus. PBTA fibers were spun from anisotropic solutions (C>Ccr) and PBTA/PI copolymer fibers from isotropic solutions (C<Ccr). The tensile strength and modulus of copolymer fibers were found to locate between those of the PBTA fiber and the PI fiber. Take-up speed was varied during the spinning process for investigating the influence of draw ratio to the mechanical properties of the spun fibers. Liquid-crystalline textures could be observed in the resultant PBTA fibers under a polarized optical microscope. The morphology of all spun fibers was also present in the study.

The concept of a self-reinforcing molecular composite of rigid-rod and flexible-coil polymer components was introduced in composite materials and polymer science(*1-3*) in the early 1980's. Since then, rigid-rod molecular composites have attracted much interest(*4-8*). This new class of high performance structural polymers has outstanding strength and modulus, as well as excellent thermal and environmental stability. In addition, the motivation behind developing such materials is based on the improvement in impact and fracture toughness, as well as dimensional stability over those of conventional composites, which are limited by the interfacial adhesion or bonding problems arising in the reinforcing solid fibers and the resin matrix.

Molecular composites development has been concentrated primarily in physical blend systems in which rigid-rod aromatic-heterocyclic polymers functioned as the

[3]Corresponding author

0097–6156/96/0632–0022$15.00/0

reinforcing elements. A brief review was made by Krause and Hwang(*9*) For physical blend systems, however, these rigid-rod molecules strongly tend toward aggregating to form micro domains with low aspect ratios, thereby leading to ineffective reinforcement and, ultimately, low mechanical properties. To prevent phase separation, many investigations have been focused on creating an intimate physical blend of rigid-rod molecules and coil-like molecules through some strategies, e.g., grafting of a matrix polymer onto the rigid polymer(*10-12*), creating ionic interactions between coil-like and rigid-rod polymers(*13-15*), and developing in-situ molecular composites(*16*).

In spite of those efforts, achieving a true molecular composite still remains difficult. The blends can only be processed from solutions of which the concentration is controlled below the corresponding critical concentration (C_{cr}), to prevent the segregation of the rigid component, and to insure the molecular dispersity of the reinforcing elements. This limitation might eventually result in a poor processibility of molecular composites due to a rather low processing concentration. An alternative approach employed to produce a molecular composite involves synthesizing a block copolymer comprised of rigid-rod molecules and flexible-coil molecules(*17*). Directly introducing a reinforcing segment in the copolymer main chain would provide several advantages over the physical blends. For instance, the rigid segment is difficult to aggregate, consequently leading to an increase of critical concentration and hence a better processibility. Another advantage is that the mechanical properties should be improved due to an elimination of a rigid/flexible molecule interface and an easier translation of stress and strain between rigid/flexible molecules. All of these advantages can be attributed to the chemical bonding created in copolymer systems instead of the physical entanglement occurring in blend systems.

The candidate systems of molecular composites developed by our laboratory was polyamide/polyimide (PA/PI) block copolymers with liquid-crystalline PA block participating in the polymer backbones. Several novel PA/PI block copolymers have been synthesized by our laboratory(*18,19*). The rigid components (PA) were verified to retain their liquid-crystalline property in static solutions and would therefore be expected to provide effective reinforcement. After the chemical architecture of block copolymers has been appropriately designed and synthesized, the primary consideration in producing a molecular composite from solution to solid state is how to achieve the finest possible dispersion of the reinforcing segments. Hwang et al. successfully developed the solution processing of the molecular composite blend systems, which was based on some guidelines regarding the solution behavior in ternary systems of a rigid-rod polymer/flexible coil polymer/solvent(*20*).

On the other hand. the processing of liquid-crystalline polymers has received considerable interests. One of the most unique aspects of liquid crystalline fluids is the rigid-rod constituents' capability to orient during processing. Furthermore, these orientations and the resultant texture could be maintained during the solidification process. Hence, the rheological properties of spinning dopes are directly correlated to the final mechanical properties of fibers.

In this article, the synthesis, characterization, rheology, solution processing, and mechanical properties of novel molecular composites are discussed. These molecular composite are block copolymers composed of a liquid-crystalline polyamide and two amorphous polyimides. Also, the rheological properties of both isotorpic and anisotropic solutions of PBTA homopolymers and PBTA/PI block copolymers are investigated before the solution processing is taken. The mechanical properties and morphologies of the spun fibers are also examined in this study.

Experimental

Materials The materials in this work included a lyotropic polyamide (PBTA) composed of terephthaloyl dichloride and 2,2'-dimethyl-4,4'-diamino-biphenyl; two amorphous polyimides respectively consisting of 3,3'4,4'-benzophenone tetracarboxylic dianhydride (BTDA) and 2,3,5,6-tetramethyl-p-phenylene diamine (TMPD), as well as 4,4'-(hexafluoroisopropylidene)-bis(phthalic anhydride) (6FDA) and 2,2'-dimethyl-4,4'-diamino-biphenyl (DMDB); and the block copolymers composed of PBTA and those two polyimides with various compositions. The PBTA/PI block copolymers and the corresponding homopolymers were prepared according to our previous work(*19*).

All the polymers were obtained as powders after synthesis, and were carefully dried at 150 °C under vacuum for at least 12 h prior to solution preparation used in rheological measurement and wet spinning. In rheological measurements and solution processing, the solvent used was NMP for PI homopolymers, NMP/4% LiCl for PBTA homopolymer, and NMP/3% LiCl for PBTA/PI block copolymers.

Rheological Measurements Polymer solutions with various concentrations were prepared in a small sample bottle by moderately heating for several days, and then used for rheological measurements. The study was carried out using a cone-and-plate rheometer (Haake CV-20N) with a shear rate range of 0 ~ 300 s^{-1} at 30 °C.

Preparation of Spinning Dopes The spinning dopes were prepared by dissolving polymer powders with the specified solvents. The dissolution was performed in a three-neck kettle by mechanically stirring and moderate heating. A steady stream of dry nitrogen gas was applied to ensure an inert atmosphere for preventing humidity up-take by the hydrogen bonding of amide groups in the polymer main chain. When the polymer powder was completely dissolved, any insoluble impurity was separated by pressure filtration, in which the filter was a circular disk made of polypropylene. The the filter pore size was ca. 20 μm and the pressure source originated from compressed nitrogen gas. The resultant solution after filtration was placed in a bottle, sealed with PTFE sealing tapes, and then directly heated with a hot plate to remove any bubbles.

Wet Spinning Aparatus and Processing Fiber spinning was carried out on a laboratory-made model of a wet-spinning apparatus. Figure 1 shows a schematic drawing of the equipment, which consists of (1) a spinning-dope reservoir with heating jacket, (2) a gear pump with speed and temperature controller, (3) a spinnerette, (4) a coagulation bath of 120 cm length, and (5) take-up and drawing devices.

The spinning dopes were poured into the reservoir, in which the temperature was maintained at 60 °C (before spinning) for further degassing. The solution was extruded via a gear pump at a flow rate of 0.21~0.4 cm^3/min through a connecting cylindrical stainless steel tube with a spinnerette in its end. A single-hole spinnerette with 200 μm diameter and 2 mm length was used in this study. The extrudate was allowed to pass downward through an air gap of 5 cm prior to entering the liquid coagulation bath, in which the coagulation medium is 7 vol % aqueous NMP solution at room temperature.

The solidifying filament was drawn through the coagulation bath and onto the winder. The tale-up speed varied from 20 m/min to 60 m/min. Next, the bobbins with collected filaments were washed in runnning water for 24 h, dried in a vacuum oven at room temperature for 4 h and then at 90 ℃ for 24 h. No postspinning treatments were performed.

Characterization Mechanical properties of fibers were determined using a Minimate tensile machine with a strain rate of 0.2 mm/min and a gauge length of 5 cm. Seven or more monofilamemts were tested for each sample and their average values are reported.

An examination of anisotropic texture for solid monofilaments was next carried out by a polarizing optical microscope (Leitz). Hitachi S-2300 scanning electron microscope (SEM) was applied to observe the morphology of either longitudinal surface or fractured transverse surface of those fibers. In the latter case, the fiber was fractured by immersion in liquid nitrogen.

Results and Discussion

Synthsis and Characterization Two series of PBTA/PI block copolymers have been prepared from two different flexible polyimides. The chemical structures are illustrated in Scheme 1.

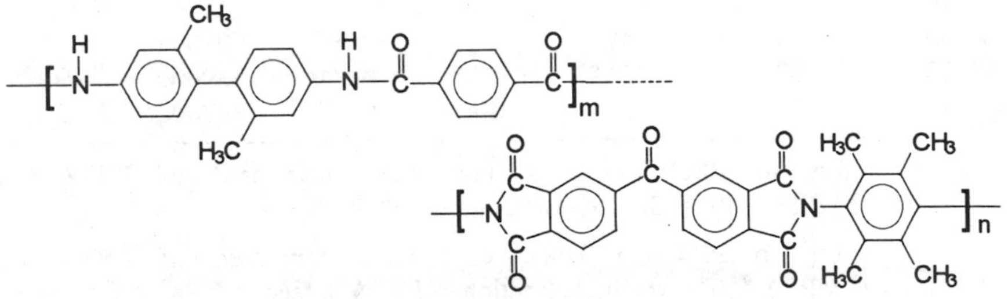

PBTA/PI Block Copolymer 1 (Co1)

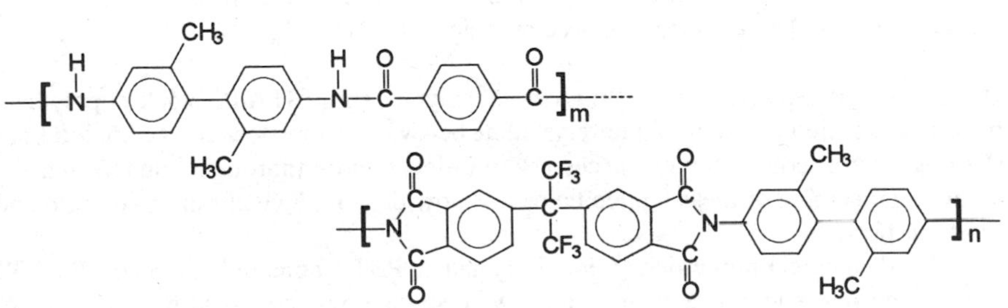

PBTA/PI Block Copolymer 2 (Co2)

Scheme 1. Chemical structures of two PBTA/PI block copolymers

The synthesis, characterization, and liquid-crystalline property of Co1 block copolymers with various rigid/flexible composition were reported elsewhere(*19*). Co2 series was prepared by the same synthetic procedure as described in that article(*19*) but instead of PI(6FDA/DMDB) as the flexible component. The characterization was also made; however, the results have not yet been reported. PI(6FDA/DMDB) exhibited superior solubility by introducing the fluoro-containing monomer and 2,2'-disubstituted biphenylene moiety. This polyimide could be dissolved not only in polar solvents like 1-methyl-2-pyrrolidone (NMP) and N,N-dimethylacetamide (DMAc), but also in common organic solvents such as tetrahydrofuran (THF), chloroform, and acetone. Therefore, Co2 series possesses better solubility, i.e., 16~19 wt% in NMP-4%LiCl, than Co1 series (13~14 wt% in NMP-4%LiCl). Table I lists the compositions, inherent viscosities, and molecular weights of the copolymers which could be processed into fibers.

Table I. Compositions, Inherent Viscosities, and Molecular Weights of Block Copolymers for Wet Spinning

copolymer	PBTA content in feed (wt%)	Exact PBTA content (wt%)[c]	η_{inh} (dL/g)[d]	Mn[e]	Mw[e]	Mw/Mn[e]
Co1-30 [a]	30	41	2.43	55000	315000	5.72
Co1-40	40	49	2.95	65000	401000	6.12
Co1-50	50	55	2.78	49000	339000	6.91
Co2-40 [b]	34	39	1.42	48000	347000	7.20

[a] "Co1" refers to PBTA/PI block copolymers composed of PBTA and PI(BTDA/TMPD), "30" is the weight fraction of PBTA in feed.

[b] "Co2" refers to PBTA/PI block copolymers composed of PBTA and PI(BTDA/TMPD), "40" is the mole fraction of PBTA in feed.

[c] Determined by NMR spectra analysis.

[d] 0.5 g/dL in NMP containing 3 % LiCl

[e] Measured by GPC, in which DMF-3%LiCl was used as solvent and poly(ethylene oxide) was applied as narrow molecular weight standard.

Rheological Properties Since PBTA homopolymer and PBTA/PI block copolymers have been verified to exhibit liquid-crystalline behavior, completely characterizing their rheological properties becomes necessary in order to more thoroughly understand the solution processing of these unique fluids and how they produce ultrahigh strength and modulus fibers.

The rheological properties of liquid-crystalline PBTA homopolymer and PBTA/PI block copolymers in a steady shear flow were studied by a cone-and-plate rheometer with a shear-rate range of 0-300 s^{-1}. The data presented in Figure 2 show the concentration dependence of zero-shear viscosity (η_0) for PBTA solutions. Interestingly, η_0 increases rapidly with the concentration reaching a maximum with the onset of liquid-crystalline order, and then decreases for higher concentration to a minimum, whereupon it begins to increase again with a further increase of concentration. This experimental phenomema coincides with the result of poly- γ -

benzyl-L-glutamate (PBLG) solutions as repoted by Kiss and Porter(*21*). The concentration at which the peak appears is generally referred to as the critical concentration (C_{cr}). For $C < C_{cr}$, the solution is in an optically isotropic state and its viscosity increases normally with the concentration. As the concentration increases higher than C_{cr}, the rigid polymer chain tends to aggregate and an anisotropic solution is formed. This formation is evidenced by the appearance of birefrigence in a static solution under a polarized microscope. In this state, polymer chains align themselves to

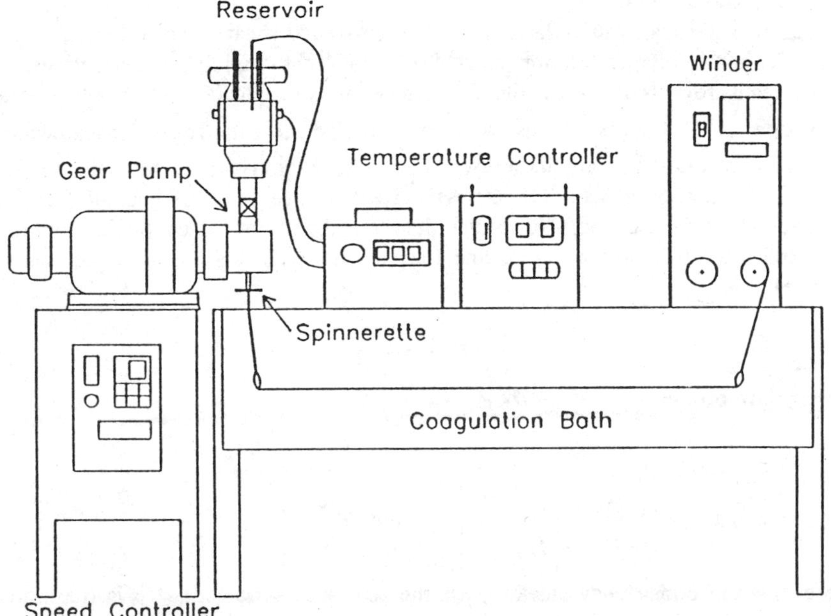

Figure 1 Schematic drawing of the wet-spinning apparatus

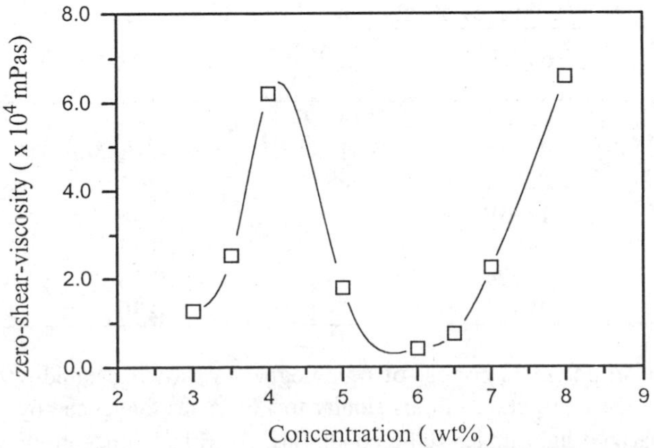

Figure 2. Zero-shear-viscosity as a function of PBTA concentration.

form various structured phases with high degree of local order and eventually result a biphasic fluids, in which highly ordered regions are suspensed in an isotropic matrix. Since the polymer chains can no longer tumble freely in the regularly aligned region (anisotropic phase), the hydrodynamic volume might be significantly reduced. Consequently, an easier slippage between polymer chains and a decrease in viscosity would occur. Moreover, increasing the concentration would not only increase the size and density of the anisotropic phases but also increase the local order. The hydrodynamic volume per polymer chain would thereby be further reduced and, ultimately, the viscosity.

Figure 3 illustrates the relationship between steady shear viscosity and shear rate for PBTA homopolymer solutions in NMP/4% LiCl with various concentrations. This figure clearly reveals the shear-thinning effect for isotropic ($C < C_{cr}$) solutions and anisotropic ($C > C_{cr}$) solutions with the most shear rate region. Meanwhile, a Newtonian plateau appears in a low shear rate region for anisotropic solutions, especially for C = 6 wt% and C = 6.5 wt%. Furthermore, the experimental data could be fitted with theoretical non-Newtonian fluid model. Among which, power-law model was applied for isotropic solutions and Carreau model (*22*) for anisotropic solutions, as shown below:

power - law model
$$\eta = m\dot{\gamma}^{n-1} \tag{1}$$

Carreau model
$$\frac{\eta - \eta_\infty}{\eta_0 - \eta_\infty} = \left[1 + (\lambda\dot{\gamma})^2\right]^{\frac{n-1}{2}} \tag{2}$$

where m is the consistency index, η_0 is the zero-shear-viscosity, η_∞ is the infinite-shear-viscosity, λ is a time constant, and n is the power-law exponent. A summary of these parameters for PBTA solutions with different concentrations is given in Table II.

Table II. Zero-Shear-Viscosity, and Power-Law and Carreau Model Parameters for Solutions of PBTA in NMP-4%LiCl at 30°C

Concentration (wt%)	η_o (mPas)	m	λ (sec)	n
3.0	12597	13566	-	0.47
3.5	25314	31033	-	0.41
4.0	62124	66811	-	0.28
5.0	17839	-	0.46	0.38
6.0	4681	-	0.13	0.55
6.5	7524	-	0.24	0.59
7.0	22423	-	0.30	0.26

The block copolymers consist of rigid segments showing liquid-crystalline order, and also could form structured fluids similar to PBTA, as evidenced by their ability to transmit a polarized light under static conditions. A critical concentration also appears in a plot of zero-shear viscosity, obtained from the measurement of a cone-and-plate rheometer, versus concentration. Figure 4 shows this critical concentration curve for Co1-40. The critical concentraions for Co1 series are listed in Table III, revealing that C_{cr} increases with an increase in PBTA content.

Figure 5 shows the flow curve of shear viscosity vs. shear rate for Co1-40. The flow behavior of solutions with concentrations below C_{cr} could be exactly fitted by the power-law model in the low shear rate region. Whereas the shear viscosities rapidly decrease for shear rate higher than ca. 100 s^{-1}. The latter phenomena might be attributed to a shear-induced orientation. In the case of the anisotropic solution (C=9 wt%) for Co1-40, the flow curve was also fitted by Carreau model in spite of a large deviation occurring in a high shear rate region. All the fitted parameters for the power-law and Carreau model are summarized in Table IV.

Table III. Critical Concentration (C_{cr}) and Processing Concentration ($C_{proc.}$) of PBTA/PI Block Copolymers

Copolymer	C_{cr} (wt%) [a]	$C_{proc.}$ (wt%)
Co1-30	9.0	8.5
Co1-40	8.5	8.0
Co1-50	9.2 [b]	8.0
Co2-40	12.0	11.5

[a] Determined by cone-and-plate viscometer.

[b] Determined by Brookfield viscometer.

Table IV. η_o, and power-law and Carreau Model Parameters for Solutions of Co3-40R in NMP-3%LiCl at 30°C

Concentration (wt%)	η_o (mPas)	m	λ (sec)	n
4.0	8389	11400	-	0.57
5.0	10056	13900	-	0.51
6.0	10223	14100	-	0.57
7.0	13220	18700	-	0.45
8.0	28258	42300	-	0.46
8.5	44556	67500	-	0.35
9.0	14754	-	0.21	0.40

Fiber Spinning and Mechanical Properties Aside from providing the critical concentrations at which liquid-crystalline behavior appears, Table III also lists the processing concentrations of the candidate block copolymers. Since rigid molecules would aggregate together to form anisotropic domains as their concentration exceeds C_{cr}, the processing of molecular composites must be controlled within this limit.

The dry-jet/wet-spin technology was applied to produce molecular composite and PBTA fibers. The spinning dope was prepared from the anisotropic solution of PBTA in NMP-4% LiCl with a concentration (6.5 wt%) higher than C_{cr}. Since the liquid-crystalline domains in these lyotropic solutions were frozen in the coagulation stage,

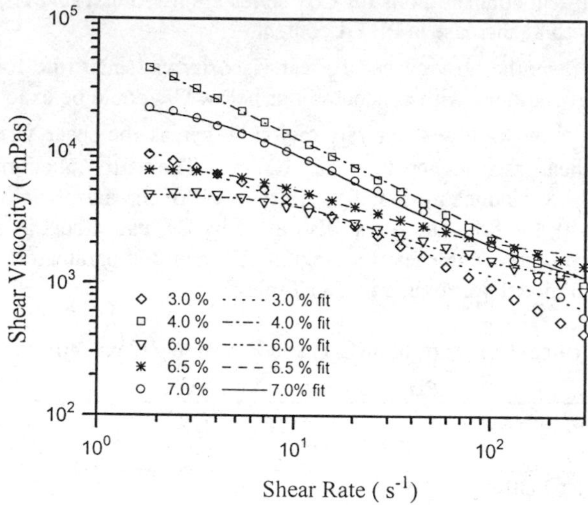

Figure 3. Shear rate dependence of viscosity for isotropic and anisotropic solutions of PBTA in NMP-4% LiCl. The lines represent the power-law or Carreau fit.

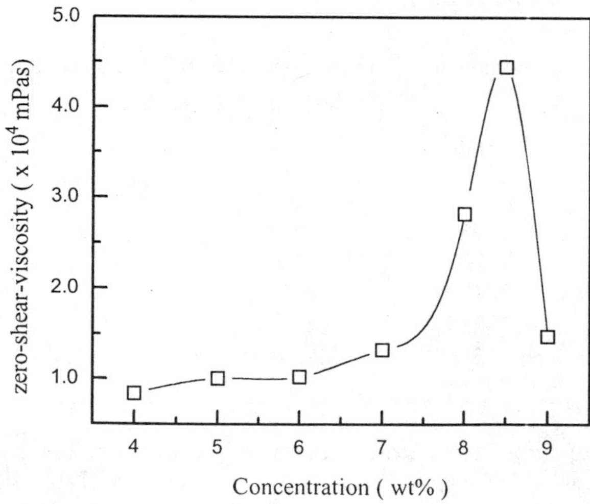

Figure 4. Zero-shear-viscosity as a function of Co1-40 concentration.

anisotropic textures could be observed in the spun PBTA fibers under a polarized light, as shown in Figure 6. PBTA/PI copolymer fibers were produced from dilute isotropic solutions to prevent phase separation. The spinning dope was prepared at concentrations below C_{cr}, as indicated in Table III. In processing these isotropic solutions of block copolymers, fine dispersions of rigid-rod PBTA segments in a flexible PI matrix would be frozen by rapid coagulation to form molecular composite fibers.

Table V makes a comparison of the mechanical properties of Co1 fibers with those of the pure PI and PBTA fibers. For the sake of possessing poor solubility, polyimide fibers have once generally spun from solutions of their precusor, polyamic acid, following a heat treatment for thermal imidization. In general, micropores and voids exist in polyamic acid fibers. Although many coagulation media have been investigated to prevent this drawback, the production of totally void-free aromatic polyamic acid fibers was not found.

Table V. Mechanical Properties of PBTA, PI, and Co1 Fibers

Fiber	diameter (μm)	tensile strength (MPa)	initial modulus (GPa)	elongation (%)
PI(BTDA/TMPD) [a]	49	24±6	1.6±0.3	2.5
Co1-30	55	117±5	5.1±0.5	5.2
Co1-40	49	162±19	7.9±0.8	5.2
Co1-50	41	224±19	12.5±1.1	4.9
PBTA	33	498±28	19.2±0.5	4.8

[a] Fiber spun from 17 % solution in NMP.

Organosoluble polyimides could be directly processed by wet spinning while the presence of voids could not be prevented. Both PI(BTDA/TMPD) and PI(6FDA/DMDB) fibers, spun from their NMP solutions, are very brittle; in addition, their tensile strength, modulus, as well as elongation are relatively low. The poor mechanical properties of PI fibers might be attributed to the void formation during consolidation stages, as evidenced by the scanning electron microscope observation. The porous crossection area is shown in Fig 7.

For copolymer fibers, the tensile strength and modulus were found to be much higher than those of the PI fibers. Furthermore, Figure 8 shows that tensile strength and modulus of PBTA/PI molecular composite fibers increase with an increase of the PBTA content in block copolymers. It is evident that introducing a liquid-crystalline polymer in molecular architecture makes considerable reinforcing effects in molecular composites.

During the spinning process, isotropic solutions of these copolymers were forced to flow through a spinnerette, thereby leading to a high shear rate and hence resulting in a shear induced alignment of the rigid blocks. As the filament went through the spinline, i.e., an air gap, a coagulation bath, and a take-up device, the elongational flow field could further orient the directors of the rigid segments in the consolidation stages. Consequently, increasing the draw ratios would give rise to a higher degree of orientation for rigid blocks and eventually result in an increase in the mechanical properties of copolymer fibers. Figure 9 illustrates this trend for Co2-40 as a representation.

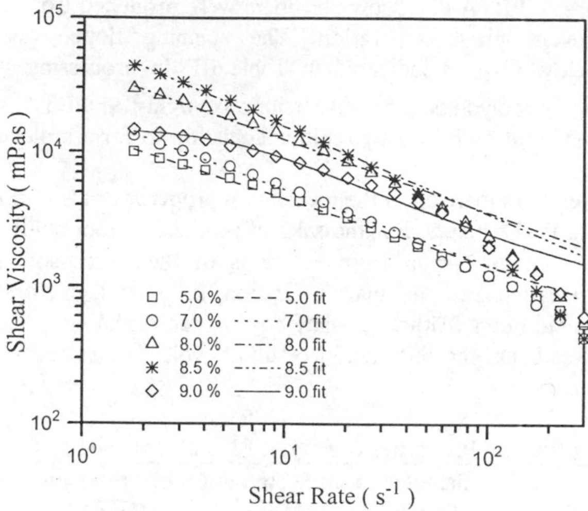

Figure 5. Shear rate dependence of viscosity for isotropic and anisotropic solutions of Co1-40 in NMP-3% LiCl. The lines represent the power-law or Carreau fit.

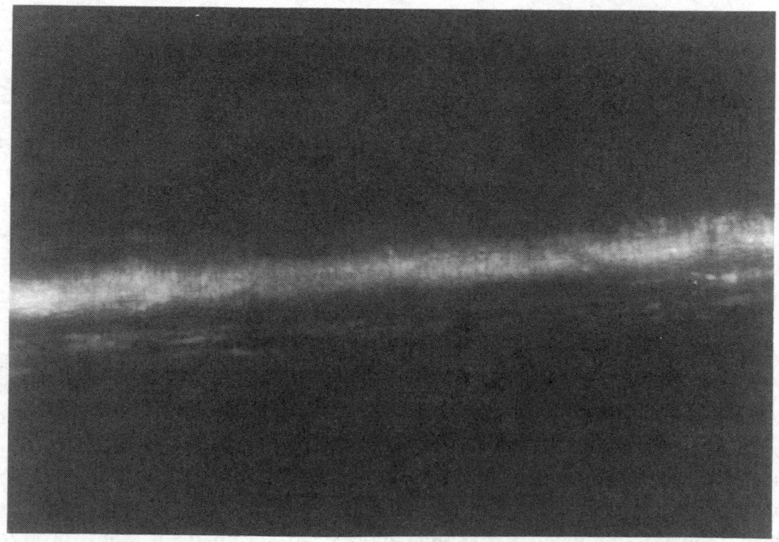

Figure 6. Optcal micrographs of liquid-crystalline texture in PBTA fibers, magnification, 400x, under polarized light.

(A)

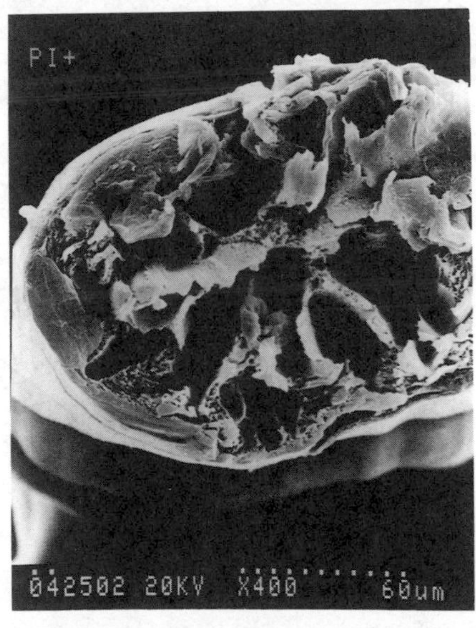

(B)

Figure 7. Scanning electron micrographs of PI fibers (A) PI(BTDA/TMPD) ,(B) PI(6FDA/DMBD).

Morhology A thorough description of morphology is deemed necessary for assessing and predicting the ultimate properties of molecular composite materials. No anisotropic domains or textures were observed in the spun copolymer fibers by an optical microscope with polarized light. This would suggest that liquid-crystalline PBTA segments did not aggregate to form large-scale phase separation, which has

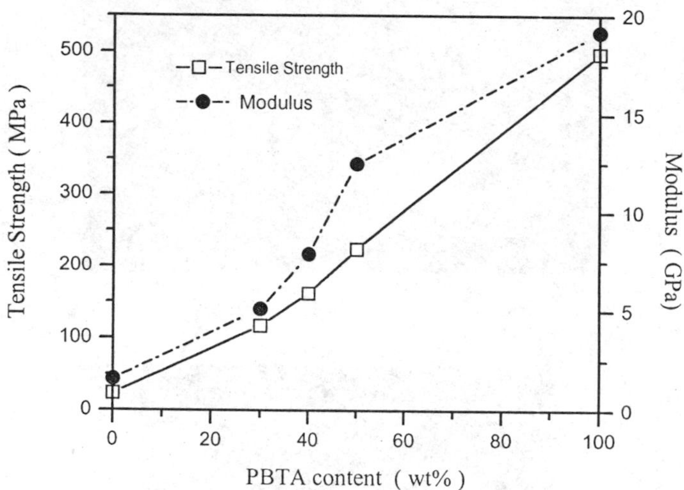

Figure 8. Tensile strength and modulus of Co1 fibers with various composition.

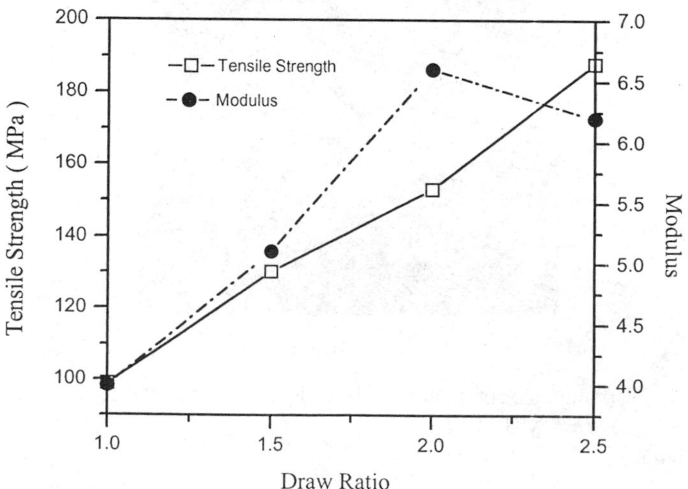

Figure 9. Tensile strength and modulus of Co2 fibers as a function of draw ratio.

many disadvantages for material performance, e.g., the reduced aspect ratio of the reinforcing phase and development of discrete interphase. The morphological features of spun PBTA and copolymer fibers were also examined by scanning electron microscopy (SEM). The resultant SEM-images are shown in Figure 10. No large scale voids were found in the fractured surface of the copolymer fibers.

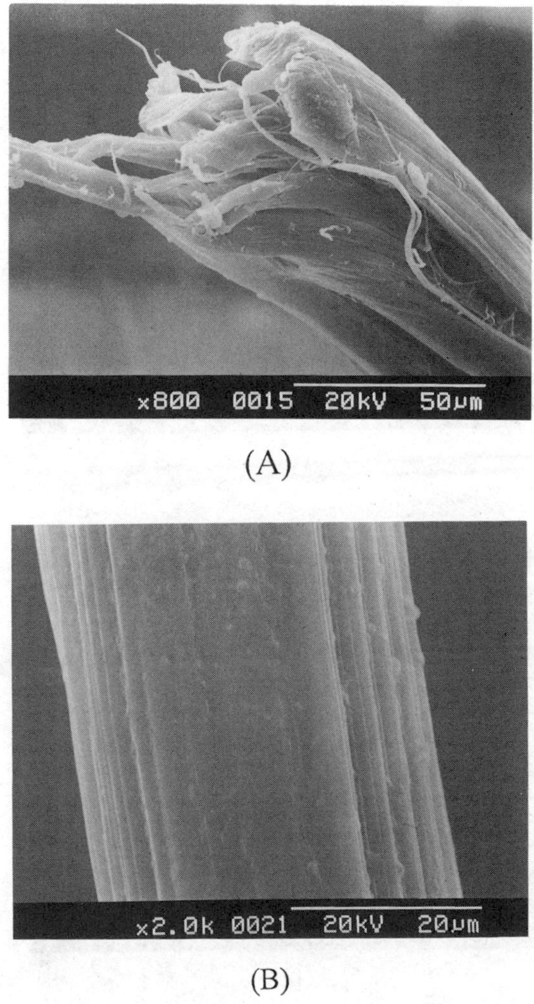

Figure 10. Scanning electron micrographs of (A), (B) PBTA fibers and (C), (D) Co1-40 fibers. Continued on next page.

(C)

(D)

Figure 10. Continued.

Conclusions

Two series of PBTA/PI block copolymers were synthesized in this study and solution processed into molecular composite fibers via dry-jet wet-spinning. The unique rheological properties of liquid-crystalline PBTA homopolymers and PBTA/PI block copolymers were studied with a cone-and-plate rheometer. For block copolymers, the critical concentration decreased with an increase in PBTA content. The flow curves of isotropic and anisotropic solutions could be described via the power-law model and Carreau model, respectively. Copolymer fibers possess tensile strength and modulus located between those of PBTA fibers and PI fibers. Moreover, the tensile strength and modulus of Co1 fibers increase with an increase in PBTA content. Besides, increasing the draw ratios would give rise to an increase in the mechanical properties of copolymer fibers

Acknowledgements

The authors would like to thank the National Science Council of ROC for its financial support under Grant no. NSC 83-0405-E-007-056.

Literature Cited

1. Helminiak T. E.; Benner C. L., Arnold F. E., and Husman G. E. *U. S. Patent* 4, 207, 407 (**1980**).
2. Hwang W-F.; Wiff D. R.; Benner C. L., and Helminiak T. E. *J. Macromol. Sci.- Phys.*, **1983**, *B22(2)*, 231.
3. Takayanagi M.; Ogata T.; M.; Morikawa and Kai T. *J. Macromol. Sci.- Phys.*, **1980,** *B17(4)*, 591.
4. Wiff D. R.; Hwang WF.; Chuah H-H, and Soloski E. *J., Polym. Eng. Sci.*, **1987**, *27 (20)*, 1557.
5. Wiff D. R.; Timms S.; Helminiak T. E., and Hwang W-F. *Polym. Eng. Sci.*, **1987**, *27 (6)*, 424.
6. Chuah H. H.; Arnold F. E.; and Tan L. S. *Polym, Eng, Sci,* **1989**, *29 (2)*, 107.
7. Yamada K.; Uchida M.; Takayanagi M., and Goto K. *J. Appl. Polym. Sci.*, **1986**, *32*, 5231.
8. Yokota R.; Horiuchi R.; Kochi M.; Soma H., and Mita I., *J. Polym. Sci., Polym. Lett.*, **1988**, *26*, 215.
9. Krause S. J. and Hwang W-F, Polymer Based Molecular Composites, Material Research Society, 1990; pp 131-140.
10. Evers R. C.; Dang T. D., and Moore D. R. *Polym. Preprints*, **1988**, *29 (1)*, 244.
11. Bai S. J.; Dotrong M., and Evers R. C. *J. Polym. Sci., Polym. Phys. Ed.*, **1992**, *30*, 1515
12. Ueta S.; Lei W-Y; Koga K., and Takayanagi M. *Polym. J.*, **1993,** *25*, 185.
13. Lee C. Y-C.; Bai S. J.; Tan L-S.; Smith B., and Huh -W *Polym. Mater. Sci. Eng.*, **1990**, *62*, 91.
14. Tan L-S; Arnold F. E., and Chuah H. H. *Polymer*, **1991**, *32*, 1376.
15. Hara M. and Parker G. J. *Polymer*, **1992**, *33*, 4650.
16. Tan L-S and Arnold F. E. *Polym. Preprints*, **1991**, *32 (1)*, 636.

17. Tsai T. T. and Arnold F. E. (a) *Polym. Preprints*, **1985**, *26*, 144; (b) *J. Polym. Sci., Polym. Chem. Ed.*, **1989**, *27*, 2839.
18. Chang K-Y. and Lee Y-D., *J. Polym. Sci., Polym. Chem. Ed.*, **1993**, *31*, 2775.
19. Chang K-Y, Chang H-M, and Lee Y-D, *J. Polym. Sci., Polym. Chem. Ed.*, **1994**, *32*, 2629.
20. Hwang W -F.; Wiff D. R.; Verschoore C.; Price G. E.; Helminiak T. E., and Adams W. W. *Polym. Eng. Sci.*, **1983**, *23 (14)*, 784.
21. Kiss G. and Porter R. S., *J. Polym. Sci. , Polym. Symp.*, **1978**, *65*, 193.
22. Bird R. B. and Carreau P. J. *Chem. Eng. Sci.*, **1968**, M.S. 600, *23*, 427.

Chapter 3

Miscibility, Structure, and Property of Poly(biphenyl dianhydride perfluoromethylbenzidine)–Poly(ether imide) Molecular Composites

Thein Kyu[1], J.-C. Yang[1], C. Shen[1], M. Mustafa[1], C. J. Lee[2], F. W. Harris[2], and Stephen Z. D. Cheng[2]

[1]Institute of Polymer Engineering and [2]Maurice Morton Institute of Polymer Science, University of Akron, Akron, OH 44325

Miscibility of segmented rigid-rod polyimide (PI), viz., biphenyl dianhydride perfluoromethylbenzidine (BPDA-PFMB), and flexible polyether imide (PEI) molecular composites was established by differential scanning calorimetry. The composite films of BPDA-PFMB/PEI were drawn at elevated temperatures above their glass transitions. Tensile moduli of the films were evaluated as a function of composition and draw ratio. Molecular orientations of polyimide were determined by birefringence and wide-angle X-ray diffraction. The crystal orientation behavior of the 80/20 BPDA-PFMB/PEI was analyzed in the framework of the affine deformation model.

Recently, various types of rigid-rod polyimide (PI) derivatives and polyether imide (PEI) molecular composites were found to be completely miscible (1). This miscibility has been ascribed to similarity of the imide structure between PI and PEI. The tensile modulus and strength of bulk molecular composites showed remarkable improvement with increase of the polyimide content. Although the molecular reinforcement of the flexible PEI matrix via incorporation of the PI rigid-rod molecules appears unequivocal, the molecular weight of the above rigid-rod polyimide derivatives was rather low as exemplified by their low inherent viscosity (2.2~2.63 dl/g) (1). Such low molecular weight systems tend to favor miscibility. Hence, we have selected a segmented rigid-rod polyimide, viz., biphenyldianhydride perfluoromethylbenzidine (BPDA-PFMB) having a higher molecular weight, (i.e., inherent viscosity of 4.9 dl/g) and a high aspect ratio to ascertain the concept of molecular composites. It is promising to achieve true miscibility between the segmented rigid-rod BPDA-PFMB and the flexible PEI because of their similar imide structure.

The highly drawn rigid-rod polymers generally show extremely high modulus and high strength close to their theoretical values. However, the compressive strength of these rigid materials is rather poor, thereby prohibiting their applications as a

0097–6156/96/0632–0039$15.00/0

structural material. The incorporation of a small amount of flexible coil molecules into the rigid-rod matrix has been sought as a route for improving the compressive strength of the highly oriented rigid-rod materials (2-9). On the other hand, a minor amount of rigid-rod polymer molecules may be dispersed in the matrix of flexible coils in order to reinforce the matrix (7). This concept is similar to the principle of the chopped fiber reinforced composites (10,11), except that reinforcement is expected to occur at a molecular level or a nanometer-scale. The reinforcement can be further enhanced via molecular drawing (i.e., orientation), but it is difficult to quantify crystal orientation by x-rays or by other techniques due to the low crystallizable rod contents in most molecular composites. We therefore focus on the characterization of molecular orientation and mechanical properties of the composite materials having high-rod contents.

In this paper, the BPDA-PFMB/PEI molecular composites were oriented by means of zone annealing/drawing slightly above the glass transition temperatures of the respective molecular composites (280 - 400 °C). The dependence of draw ratio on tensile modulus, crystal orientation, and birefringence was determined as a function of composition. The relationship between structure (crystal orientation) and tensile property (modulus) of drawn films has been examined by comparing crystal chain orientations with the prediction of affine deformation (12-16).

Experimental

Sample Preparation

Polyether imide, commercially known as Ultem 1,000, was kindly supplied by GE. The segmented rigid-rod BPDA-PFMB was synthesized at the University of Akron. It has an intrinsic viscosity of 4.9 dl//g. The BPDA-PFMB/PEI molecular composites were prepared by dissolving in m-cresol at a polymer concentration of 3 wt%. Molecular composite films were cast from m-cresol solution at 110 °C in an air oven. The cast films were soaked overnight in a boiling methanol bath to extract the residual m-cresol, then dried at 250 °C in a vacuum oven over 24 hours. The BPDA-PFMB/PEI molecular composite films were stretched on a heated roller by zone-drawing at elevated temperatures (280 - 400 °C) above the glass transition temperatures of the respective molecular composites. After zone annealing/drawing, the composite films were air-quenched to room temperature to freeze the structure and molecular orientation.

Characterization

Differential scanning calorimetry (DSC) scans were acquired on a Du Pont thermal analyzer (Model 9900) with a heating module (Model 910). The heating scans were carried out from ambient temperature to 330 °C in a circulating dry nitrogen environment. Indium standard was used for temperature calibration. The heating rate was 20 °C/min unless indicated otherwise.

A Seiko thermal mechanical analyzer (TMA/SS 100) was utilized to measure tensile modulus of the composite films. The gauge length of measurement was 10 mm. A force mode with a ramp rate of 5 g/min up to the maximum force of 40 g was utilized. Each data point of tensile modulus and birefringence represented an average value of 4 measurements.

Birefringence of the drawn films was measured by using a Leitz polarized light microscope with crossed polarizers and a Berek compensator. The orientation of the drawn films was examined by wide-angle X-ray diffraction (WAXD) film patterns using a Siemens X-100 area detector. The rotating anode (12 kW Rigaku) with CuKα radiation having a wavelength of 0.1541 nm was used. The rotating anode generator was operated at 150 mA and 40 kV. The sample-to-film distance was 5.3 cm.

Floating Rod Model

The orientation behavior of fibers may be analyzed by defining an orientation factor, $f(\psi)$, of a fiber molecule with respect to a reference axis (e.g., the direction of spinning or drawing) assuming a uniaxial symmetry around the fiber axis (12), i.e.,

$$f(\psi) = \frac{3 < \cos^2 \psi > -1}{2} \quad (1)$$

where, ψ represents an angle that the fiber axis makes with the reference axis. The average directional cosine of the radial fiber direction, $< \cos^2 \psi >$, is defined by

$$< \cos^2 \psi > = \frac{\int_0^1 N(\psi) \cos^2 \psi \, d\cos\psi}{\int_0^1 N(\psi) \, d\cos\psi} \quad (2)$$

where $N(\psi)$ is an orientation distribution function, which is a probability of finding the fiber molecule within the solid angle ψ.

The molecular composite system under consideration may be envisaged as a floating rod model proposed by Kratky (13) in which the fiber crystal molecules are imbedded in the matrix of flexible amorphous matrix. The orientation distribution function of the molecular composite subjected to an external deformation may be expressed by assuming affine deformation (14-16) as follows:

$$N(\psi) = \frac{N_0}{4\pi(\lambda^{-2} \cos^2 \psi + \lambda \sin^2 \psi)^{3/2}} \quad (3)$$

where N_0 is the distribution function in the unreformed state. Substituting eq. 3 into eq. 2, one obtains

$$< \cos^2 \psi > = \frac{\lambda^3}{\lambda^3 - 1}\left[1 - \frac{\tan^{-1}(\lambda^3 - 1)^{1/2}}{(\lambda^3 - 1)^{1/2}}\right] \quad (4)$$

where λ is the draw ratio. From eq. 4, it is interesting to note that the average directional cosine or the orientation factor is no longer a function of N_0 and ψ, i.e., it can be simply expressed as a function of elongation. The floating rod model is based on the affine deformation that assumes the microscopic local deformation is the same as that of the macroscopic bulk deformation and the total volume remains constant during deformation. Although the floating rod model appears to be too simplistic to account for the deformation mechanisms of semicrystalline polymers (15,16), it is worthy to test with the present BPDA-PFMB/PEI systems because of the close resemblance of the rod/coil structure between the floating rod model and the rigid-rod molecular composite.

Results and Discussion

DSC traces of some representative BPDA-PFMB/PEI composite films are depicted in Figure 1. The glass transition (Tg) of the neat PEI is evident at 220 °C. The neat BPDA-PFMB shows no discernible Tg or melting transition in the DSC scans as typical for rigid-rods. The DSC traces of low PI compositions reveal a distinct single glass transition, but the transition becomes broader while the Tg of the mixture shifts progressively to a higher temperature with increasing BPDA-PFMB content. These DSC results are reproducible in the repeated heating and cooling cycles tested (c.f. up to fourth cycle), suggesting that BPDA-PFMB and PEI are intimately mixed at least in their amorphous phase. Figure 2 shows the glass transition temperature as a function of composition for BPDA-PFMB/PEI molecular composites. The non-linear curvature of the Tg vs composition data does not permit unambiguous determination of the glass transition temperature of the neat PI from the DSC experiments. However, the glass transition temperatures of the mixtures are seemingly lower than that of the linear additivity rule, implying lack of specific interaction between the PI and PEI. Such a shift of Tg with composition confirms our previous claim that rigid-rod polyimide derivatives and flexible PEI are completely miscible in their amorphous state. This conclusion certainly contradicts the predicted unfavorable thermodynamics of mixing between rigid-rod and flexible polymer blends (17,18). Hence, the present observation may be taken as an exception rather than the rule.

Now that the molecular dispersion of rigid-rods in the flexible coil matrix has been established, it is interesting to find out the reinforcement effect of the flexible matrix by the rigid-rods. Figure 3 shows the increasing trend of tensile modulus with increasing rod content at very low-rod compositions in the bulk as well as in their fiber form. The x-ray studies of those fibers show no detectable level of crystallinity, and thus the crystal orientation cannot be determined. In the unoriented bulk composites, the incorporation of 5 wt% of rod molecules to the PEI matrix increases the modulus value for about 10 %, but when it is highly oriented in the fiber state, a two-fold improvement of modulus can be observed. There are three possible causes to improve the mechanical properties in general, and modulus in particular; (i) reinforcement of flexible coils by rod molecules, (ii) increase in crystallinity, and (iii) crystal orientation. The development of crystallinity in high-rod composites due to

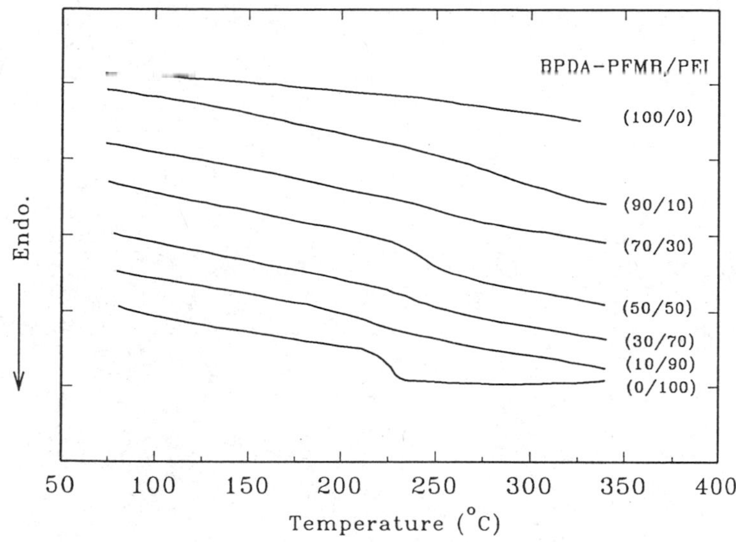

Figure 1. DSC thermograms of the PI/PEI molecular composites, exhibiting the variation of a single glass transition as a function of composition.

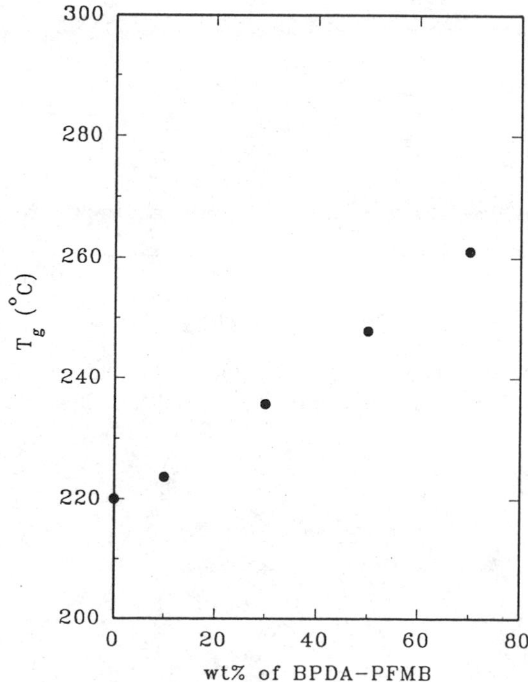

Figure 2. Variation of glass transition temperature with composition for the PI/PE1 molecular composites.

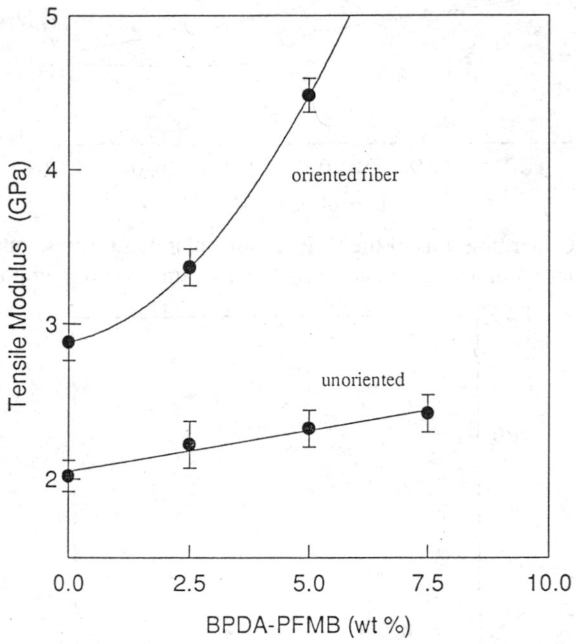

Figure 3. The effect of rod content on tensile modulus of the unoriented PI/PEI molecular composites in comparison with that of their oriented fibers.

increasing rod-alignment (chain orientation) during drawing or spinning casts some doubt about the concept of molecular composite. The molecular composite is often called "nano-composite" to reflect more accurately on the state of phase segregation of rigid-rod crystals from the rod/coil matrix. It should be pointed out that the rod and coil molecules are intimately mixed in their amorphous state. The presence of crystallites should not deter the concept of molecular composites as long as the reinforcement occurs at the length scale smaller or equal to the radius of gyration of the rod and/or coil molecules; the crystallite size is certainly smaller than the radius of gyration of the amorphous chains.

Figure 4 shows a plot of the tensile modulus as a function of draw ratio for the (80/20) BPDA-PFMB/PEI drawn films. The undrawn film shows a relatively low tensile modulus of about 2 ~ 2.5 GPa. When the draw ratio increases, the tensile modulus of the 80/20 film exhibits a gradual increase at low draw ratios, but it rapidly increases at a higher draw ratio ($2.5 < \lambda \leq 4$). An average tensile modulus value of 21 GPa was obtained at the draw ratio $\lambda = 4$. A similar trend was also observed in other compositions containing high rigid-rod contents.

Figure 5 depicts the composition-dependence of the tensile modulus for various BPDA-PFMB/PEI drawn films at the draw ratio $\lambda = 4$ in which the (50/50) BPDA-PFMB/PEI drawn film has a tensile modulus of 7.6 GPa. The tensile modulus of BPDA-PFMB/PEI molecular composites shows improvement with further increase of the BPDA-PFMB content. At 80 PI wt%, the modulus value has increased to 21 GPa.

The average over-all orientation of the composite system was estimated by birefringence measurement. The drawn BPDA-PFMB/PEI films could be identified as positively birefringent. Figure 6 illustrates the variation of birefringence with draw ratio. The birefringence of drawn film shows a drastic increase at a low draw ratio and a smaller increase at a higher draw ratio. As shown in Figure 7, the birefringence measurement shows a similar increasing trend with composition. At the draw ratio $\lambda = 4$, the birefringence (Δn) is 0.08 for the (50/50) composition and it rapidly increases to 0.19 at the (70/30) BPDA-PFMB/PEI composition. Subsequently, the birefringence of a film levels off gradually and only shows a subtle increase at the (80/20) composition ($\Delta n = 0.20$). This birefringence behavior is considerably different from the increasing trend of the modulus versus composition and draw ratio, shown in Figures 4 and 5. In elastomeric networks, stress and birefringence have been shown to have the same origin and therefore behave similarly during deformation (15). This situation is obviously not the case for the present PI/PEI molecular composites. This discrepancy may be attributed to the possible crystallization of the PI rod molecules due to (i) thermodynamic effects (15) and (ii) stress-induced crystallization (19). When the rod content increases beyond the critical concentration, crystallization may occur because of the entropic effect of the anisotropic rod molecules. With increasing draw ratio, the rods tend to align increasingly which eventually leads to crystallization of the BPDA-PFMB molecules due to the orientation-induced crystallization. Perhaps, birefringence of the PI/PEI molecular composites is more sensitive to the stress-induced crystallization at small strains than the corresponding tensile measurement.

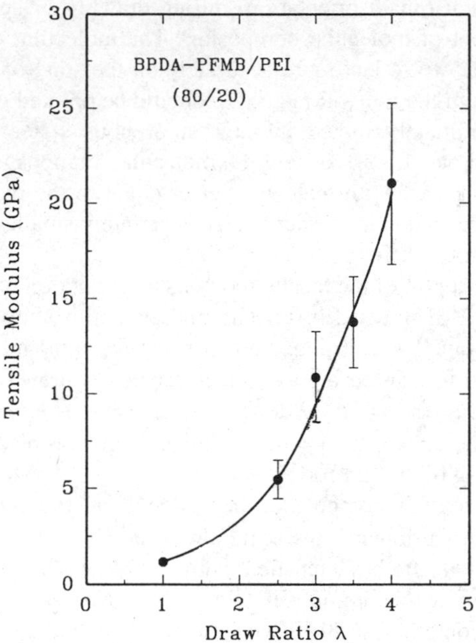

Figure 4. Variation of tensile modulus as a function of draw ratio for the 80/20 PI/PEI molecular composites.

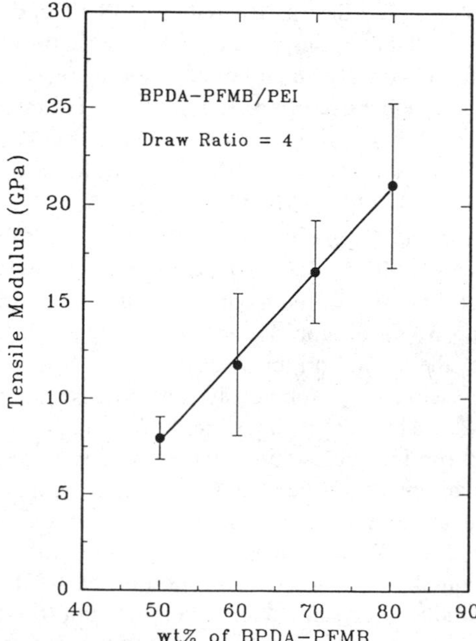

Figure 5. Variation of tensile modulus as a function of composition for the PI/PEI molecular composite at the draw ratio of 4.

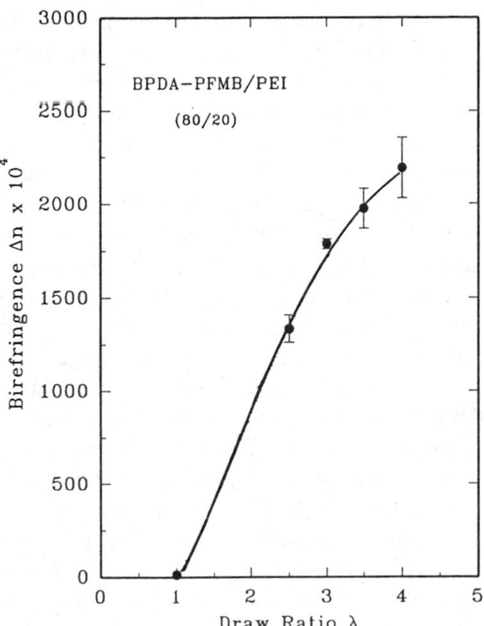

Figure 6. Variation of birefringence as a function of draw ratio of the 80/20 PI/PEI molecular composites.

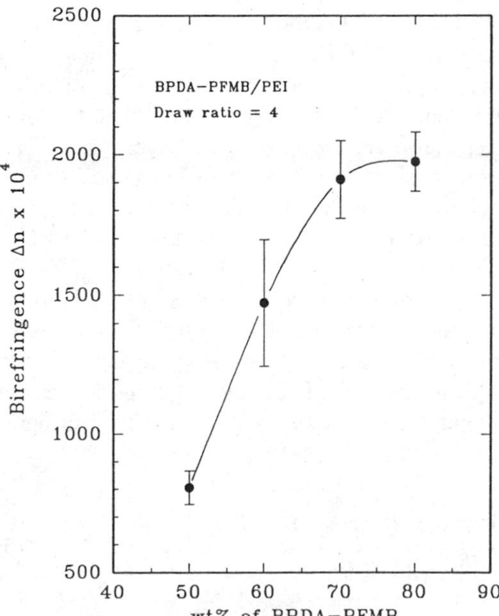

Figure 7. Variation of birefringence as a function of composition of the PI/PEI molecular composites at the draw ratio of 4.

In principle, the total birefringence of the composites depends on various factors such as intrinsic birefringence of the constituents, orientation and composition in their amorphous state. The intrinsic birefringence of the crystalline phase of the rod molecules, the crystal orientation factor, and the crystallinity further contributes to the total birefringence. It is extremely difficult, if not impossible, to evaluate the individual contributions to the total birefringence since their intrinsic birefringence values are not known at this writing and the determination of intrinsic birefringence values of the individual phase is beyond the scope of this work. The crystal orientation may be determined by wide angle x-ray diffraction (WAXD) for composites with high rod contents, but the low rod content materials show diffuse scattering and thus not suitable for determining the crystal orientation.

The development of crystalline orientation during drawing of PI/PEI molecular composites may be qualitatively examined from the wide-angle X-ray diffraction (WAXD) film patterns. Figure 8 illustrates the change of WAXD patterns of the 80/20 BPDA-PFMB composite as a function of draw ratio. For the undrawn sample, a diffuse ring appears because the crystallites are randomly oriented. When the crystallites are oriented progressively along the drawing (reference) direction, the circular rings transform into the corresponding arcs. When the draw ratio increases further, the arcs become shorter, but sharper. The extend of crystal orientation can be determined by analyzing the integrated intensities of the arcs. Figure 9 shows the Bragg's scans of the x-ray intensity along the equatorial (0°) and meridional (90°) directions. The crystal structure of the oriented neat BPDA-PFMB film is known to be monoclinic with a unit cell dimension of crystal axis a = 1.54 nm, b = 0.99 nm, c = 2.02 nm, and angle $\gamma = 56.2^{o}$ (20). On the basis of the above monoclinic structure of BPDA-PFMB, the arc at the 2θ angle of 16.2^{o} in the equatorial direction can be indexed as the (310) crystal plane and the arc at 13.6^{o} in the meridional direction can be ascribed to the (003) plane. It should be pointed out that the direction of the plane normal of the (003) reflection corresponds to the chain axis. The change from a diffuse circular pattern to sharp arcs of the (003) plane with drawing is the manifestation of the increase of orientation of the crystal c-axis.

The crystal orientation factors of each hot drawn BPDA-PFMB/PEI samples may be evaluated quantitatively from the WAXD results. The (003) diffraction plane was chosen to calculate the orientation factor as it corresponds to the chain direction of the crystals. Experimentally the orientation factor of the chosen diffraction plane can be calculated from the integrated intensity (over the Bragg angles), $I(\psi)$, in accordance with the Wilchinski method (21) by assuming the orientation distribution function, $N(\psi)$, to be equal to the experimental intensity distribution ($I(\psi)$). Eq. 2 may be re-written as:

$$\left\langle \cos^2 \psi \right\rangle = \frac{\int_0^1 I(\psi)\cos^2 \psi d(\cos \psi)}{\int_0^1 I(\psi) d(\cos \psi)} \qquad (5)$$

The observed intensity of the (003) plane in the meridional direction was slightly overlapped with that of the broad (310) peak approximately at 16° in the

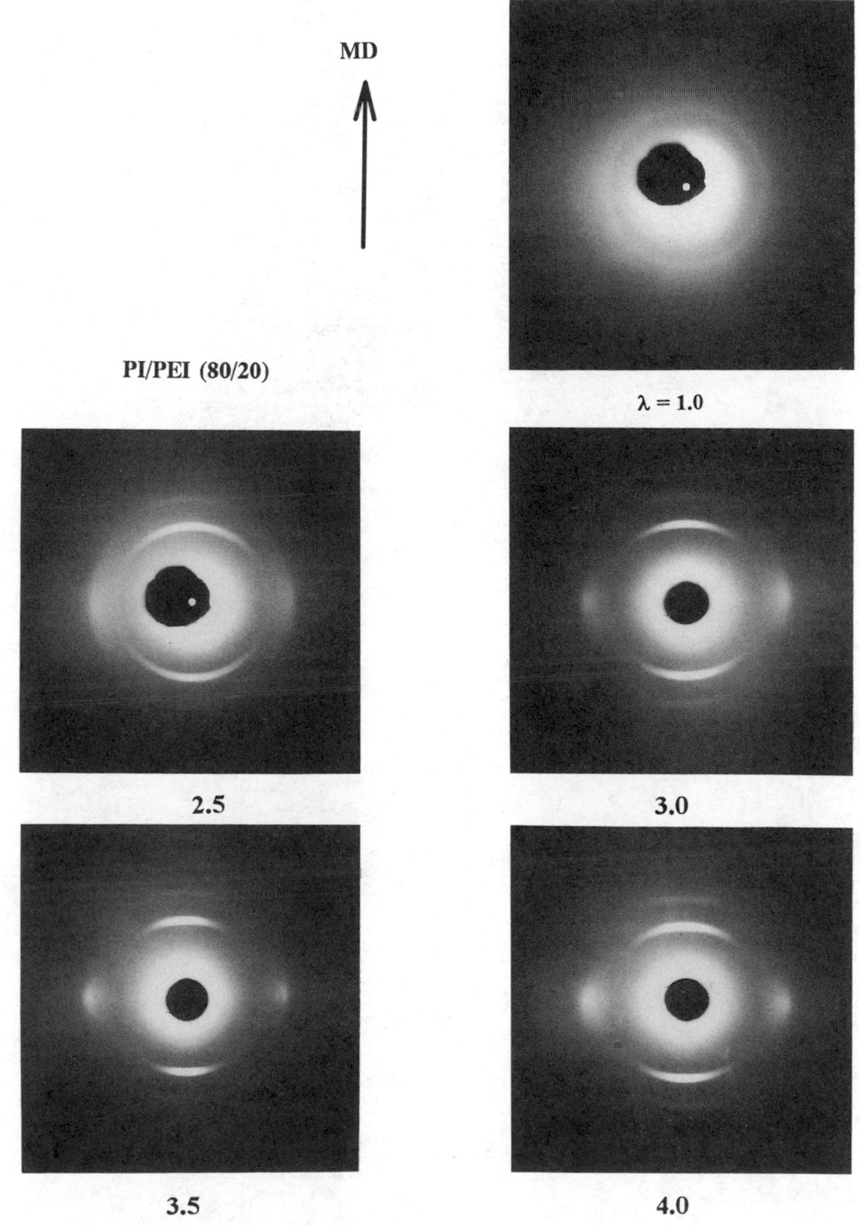

Figure 8. Evolution of WAXD patterns of the 80/20 PI/PEI molecular composite as a function of draw ratio.

(a)

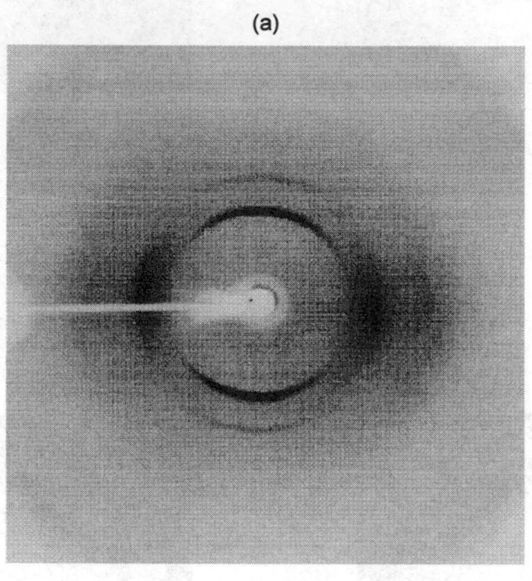

(b)

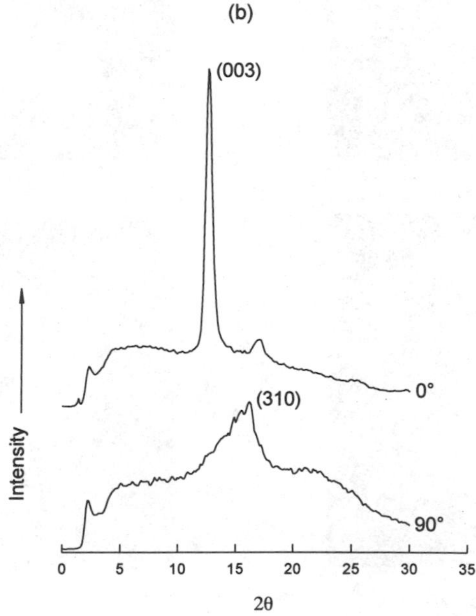

Figure 9. (a) 2-D WAXD patterns of the drawn 80/20 PI/PEI molecular composite $\lambda = 3.5$, and (b) Bragg's 2θ scans in the equatorial (azimuthal angle of 90°) and the meridian direction (0°). The draw direction is vertical.

equatorial direction (Figure 10). The background scattering was taken as the minimum intensity level of the two diffraction planes along the azimuthal direction. Following the background subtraction, the two overlapping peaks were deconvoluted with maxima at 90 and 0°, respectively. The orientation factors of the (003) plane normal for each drawn sample were calculated in accordance with eq. 5.

The crystal orientation factor of the 80/20 BPDA-PFMB/PEI is plotted as a function of draw ratio in Figure 11 in comparison with that calculated according the floating rod model. Although the trend is strikingly similar, the experimental values are slightly higher than the model prediction. This discrepancy may be ascribed to the orientation-induced crystallization which is not accounted for in the model. In the present 80/20 PI/PEI composition, the rigid-rod PI crystalline molecules are the continuum in which the flexible PEI coils are mixed intimately with non-crystalline PI molecules, thus it is just the reverse case of the floating rod model. The affine deformation could have been a reasonably good approximation in describing the crystal orientation behavior of the PI/PEI molecular composites if additional effect such as orientation-induced crystallization did not arise.

Conclusions

The miscibility of the BPDA-PFMB/PEI mixture was established by the systematic variation of the single glass transition with composition. The tensile modulus of molecular composites shows appreciable improvement with the addition of small amount of rod molecules. Such improvement is more dramatic upon increasing the

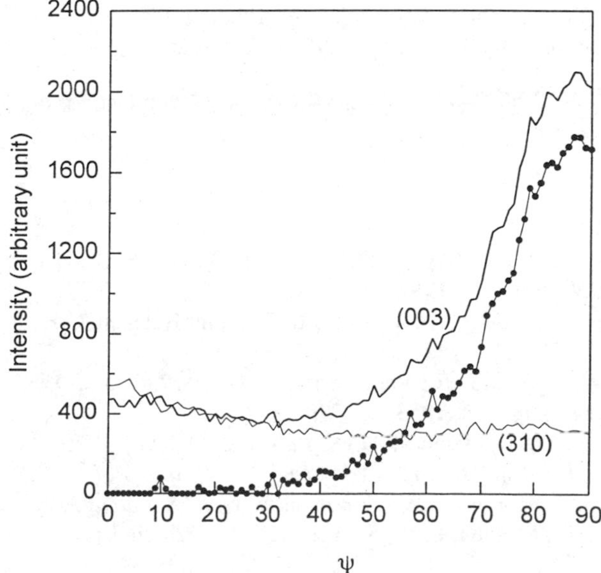

Figure 10. Polar angle (ψ) dependence of the (003) and (310) planes of the 80/20 PI/PEI molecular composites at the draw ratio of 3.5. The line with dots represents the (003) plane after background subtraction.

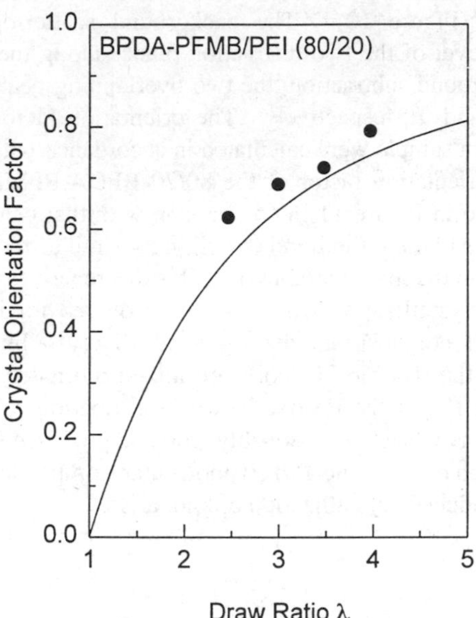

Figure 11. The observed crystal orientation factor (filled circles) of the 80/20 PI/PEI molecular composite in comparison with the prediction of the floating rod model (the solid line).

draw ratio or molecular orientation. The stress-induced crystallization of the rigid rod PI presents additional complications to the floating rod model in describing the orientation behavior of the BPDA-PFMB/PEI molecular composites.

Acknowledgment: This work was supported by the Edison Polymer Innovation Corporation (EPIC).

Literature Cited

1. Fukai, T.; Yang, J. C.; Kyu, T.; Cheng, S. Z. D.; Lee, S. K.; Hsu, S. L. C.; Harris, F.W. *Polymer*, **1992**, 33, 3621.
2. Helminiak, T.E.; Benner, C.L., Arnold, F.E., and Husman, G.E. *U.S. Patent* 4,207,407; **1980**.
3. Hwang, W.F.; Wiff, D.R.; Verschoore, C. *Polym. Eng. Sci.*, **1983**, 23, 784; ibid., *Polym. Eng. Sci.*, **1983**, 23, 789
4. Takayanagi, M.; Kajiyama, T. *U.K. Patent* 2,008,598, **1978**
5. Hiraguchi, K.; Kajiyama, T.; Takayanagi, M. *J. Appl. Polym. Sci.* **1979**, 27,424
6. Kwolek, S.L.; Morgan, P.W.; Sorenson, W.R. *U.S. Patent* 3,063,966, **1962**; Kwolek, S.L.; Morgan, P.W.; Schaetgen, J.R.; Gulrich, L.W. *Macromolecules*, **1977**, 10, 1381.
7. Krause, S.J.; Haddock, T.; Price, G.E., Lenhert, P.G.; O'Brien, J.F.; Helminiak, T.E.; Adams, W.W. *J. Polym. Sci. B, Polym. Phys.* **1986**, 24, 1991.

8. Guadiana, R.A., Adams; T., Guarrere, D.; Roy, S.K.; Stein, R.S.; Sethumadhavan, M. Macro. Report, **1994**, A31, 747.
9. Wiff , D. R.; Lenke G. M.; Flemming III, P.D. *J. Polym. Sci., B, Polym. Phys.*, **1994**, 32,2555; Wiff, D.R.; Lenke, G.M., *Mat. Res. Soc. Symp. Proc.* **1993**, 305, 161.
10. Nielsen, L.E., "Mechanical Properties of Polymers and Composites," Marcel Dekker, New York, **1975**.
11. Van Krevlen D. W.; Hoftyzer, P. J. "*Properties of Polymers*", Elsevier, New York., **1976**, Ch. 14
12. Hermans, P.H., "*Physics and Chemistry of Cellulose Fibers*," Elsevier, New York, **1949**, Ch. 5.
13. Kratky, O.; Kolloid Z., **1933**, 64, 213; ibid., **1935**, 70, 14.; ibid., **1938**, 84, 149.
14. Stein, R. S. *J. Polym. Sci.*, **1958**, 31, 327; ibid., *J. Polym. Sci.*, **1959**, 34, 709.
15. Sasaguri, K.; Hoshino, S.; Stein, R.S. *J. Appl. Phys.*, **1964**, 35, 47.
16. Oda, T.; Nomura, S.; Kawai, H. *J. Polym. Sci. A*, **1965**, 3, 1993. Nomura, S.; Kawai, H. *J. Polym. Sci., A-2*, **1966**, 4,797.
17. Abe, A.; Flory, P.J. *Macromolecules*, **1978**, 11, 1122; ibid. *Macromolecules*, **1978**, 11, 1141.
18. Chuah, H.H.; Kyu, T.; Helminiak, T.E. *Polymer*, **1987**, 28, 2129; ibid., *Polymer*, **1989**, 37, 201.
19. Miller, R.L. Ed. "*Flow Induced Crystallization in Polymer Systems*," Midland Macromolecular Monographs, Gordon and Breach, New York, **1979**, vol 6.
20. Cheng, S. Z. D.; Arnold, F. A.; Zang, A. Q.; Hsu, S. L. C.; Harris, F. W. *Macromolecules*, **1991**, 24, 5856.
21. Wilchinski, Z.W.; *J. Appl. Phys.* **1959**, 30, 792; ibid. *J. Appl. Phys.*, **1960**, 31,1969.

Chapter 4

Molecular Composites via Ionic Interactions and Their Deformation–Fracture Properties

G. Parker, W. Chen, L. Tsou, and M. Hara[1]

Department of Mechanics and Materials Science, Rutgers University, Piscataway, NJ 08855–0909

Molecular composites have been made from three types of ionic PPTA's and various polar polymers (PVP, S-AN, PVC, PEO), in which a good dispersion of rod molecules is achieved via ionic (ion-dipole) interactions. Optical/thermal testing and morphological observations by electron microscopes have indicated good dispersion of the rigid-rod PPTA molecules. Molecular composites based on amorphous matrix polymers are all transparent and show no phase separation upon heating; therefore, they are melt-processable. The deformation mode of the matrix polymer is modified significantly with the addition of rod molecules: e.g., while crazing is the only deformation mechanism of PVP and S-AN (30% AN), the addition of ionic PPTA molecules into these amorphous polymers induces shear deformation. This is due to interactions between rod and coil molecules at the molecular (or microscopic) level, unlike the situation in conventional fiber composites, in which interactions between fiber and matrix polymer occur only at the interface during load transfer. The observed deformation modes suggest that fracture properties of these molecular composites should be enhanced. Mechanical tests made on three different composite systems do show enhanced ductility (toughness) in addition to increased stiffness/strength for the molecular composites, having either an amorphous or semi-crystalline polymer matrix.

Conventional fibers for advanced composites, such as carbon fiber and Kevlar fiber, are aggregates of fibrils and microfibrils, and therefore they contain many inherent defects that can initiate cracks and lead to premature failure of the composite. The idea of "molecular (level) composites" is based on the fact that an individual rigid-rod molecule, such as a poly(p-phenylene terephthalamide) (PPTA) molecule, has no defect; therefore, the theoretical strength due to covalent bonds in the

[1]Corresponding author

0097–6156/96/0632–0054$15.00/0

backbone chain may be used for reinforcement of matrix materials (*1-6*). Molecular composites are also envisaged to be polymer blends in which rigid-rod molecules are dispersed in a matrix of a flexible coil polymer such that the rods act as reinforcement.

Although the idea of molecular composites is promising, most molecular composites developed to date have a major drawback: i.e., these molecular composites are not in a thermodynamically miscible state, since entropy of mixing is very small as demonstrated for polymer blends (*7,8*), and since rod molecules have a strong tendency to segregate (*9*). One way to circumvent this problem is a rapid coagulation from a ternary solution (rod polymer/flexible polymer/solvent) to "freeze" the miscible rod/coil state, thereby overcoming the unfavorable thermodynamic driving force to phase separation. However, for many molecular composites made by this method, the homogeneous phase morphology is only temporary; for example, phase separation is found to occur after heating (*3*). Another way to overcome the problem is using block or graft copolymers that connect rigid-rod and flexible-coil components by covalent bonds; the covalent bonds can force rod and coil components to be in close proximity. Although mechanical properties of these molecular composites are improved compared with simple blends of component polymers, presumably due to improved dispersion of rod molecules (*10-12*), this usually leads to microphase separation or microfibril formation of rigid rods, again due to inherent thermodynamic immiscibility of the component polymers, as is well known for block/graft copolymers (*13,14*). After reviewing the work on molecular composites, Hwang and Helminiak (*15*) pointed out that a phase-separated rigid-rod/thermoplastic blend is no better in its physical/chemical properties than a corresponding fiber composite and that *enhanced and desirable properties can only be realized in a true molecular composite*; therefore, the key to the success of rigid-rod molecular composite technology lies in having good molecular dispersity of the rigid rods in a matrix material.

One promising approach to producing a true molecular composite is to make rod and coil components thermodynamically miscible by introducing attractive interactions, such as hydrogen bonds (*16-18*), between them. This method has proven useful for enhancing miscibility in flexible-flexible blends. Even more useful (stronger) interactions may be ionic interactions, such as ion-ion and ion-dipole interactions: various studies on ionomer blends have demonstrated that ionic interactions can enhance the miscibility of otherwise immiscible polymer pairs (*19*). Polymers studied include polystyrene, poly(ethyl acrylate), poly(ethyleneimine), nylon, and poly(ethylene oxide) (*20-22*).

Recently, several studies have been reported on molecular composite formation using ionic interactions (*23-27*). These studies are beginning to indicate the potential usefulness of ionic bonds in creating homogeneous, melt-processable molecular composites. Although these results show some miscibility enhancement, as investigated by several widely used techniques for flexible-flexible blends, little information has been presented on mechanical properties and their relationships with the degree of miscibility (dispersity). Because of their potential as high-performance polymers, much work needs to be done to fully exploit the effective use of ionic bonds in molecular composites. We have developed various molecular composites via *ion-dipole interactions* based on three types of ionic PPTA's and four polar polymers, such as poly(4-vinylpyridine) (PVP) and poly(ethylene oxide)(PEO)

(*24,25*). In this article, we focus on general conclusions drawn to date from studies on various ionic PPTA/polar polymer molecular composites by showing representative data. Many of them are PVP-matrix composites, which have been most widely studied in this laboratory. Detail accounts will be reported in separate articles (*28-30*).

Experimental

As ionic rigid-rod molecules, three types of ionic PPTA's have been synthesized: the first one has ionic groups right on the backbone chain (i.e., PPTA anion), the second one has ionic groups on the side chains (e.g., potassium propanesulfonate) (i.e., PPTA-PS) (*31*), and the third one has ionic groups attached to phenylene rings (i.e., S-PPTA) (*32,33*) (see Figure 1). PPTA-PS was made from PPTA anion by reacting with 1,3-propane sultone (*34*). S-PPTA was prepared by solution step polymerization of p-phenylenediamine and terephthaloyl chloride, where proper amounts of 2,5-diaminobenzenesulfonic acid was added to control acid (ion) content. Details on synthesis conditions are described elsewhere (*28-30*). Since too many side groups may severely disturb the rod-like conformation of PPTA (*35*), we have kept the number of propanesulfonate groups in PPTA-PS to the minimum needed for solubility. Usually, this is ca. 30 mol%. As a flexible coil (polar) polymer, PVP, PEO, poly(styrene-co-acrylonitrile)(S-AN), and poly(vinyl chloride)(PVC) were used (see Figure 2). An ionic PPTA and a polar polymer were dissolved in DMSO separately, followed by mixing under stirring. The transparent yellow (or orange) solution was then precipitated into nonsolvent, ether. The polymer was then dried under vacuum at high temperature, followed by compression molding for making specimens. It should be added that, although the PPTA anion is known to be sensitive to moisture (H_2O) and converts back to PPTA (*34*), the PPTA anion/PVP composite, once formed, seems to be stable, probably because the ionic groups are surrounded by PVP molecules.

Differential scanning calorimetry (DSC) and thermogravimetric analysis (TGA) were conducted with a 910 Differential Scanning Calorimeter and a 951 Thermogravimetric Analyzer, both of which were controlled by a TA Instruments 2100 Thermal Analyst (DuPont Instruments). Dynamic mechanical thermal analysis (DMTA) was conducted with a Piezotron (Toyoseiki). A typical sample size for DMTA measurements was 0.25 mm x 3.0 mm x 12 mm. The heating rate was 4 °C/min and the frequency was 1 Hz. A Leitz microscope coupled with a hot stage (Mettler F52) was used for polarizing microscopy. Tensile tests of small specimens were carried out on a Minimat Materials Tester (Polymer Laboratories). Films for deformation mode testing were made by casting on microscope slides; after cutting the film into 2 mm x 2 mm sections, they were floated off the glass slides onto the surface of distilled water and picked up on a ductile copper grid for TEM observation. After vacuum drying either at room or elevated temperatures, the microstructure of these strained thin films was observed by a transmission electron microscope (JEM-100 CX II) operating at 100 kV. Details concerning TEM work are described elsewhere (*36*).

PPTA-PS

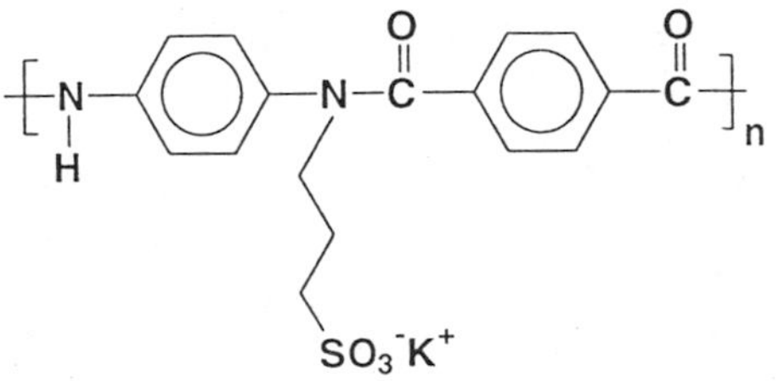

PPTA anion

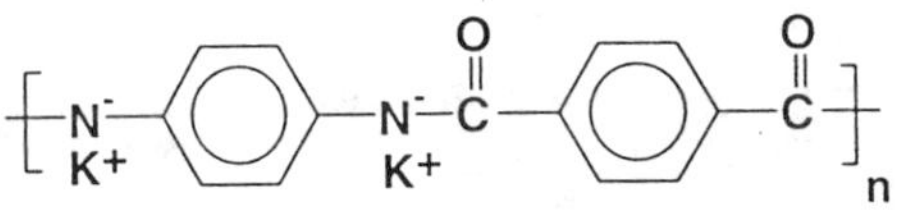

S-PPTA

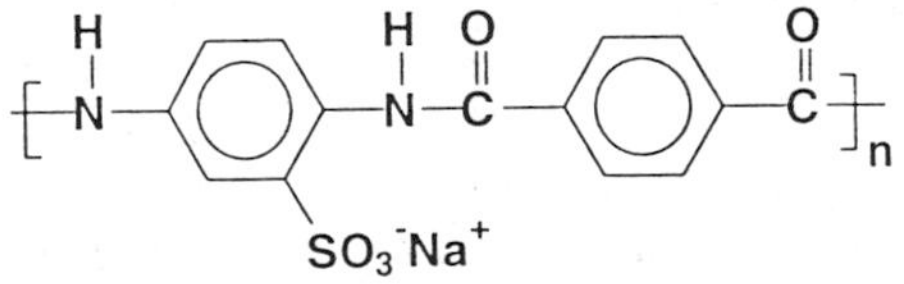

Figure 1. Chemical structures of ionic PPTA's.

Results and Discussion

Ionic PPTA's. Generally, ionic PPTA's show liquid crystalline behavior in solution, as we observe birefringence in concentrated solutions. It is also reported that PPTA anion solutions show birefringence at ca. 5 wt.% for Na salt and ca. 12 wt.% for K salt (*34*). Moreover, as Figure 3 indicates, a polarized microscope picture of the PPTA anion shows anisotropy in the solid state, indicating retention of the rod-like structure of PPTA anion molecules (*34*). The PPTA-PS and S-PPTA samples also show birefringence, although in different scales than for the PPTA anion, presumably reflecting the different rigidity of the respective polymer chains.

Figure 4 shows TGA data obtained on ionic PPTA's and unmodified PPTA. Although the thermal stability of the ionic PPTA's is reduced to some extent, they show that no weight loss occurs at least up to 450 °C. S-PPTA shows a slightly higher stability than PPTA-PS, presumably because the former is more rigid than the latter. It is also observed that the salt-form PPTA's are significantly more stable than the corresponding acid-form PPTA's; e.g., S-PPTA (acid) shows a significant weight loss at ca. 200 °C, but S-PPTA (Na salt) only shows a significant weight loss at temperatures above 500 °C.

Molecular Composites. Optical clarity, polarized microscopy, TEM, and T_g measurements (both by DSC and DMTA) of the ionic PPTA/PVP composites all indicate the formation of a homogeneous molecular composite that shows no phase separation upon heating, suggesting melt-processability of these materials. For example, Figure 5 shows the optical clarity of samples of ca. 1 mm thickness, which were made by compression molding. While the unmodified-PPTA/PVP sample is opaque already at 2 wt.% PPTA level, PPTA anion/PVP samples are transparent at least up to 5 wt.% of ionic PPTA, with light orange color. PPTA-PS/PVP and S-PPTA/PVP systems show transparency at least up to a 15 wt.% ionic PPTA level. Polarized microscope pictures of ionic PPTA/PVP composites show no second phase with anisotropy, while the unmodified-PPTA /PVP system indicates the existence of a phase-separated PPTA phase. These optical results indicate miscibility on a scale down to 0.1 μm (*7*). The unmodified PPTA/PVP sample was made as follows: first a solution mixture of PPTA anion and PVP was prepared and this was followed by precipitation into acidic ether, which converts PPTA anion to PPTA (*34*). The samples were then compression molded after drying at high temperature.

In addition, T_g criteria determined by DSC and DMTA measurements show miscibility on a scale down to 50-100 Å (*7*). For example, Figure 6 shows storage modulus vs. temperature curves for PVP and for the PPTA anion/PVP (2/98) molecular composite. For both, there is a single T_g, which is ca. 15 °C higher for the composite than for PVP. DSC data also show only a single T_g up to 400 °C, which increases with increasing the ionic PPTA content (see Figure 7).

Finally, TEM micrographs for the ionic PPTA/PVP composites show no phase-separated morphology, as is indicated for a PPTA-PS/PVP blend with high ion content (*24*). It should be stressed that heat treatment has not developed any sign of phase separation for all ionic PPTA/PVP composites, which is usually noted for many molecular composites (*3*).

As mentioned in the Introduction, strong attractive interactions are needed to render miscible otherwise immiscible polymer mixtures, especially when one of the

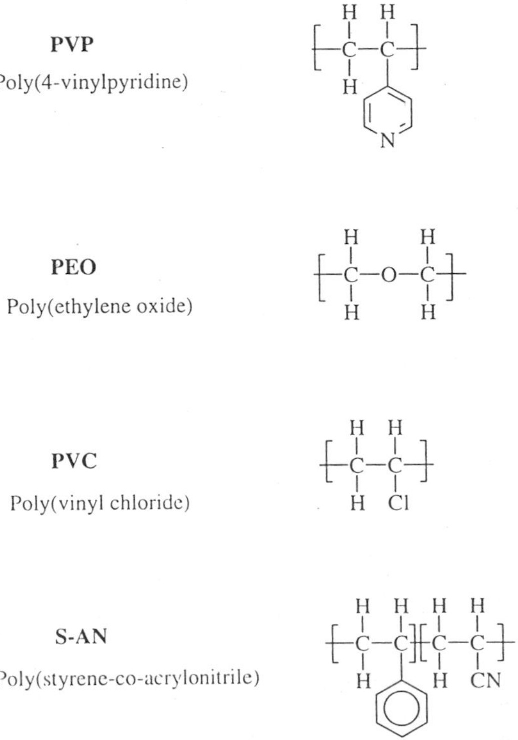

Figure 2. Chemical structures of polar matrix polymers.

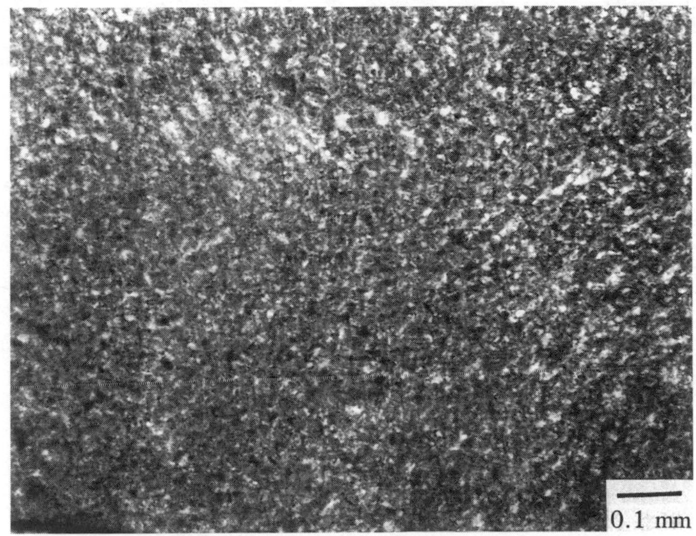

Figure 3. Polarized optical micrograph of PPTA anion film.

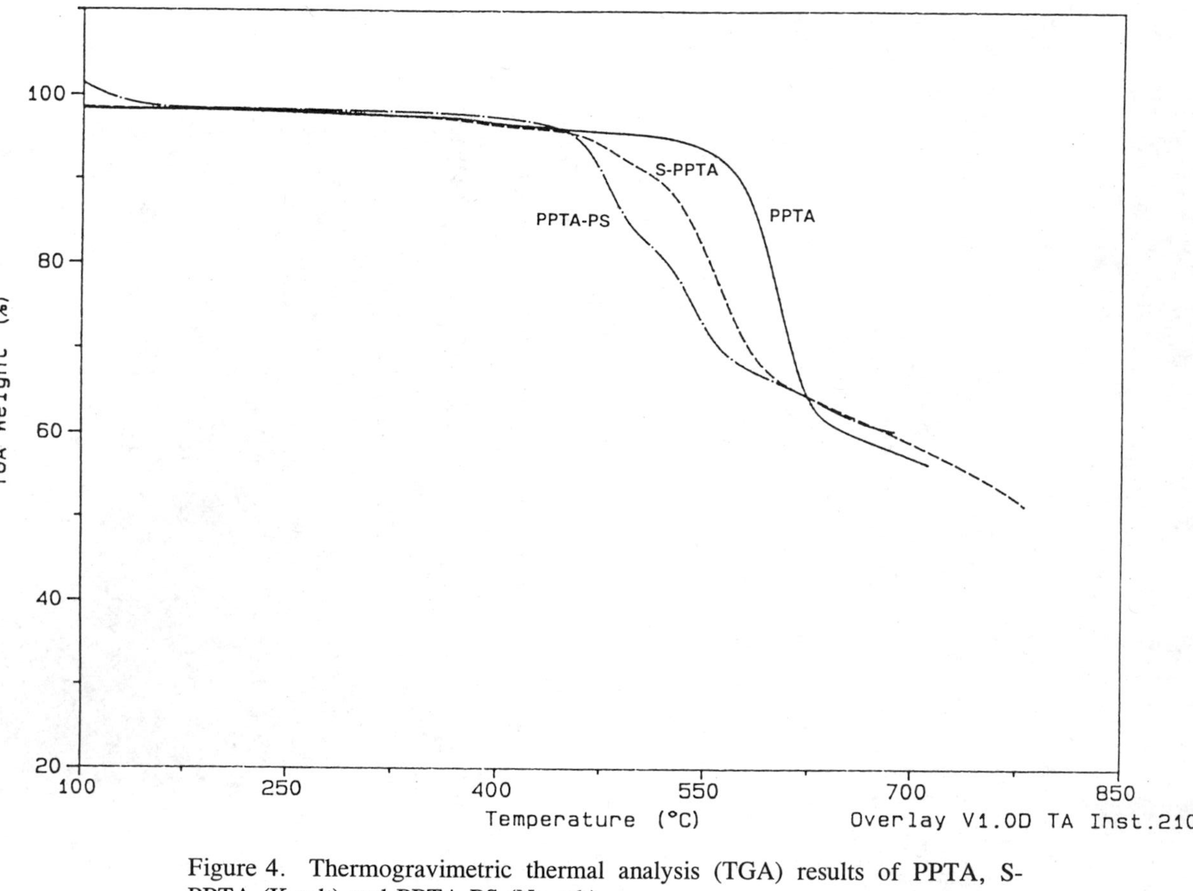

Figure 4. Thermogravimetric thermal analysis (TGA) results of PPTA, S-PPTA (K salt),and PPTA-PS (Na salt).

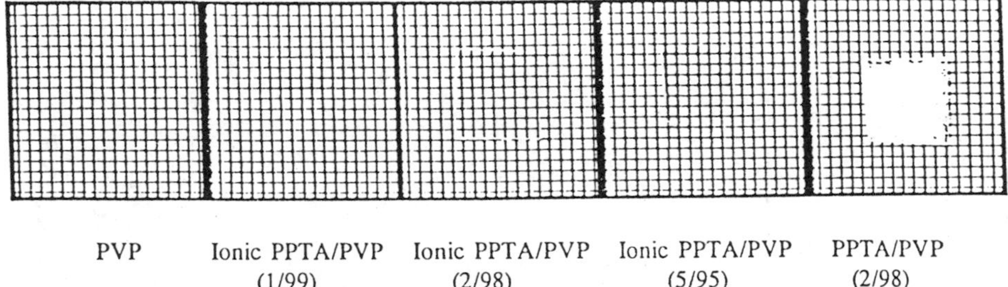

Figure 5. Optical clarity of PVP and PPTA anion/PVP composites with various PPTA anion content.

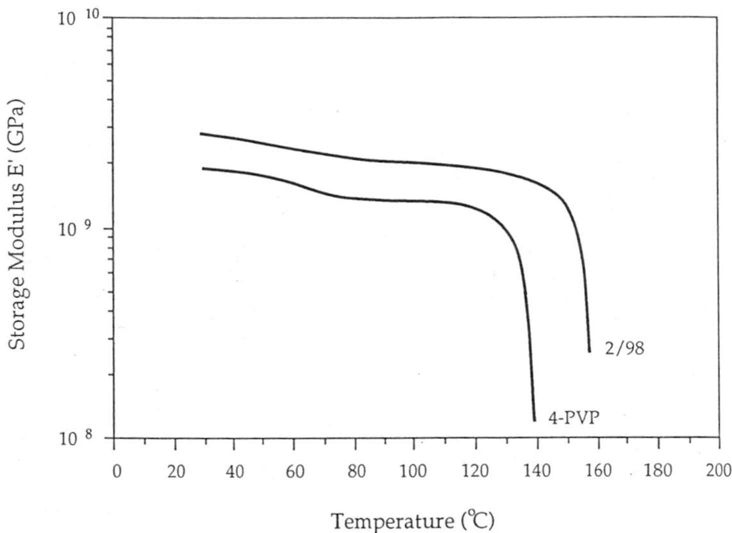

Figure 6. DMTA data for the PPTA anion/PVP (2/98) composite and for PVP.

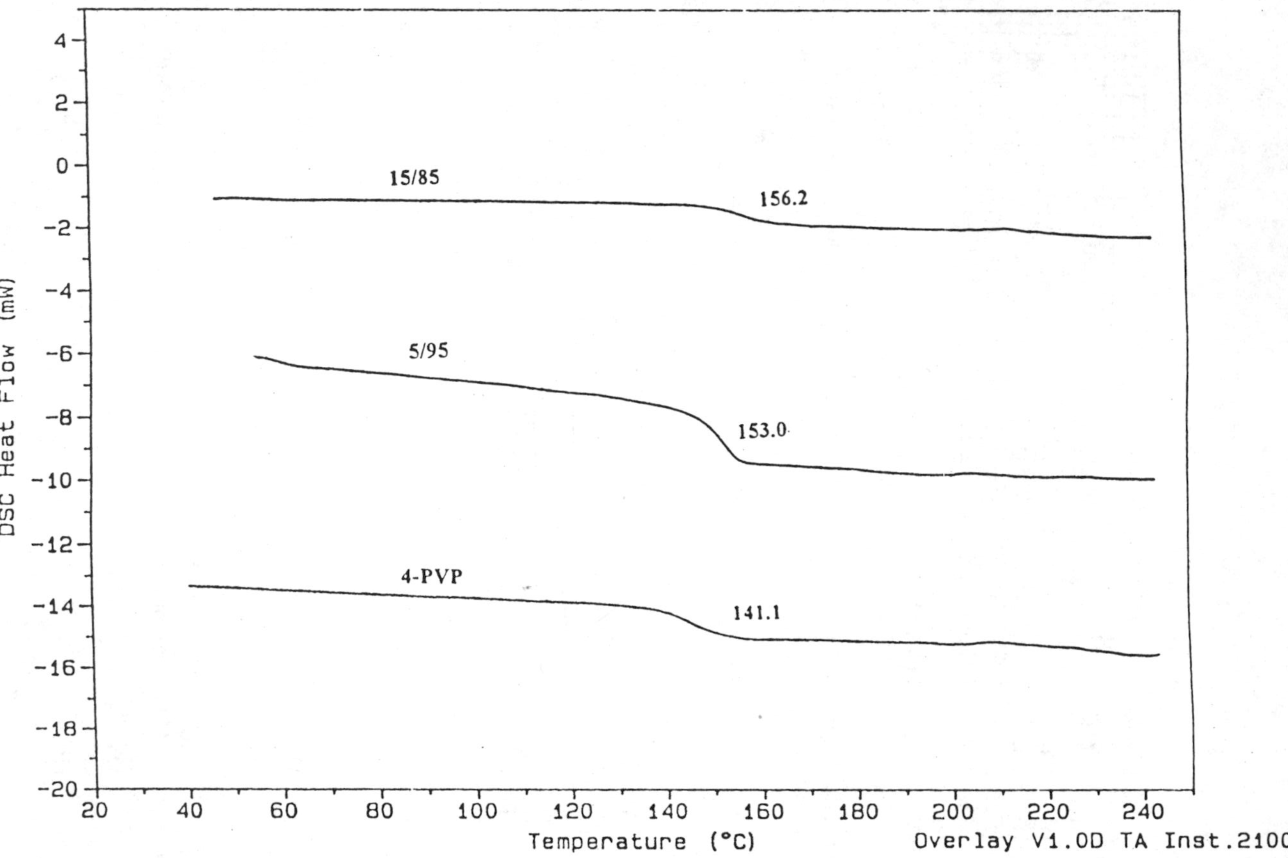

Figure 7. DSC data for the PPTA-PS/PVP composites and for PVP.

components has a rigid-rod conformation. In our case, it is expected that ion-dipole interactions are responsible for the miscibility enhancement, since the functional groups involved are ion pairs in ionic PPTA (e.g., sodium sulfonate in S-PPTA) and ionic dipoles in PVP molecules. Ion-ion interactions arising from proton transfer to N atoms are inconceivable for this system. Also, hydrogen bonds are not strong enough to achieve dispersity as indicated by the phase-separated morphology of the unmodified-PPTA/PVP system, where hydrogen bonds can be formed between amide groups of PPTA and pyridine groups of PVP. To make ion-dipole interactions effective as miscibility enhancers, the polar polymer should have many dipoles (*19*), as is the case for PVP that has an ionic dipole at every repeat unit. Actually, when the number of dipoles of the matrix polymer (S-VP) is reduced to half, a sign of immiscibility appears (*24*).

In addition to these miscibility studies, mechanical studies have been conducted. First, the deformation pattern of a strained thin film of an ionic PPTA/PVP composite is shown in Figure 8a. The TEM micrograph shows that shear deformation zones, with no fibrillation, have been developed. In contrast, fibrillated crazes are the only deformation microstructure observed in PVP (see Figure 8b). A similar change in deformation mode is also seen for PPTA-PS/S-AN systems; S-AN (30% of AN) deforms only by crazing, while the molecular composites develop shear deformation in addition to crazing. The development of shear deformation, at the expense of crazing, is an effective way to enhance fracture properties of polymers, as has been demonstrated for various polymers including miscible flexible-flexible blends (*37*). This is also apparent when we compare the deformation at room temperature of relatively ductile polycarbonate, which deforms by shear deformation, with brittle polystyrene, which deforms by crazing.

Kramer and coworkers have shown that in amorphous polymers, including homopolymers, random copolymers, and miscible blends, a determining factor governing the deformation mode is *strand density* (*37*). The strand density is defined as the number of strands (chain segments bounded by crosslinks or entanglement points) per unit volume; here, polymer is considered to consist of a network of molecular strands (*38*). With increasing strand density of the polymer, the deformation mode changes from crazing only, to crazing plus shear deformation, to shear deformation only. This is because an increased number of strands raise the crazing stress, but have little effect on yield stress, thereby suppressing crazing in favor of shear yielding. It is also known that (low-degrees of) covalent cross-links (*38*) and ionic cross-links (although less effective than covalent cross-links) (*39*) can increase strand density and modify the deformation mode. In view of these results, our current results can be explained in terms of an increased strand density of the molecular composites. Two causes may be considered: First, as indicated by studies on the deformation modes of ionomers that have ionic cross-links (*39*), ion-dipole interactions may work like ionic cross-links and increase the strand density. Second, rod molecules (PPTA) are more effective to create physical entanglements with coil polymers because of their more extended conformation in the solid. Nevertheless, it is clear that ionic rod molecules, dispersed at the molecular (or microscopic) level, interact directly with a matrix coil polymer, PVP for example, and modify the deformation mode. This is in contrast to the situation of conventional macro-fiber composites, in which fibers can only interact with the matrix polymer at the interface

(a)

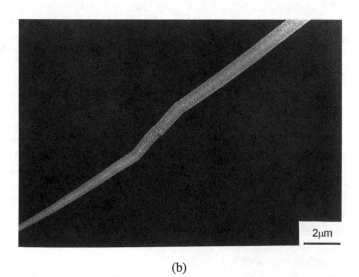

(b)

Figure 8. TEM micrograph for (a) the PPTA-PS/PVP (15/85) composite, showing shear deformation zones and (b) PVP, showing a craze.

during load transfer, and thereby have little influence on the deformation mode of the matrix polymer.

The modulus change occurring in these composites is found to be significant: for example, the addition of only 2 wt.% of PPTA anion to PVP enhances the modulus by 50 %, as seen in Figure 6. Moreover, initial mechanical data indicate a significant increase in tensile properties. Figure 9 shows stress-strain curves for PPTA-PS/PVP (5/95) and PVP. It is seen that the molecular composite is not only stiffer and stronger but also show an increase in ductility (hence also in toughness or energy to fracture). Although more systematic work on other composites is needed to substantiate these results, it seems that the appearance of the yield point and subsequent enhanced ductility are closely related to the observed changes in deformation mode of the matrix polymer, as discussed above. Somewhat similar results have been obtained for a molecular composite in which the matrix polymer is semi-crystalline. For example, Figure 10 shows stress-strain curves for PPTA anion/PEO composites and for PEO; again, both stiffness/strength and ductility/toughness are higher for the molecular composites, and these values increase with an increase in the ionic PPTA content. Similar results have also been obtained for PVC matrix composites.

Conclusions

Molecular composites have been made from ionic PPTA's and polar polymers, in which a good dispersion of rod molecules is achieved via ionic (ion-dipole) interactions. Optical clarity, polarized microscopy, T_g measurements (both by DSC and DMTA), and TEM observation of ionic PPTA/polar polymer composites have indicated good dispersion of the rigid-rod PPTA molecules. Molecular composites based on amorphous matrix polymers are all transparent and show no phase separation upon heating; therefore, they are melt-processable. The deformation mode of the matrix polymer is modified significantly with the addition of rod molecules: e.g., an addition of ionic PPTA molecules into amorphous polymers induces shear deformation. This is due to the interactions between rod and coil molecules at the molecular (or microscopic) level. In conventional fiber composites, interactions between fiber and matrix polymer occur only at the interface during load transfer. These results on deformation modes suggest that the fracture properties of these molecular composites should be enhanced. Actually, mechanical testing shows enhanced ductility (toughness) in addition to increased stiffness/strength for the molecular composites, based on either an amorphous or semi-crystalline polymer matrix. Work is under way to elucidate the degree of molecular dispersity of ionic PPTA by use of SAXS/SANS techniques and to correlate these findings with results of studies of mechanical properties.

Acknowledgment

Acknowledgment is made to ARO, ACS-PRF, and Hoechst Celanese for support of this research. We also thank Dr. Sauer for useful discussions and Drs. Newman and Scheinbeim for making their Piezotron available to us.

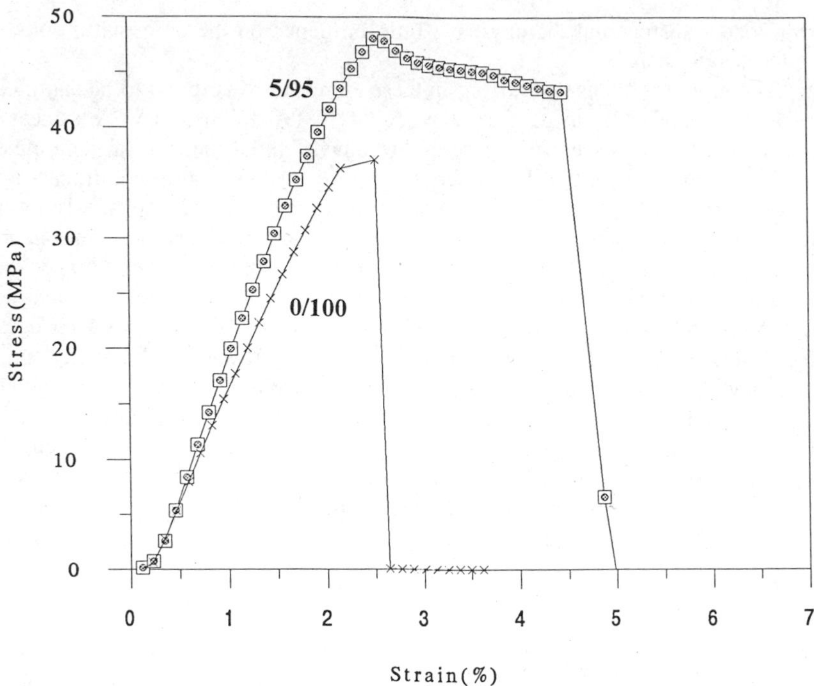

Figure 9. Stress-strain curves for the PPTA-PS/PVP (5/95) composite (⊠) and for PVP (×).

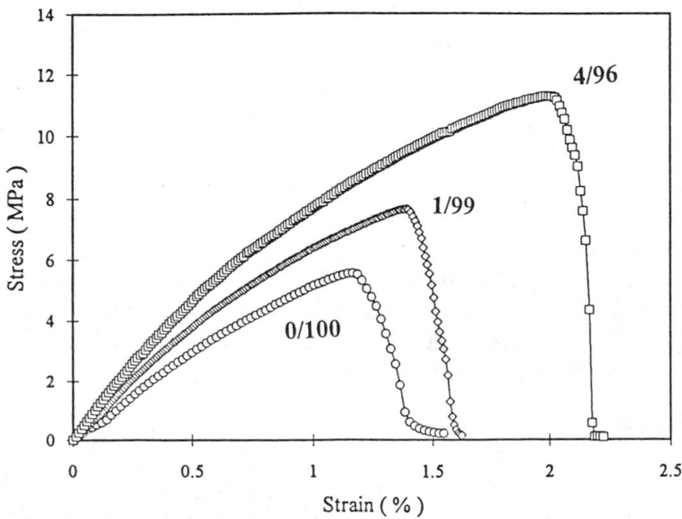

Figure 10. Stress-strain curves for PPTA anion/PEO composites and for PEO. ○: PEO, ◇: 1/99, □: 4/96.

Literature Cited

1. *The Materials Science and Engineering of Rigid-Rod Polymers;* Adams, W.W.; Eby, R.K.; McLemore, D.E., Eds.; Mat. Res. Soc. Symp. Proc. Vol. 134; Materials Research Society: Pittsburgh, 1989.
2. *Polymer Based Molecular Composites;* Schaefer, S.W.; Mark, J.E., Eds.; Mat. Res. Soc. Symp. Proc. Vol. 171; Materials Research Society, Pittsburgh, 1990.
3. Hwang, W.F.; Wiff, D.R.; Benner C.L.; Helminiak, T.E. *J. Macromol. Sci.-Phys.*, **1983**, *B22*, 231.
4. Takayanagi, M. *Pure Appl. Chem.*, **1983**, *55*, 819.
5. Hwang, W.F.; Wiff, D.R.; Verschoore, C.; Price, G.E.; Helminiak, T.E.; Adams, W.W. *Polym. Eng. Sci.*, **1983,** *23*, 784.
6. Pawlikowski, G.T.; Dutta, D.; Weiss, R.A. *Annu. Rev. Mater. Sci.*, **1991,** *21*, 159.
7. Paul D.R.; Newman, S. *Polymer Blends*; Academic Press: New York, 1978.
8. Olabisi, O.; Robeson, L.M.; Shaw, M.T. *Polymer-Polymer Miscibility*; Academic Press: New York, 1979.
9. Flory, P.J. *Macromolecules*, **1978**, *11*, 1138.
10. Takayanagi, M.; Goto, K. *J. Appl. Polym. Sci.*, **1984,** *29*, 2547.
11. Moore, D.R.; Mathias, L.J. *J. Appl. Polym. Sci.*, **1986,** *32*, 6299.
12. Bai, S.J.; Dotrong, M.; Evers, R.C. *J. Polym. Sci., Polym. Phys.*, **1992,** *30*, 1515.
13. *Recent Advances in Polymer Blends, Grafts and Blocks;* Sperling, L.H., Ed.; Plenum Press: New York, 1974.
14. *Developments in Block Copolymers;* Goodman, I., Ed.; Applied Science: London, 1982.
15. Hwang, W; Helminiak, T.E. p.507 of ref. 1.
16. Painter, P.C.; Tang, W.; Graf, J.F.; Thomson, B.; Coleman, M.M. Macromolecules ***1991,*** *24*, 3929.
17. Dai, Y.K.; Chu, E.Y.; Xu, Z.S.; Pearce, E.M.; Okamoto, Y.; Kwei, T.K. *J. Polym. Sci., Polym. Chem.*, **1994,** *32*, 397.
18. Chen, T.I.; Kyu, T. *Polym. Commun.* ***1990,*** *31*, 111.
19. Smith, P.; Hara, M.; Eisenberg, A. In *Current Topics in Polymer Chemistry*; Ottenbrite, R.H.; Utracki, L.A.; Inoue, S., Eds.; Hanser Publ.: Munich, 1987, pp 256.
20. Lu, X.; Weiss, R.A. Macromolecules, **1991,** *24*, 4381.
21. Molnar, A.; Eisenberg, A. *Macromolecules*, **1992,** *25*, 5774.
22. Douglas, E.P.; Sakurai, K.; MacKnight, W.J. *Macromolecules*, **1991,** *24*, 6776.
23. Tan, L.; Arnold, F.E.; Chuah, H.H. *Polymer*, **1991**, *32*, 1376.
24. Hara, M.; Parker, G. *Polymer*, ***1992,*** *33*, 4650.
25. Parker, G.; Chen, W.; Hara, M. *Polym. Mater. Sci. Eng.*, **1995,** *72*, 544.
26. Weiss, R.A.; Shao, L.; Lundberg, R.D. *Macromolecules*, ***1992,*** *25*, 6370.
27. Eisenbach, C.D.; Hofmann, J.; MacKnight, W.J. *Macromolecules*, ***1994,*** *27*, 3162.
28. Parker, G.; Hara, M. *Polymer*, submitted,
29. Chen, W.; Hara, M. *Macromolecules*, to be submitted.

30. Tsou, L.; Hara, M. *Macromolecules*, to be submitted.
31. Gieselman, M.B.; Reynolds, J.R. *Macromolecules*, **1990,** *23*, 3118.
32. Salamone, J.C.; Li, C.K.; Clough, S.B.; Bennett, S.L.; Watterson, A.C. *Polym. Prepr.*, **1988**, *29(1)*, 273.
33. Vandenberg,E.J.; Diveley,W.R.; Filar, L.J.; Patel, S.R.; Barth, *H.G. Polym. Mater. Sci. Eng.*, **1987,** *57*, 139.
34. Burch, R.R.; Sweeny, W.; Schmidt, H.; Kim, Y.H. *Macromolecules*, **1990,** *23*, 1065.
35. Kim, Y.H.; Calabrese, J.C. Macromolecules, **1991,** *24*, 2951.
36. Hara, M.; Jar, P. *Macromolecules*, **1988,** *21*, 3187.
37. Kramer, E.J. *Adv. Polym. Sci.*, **1983,** *52/53*, 1.
38. Henkee, C.S.; Kramer, E.J. *J. Polym. Sci., Polym. Phys. Ed.* ***1984,*** *22*, 721.
39. Hara, M.; Sauer, J.A. *J. Macromol. Sci., Rev. Macromol. Chem. Phys.* ***1994,*** *C34*, 325.

Self-Reinforced Composites

Chapter 5

Compatibilization of Thermotropic Liquid-Crystalline Polymers with Polycarbonate via Transesterification by In Situ Reactive Blending

M. J. Stachowski[1] and A. T. DiBenedetto

Polymer Science Program, Institute of Materials Science, University of Connecticut, U–136, Storrs, CT 06269–3136

Aromatic thermotropic liquid crystal polyesters (Ar-TLCP's) and TLCP's containing aliphatic linkages can be compatibilized as binary-TLCP blends by transesterification. The morphology and physical properties of the resultant binary-TLCP blend are dependent on the blockiness, composition and viscosity ratios of the two TLCP components. Polycarbonate (PC) can also be blend compatibilized with either TLCP's or binary-TLCP blends, by transesterification of aliphatic linkages from the TLCP's into the PC. In this work, the degree of selective transesterification is quantified and its effect on TLCP blend compatibility is described.

One of the fastest growing areas of materials research is the development of new materials and properties be blending of existing and largely available polymers. A miscible blend of two polymers with different glass transition temperatures (T_g) is characterized by a single glass-transition temperature due to solubility of one polymer within the other. A well known miscible polymer pair is polystyrene and polyphenylene oxide, where a single T_g is seen. The blend is optically clear since both components are amorphous. Optical clarity can not be used as an absolute measure of miscibility, however, since it is possible to have a miscible blend in which crystallization occurs. If the polymer blend exhibits the same two glass-transition temperatures as its components, it is considered to be a completely immiscible blend. Note, however, that an immiscible blend can be optically clear if the two components have the same indices of refraction. Partial miscibility is characterized by multiple glass transitions that are shifted relative to those of the pure components and indicative of a two phase material.

[1]Current address: Thomson-Gordon Group, 3225 Mainway, Burlington, Ontario L7M 1A6, Canada

0097–6156/96/0632–0070$15.00/0

The compatibility of two components in a partially miscible blend is a function of the extent to which the blend properties are altered as compared to a completely immiscible pair. The compatibility can be characterized by a fine dispersion of one phase within the other, adhesion of the dispersed phase to the matrix, shifting of glass transition temperatures relative to the pure components, reduced interfacial energy between the phases, and molecular interchange [*1*].

Compatibilization can be promoted by molecular interchange reactions between blend components, such as transesterification in ester-containing polymers pairs [*2*]. Transesterification involves the interchange of one ester group within a chain with another ester group, usually in another chain. It has been used *in-situ* to compatibilize polyester thermotropic liquid crystal polymers (TLCP's) with thermoplastic polyesters or polycarbonates [*3*]. When transesterification occurs during blending, block copolymers are initially formed and then eventually randomized [*1*]. Complex and competing reactions and rates occur, resulting in varying chain microstructure. Chains in which transesterification has occurred can enhance the reaction of unreacted chains, or enhance miscibility of the unreacted components. The degree of transesterification is highly dependent on the reaction conditions, namely: material preparation, temperature and time of mixing, shear rate, and the presence of catalytic agents. Transesterification occurs in TLCP's during their synthesis, but little has been studied in the area of induced blend transesterification with Ar-TLCP's [*2,3*].

Poly(ethyleneterephthalate-hydroxybenzoic acid), p(ET/HBA), has been blended with polycarbonate [*2,4,5*], poly(ethyleneterephthalate) [*2,4,6,7,8,9*], polyhexamethylene terephthalate [*2,4,10*], and Ar-TLCP's [*3,7,8,9,11*]. In these cases, a degree of compatibilization manifests itself in the form of good bonding between the components. This and other TLCP's containing both aliphatic and aromatic groups may serve as a coupling agent between a thermoplastic and an Ar-TLCP providing an improved interphase, thus improving the mechanical properties relative to an incompatible thermoplastic / Ar-TLCP blend [*12*].

Highly aromatic ester-based main-chain TLCP's are useful because of their high strength and modulus, and processability. A limiting factor in the use of these Ar-TLCP's as a reinforcement with commercially available thermoplastic polymers is their incompatibility, as expressed by poor interfacial adhesion [*1,4,13,14*]. This could be overcome by the utilizing intermediary phases as compatibilizers [*5,11*]. A binary-TLCP blend can provide for this.

Various results are reported on binary-TLCP blends that use p(ET/HBA) or the Ar-TLCP p(HBA/HNA) (where HNA hydroxynaphthoic acid) is as a component. In some cases, miscibility is observed [*11,15,16,17,18*]. In others, the components are shown to have a degree of compatibility [*3,7,8,9,11*]. In other blends, the components are completely incompatible [*3,19,20*]. These studies note the dependence on compositions and process conditions. Main-chain (MC) and side-chain (SC) binary-TLCP blends have shown both miscibility and compatibility [*21*,*22*,*23*].

EXPERIMENTAL

1. Blends of p(ET/HBA) Liquid Crystal Polymer with Polycarbonate

Methods and Materials - Rodrun™ LC-3000 p(ET/HBA) TLCP, [Unitika. Ltd. of Japan] in 40:60% molar ratio, is tested for compatibility with Polycarbonate. Recent studies have shown this TLCP to be more random in nature than a previous p(ET/HBA) TLCP [*3*,*24*]. The influence of the more random form of p(ET/HBA) copolymer on its compatibility with polycarbonate is here investigated.

This p(ET/HBA) TLCP was mixed with bisphenol-A PC [Sinvet™ 251, Enimont America, Inc.] in 50:50% wt., blended at 300°C for 30 mins. at 40 r.p.m. in a 30 ml Brabender mixer. Additional 50:50% PC:p(ET/HBA) blends with a catalyst, tetrabutylorthotitanate (TBOT) in the amounts of 0.15% and 0.30% wt., were produced under the same conditions. Pure components were processed at the same conditions to determine if processing changed their initial physical properties. All materials were vacuum dried at a minimum temperature of 100°C for 24 hours before processing.

Samples of extrudates were pressed at 250°C and specimens were cut to necessary size. Fracture surfaces were formed after immersion in liquid nitrogen and were analyzed using an Amray SEM Scanning Electron Microscope (SEM). Dynamic mechanical thermal analyses (DMTA) were performed on a Polymer Sciences DMTA in the tensile mode at 1 Hz and 3°C/min. heating rate. Proton (^{1}H) Fourier Transform Nuclear Magnetic Resonance (FT-NMR) was performed on pure PC using a Bruker 270 NMR Spectrometer in deuterated chloroform (CDCL). Similarly, the soluble PC portions of the blends were extracted and filtered for analyses. Integration of ^{1}H spectra were performed on the peaks for absolute ^{1}H content representing a particular group, and ratios of component peaks relative to the total peak area (weight fractions) were calculated.

Results and Discussion - There was little or no evidence of compatibility of the PC:p(ET/HBA):0%TBOT blends, as determined from SEM micrographs [Figure 1]. There is no indication of adhesion between the PC and the dispersed p(ET/HBA) phase of 3-6 μm particles. The p(ET/HBA) particles are somewhat agglomerated, as opposed to being evenly dispersed as in PC:[X-7G™ p(ET/HBA)] blends [*3*]. However, when the TBOT catalyst is used, either in the amount of 0.15% or 0.30% wt., the phase dispersion becomes very even and fine, and the size decreases to 0.5-2.5 μm, with an that appears to display good interfacial adhesion [Figure 2 a,b].

NMR proton spectra of the extracted portions (polymeric and oligomeric) of the uncatalyzed PC:p(ET/HBA) blends show peaks of ethylene (Et) at 4.66 ppm and

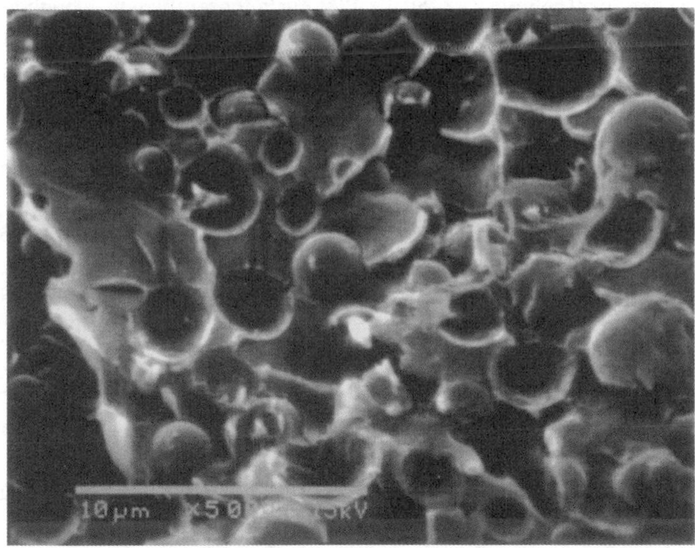

Figure 1. SEM micrograph of fracture surface of 50:50% wt. PC:p(ET/HBA)

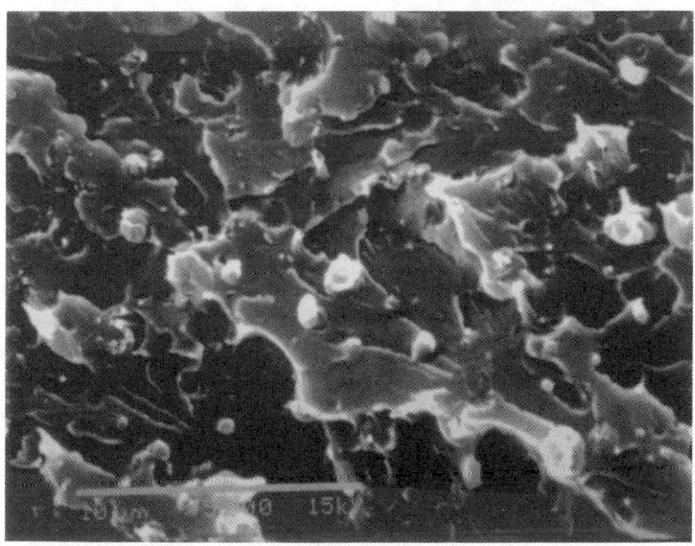

Figure 2 . SEM micrograph of fracture surface of 50:5:0:3 % wt. PC:p(ET/HBA):TBOT

terephthalate (TA) at ca. 8.1 ppm, indicating that the thermal treatment has slightly modified the chemical composition of the PC, but apparently not enough to generate adhesion between components. However, the TBOT catalyzed blends display peaks of enhanced Et and TA as well as diethylene glycol (DEG), originating from the ET block of the p(ET/HBA) TLCP [Figure 3]. It appears that enhancement of the interchange reaction has compatibilized the blends. Evidence of HBA block transesterification is difficult to ascertain as there may be peak overlap of the HBA block with the phenyl ring of the bisphenol-A of the PC. Integration of the ^{1}H peaks of PC:p(ET/HBA):*x*TBOT indicates that the extracted PC contains 37% wt. ET blocks when 0.30% TBOT is used, 22% wt. ET when 0.15% TBOT is used, and 12% wt. ET without TBOT. Thus, the catalyst increases the transesterification of the components in these 50:50% wt. PC:p(ET/HBA) TLCP blends, and the transesterification increases compatibilization. Previous work has shown the X-7G™ p(ET/HBA) [Tennessee-Eastman] is slightly blocky and generally compatible in blends with ester-containing thermoplastics, and did not require a catalyst to produce compatible blends in which ca. 10% ET is transesterified [*2, 4-10*].

DMTA spectra of the PC:p(ET/HBA):*x*TBOT blends exhibit tangent delta (tan $\delta = E'' / E'$) peaks representing glass-transition (T_g) shifts consistent with the SEM and NMR data. The p(ET/HBA) exhibits a single, broad T_g in the same 60-105°C range and the PC has a T_g ca. 164°C [Figure 4]. The p(ET/HBA) does not show a shift of its T_g in the uncatalyzed PC: p(ET/HBA) blend, though the range does narrow slightly toward the low end, ca. 65°C, and the T_g of the PC shifts to ca. 140°C. In the catalyzed blend of PC:p(ET/HBA):0.30%TBOT, there is a shift in the p(ET/HBA) T_g to ca. 75°C [Figure 4].

2. The Blending Of an Ar-TLCP, p(HBA/HNA/TA/HQ), With Polycarbonate, using p(ET/HBA) As a Compatibilizing Agent

Introduction. A binary-TLCP (Bin-TLCP) blend of an Ar-TLCP with the p(ET/HBA) TLCP is prepared over varying composition. The potentially transesterifiable ET blocks of p(ET/HBA) are blended with the Ar-TLCP to induce a small amount of aliphaticity in the Ar-TLCP. The Bin-TLCP will then be blended with PC to form a ternary blend, with and without the TBOT catalyst. It is anticipated that the TBOT will initiate a block transesterification of ET with the PC, thereby compatibilizing the components of the ternary blend.

Method and Materials. Bin-TLCP blends of Ar-TLCP Vectra™ RD-501 p(HBA/HNA/TA/HQ) with Rodrun™ p(ET/HBA) were produced with the following compositions: 30/70, 50/50, 70/30 % wt. The LCP blocks in the binary blend are: hydroxybenzoic acid (HBA); hydroxynaphthoic acid (HNA); terephthalic acid (TA); hydroquinone (HQ); ethyleneterephthalate (ET). The Bin-TLCP blend of 70/30 RD-501/Rodrun (abbreviated "RDRR" in the figures), was then used in ternary

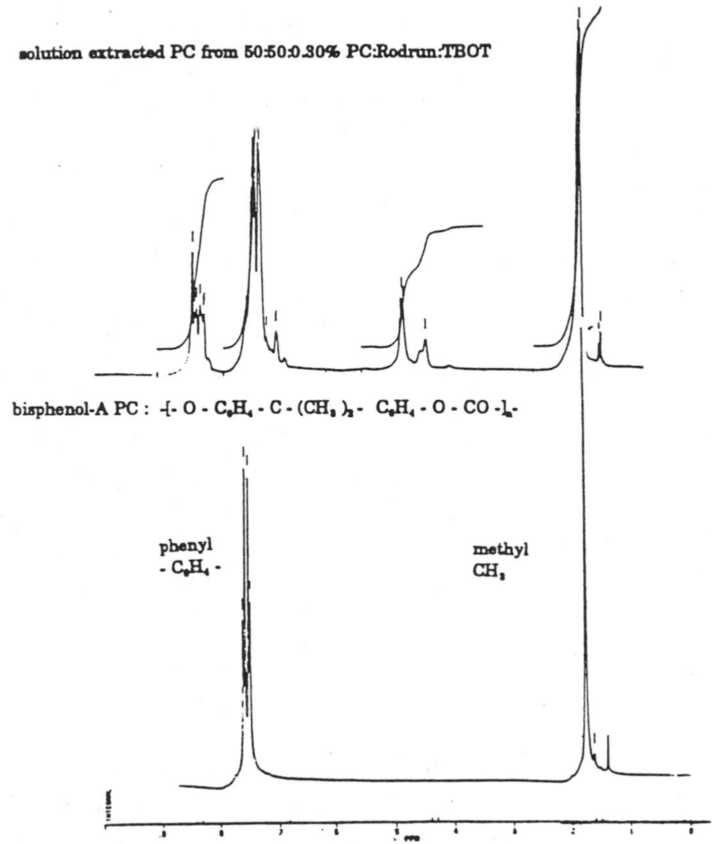

Figure 3 . NMR spectra of the extracted portion of polycarbonate (PC) from 50:50:0.3% wt. PC:p(ET/HBA):TBOT, compared to the spectra of pure PC

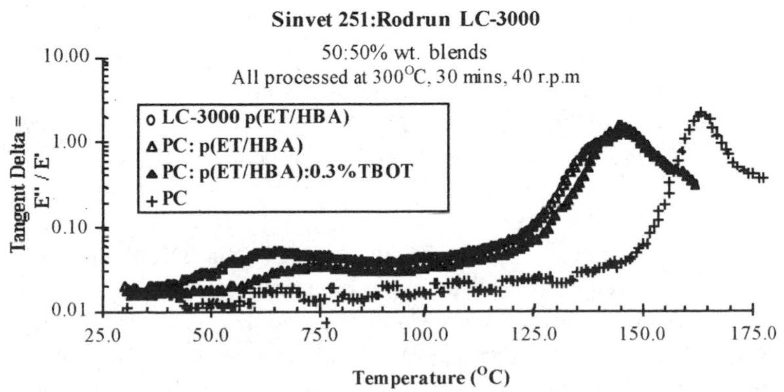

Figure 4. DMTA spectra of the 50:50:*x* wt% PC:p(ET/HBA):*x*TBOT

blends of 50:50% wt. PC:Bin-TLCP, with 0 or 0.30 % wt. TBOT. A blend of 50:50% wt. PC:Ar-TLCP was used as a reference. All materials were dried at a minimum temperature of 100°C in vacuum for at least 24 hours before blending. All blends were mixed for 30 mins. at 40 r.p.m. in a Brabender at 300°C.

Samples for morphological and thermal studies were pressed at 300°C (for Bin-TLCP blends) or 250°C (for blends with PC), and cut to size. Specimens were cold-fractured after liquid nitrogen immersion, and the fracture surfaces were analyzed for phase size, dispersion, and interfacial adhesion using SEM. The blend T_g's were measured using tensile-mode DMTA at 1 Hz and a 3°C/min. heating rate.

Extraction of p(ET/HBA) from the Bin-TLCP blends was attempted with a p(ET/HBA) solvent, trifluoroacetic acid (TFAA), then filtered and put into solution with *d*-TFAA for NMR analysis. NMR spectra of pure Ar-TLCP in *d*-TFAA was also attempted. PC was extracted from its blends in *d*-chloroform (CDCL). High-resolution proton (^{1}H) Fourier Transform Nuclear Magnetic Resonance (FT-NMR) was performed using a Bruker 270 NMR Spectrometer on the extracted polymeric and oligomeric portions of the soluble component of the blends, and compared to pure references. Quantification of the ^{1}H spectra was possible by integration of the peaks for absolute hyrogen content representing chemical groups specific to the polymers. NMR spectra indicate that the pure TLCP's and the Bin-TLCP blends are insoluble in CDCL, and it is possible to measure for any possible extent of transesterification of TLCP's in spectra of the soluble PC extract.

Results and Discussion - Bin-TLCP blends. The Bin-TLCP blends show no dispersed phase in SEM micrographs, but a fibrous texture characteristic of the Ar-TLCP [Figure 5]. DMTA speçtra exhibit a broad T_g for each blend composition indicating miscible blends. The T_g's of the blends are intermediate of the two pure TLCPs' T_g's and correlate with the blend composition [Figure 6]. Evidence of interactions between components of the blends is seen by using a TA-2200C Differential Scanning Calorimetry (DSC) [Figure 7-8]. These data show a non-linear enthalpy-of-recrystallization, during a slow cool after the first heat, indicative of a transesterification of the block structure [*19*]. The 50/50 and 70/30 ("RDRR") blends were completely insoluble in TFAA, indicating that the p(ET/HBA) is chemically attached to or physically imbibed by the Ar-TLCP. NMR spectra of the 30/70 Bin-TLCP blend showed indications of partial transesterification of Ar-TLCP blocks into the p(ET/HBA). The 70/30 % wt. Bin-TLCP ("RDRR") was then blended with polycarbonate (PC) to form a ternary PC:Bin-TLCP blend.

Results and Discussion - PC:Ar-TLCP and PC:Bin-TLCP blends. SEM micrographs cold-fractured, as well as solution etched, 50:50% wt. PC:Ar-TLCP indicate that the Ar-TLCP is the matrix while PC is the dispersed phase. Phase sizes are on the order of 20-100 μm. The appearance of both cavitation and adherence of the phases indicates that adhesion of the phases is not very strong. Dispersion of the

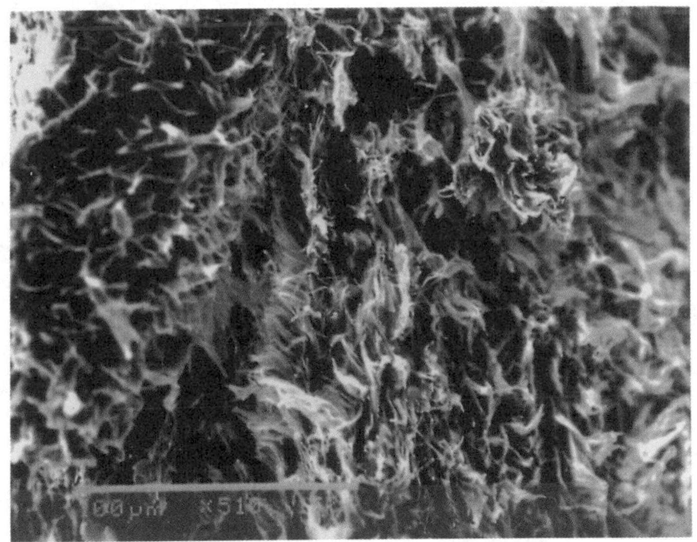

Figure 5. SEM micrograph of the Bin-TLCP blend, in 70 / 30 % wt.

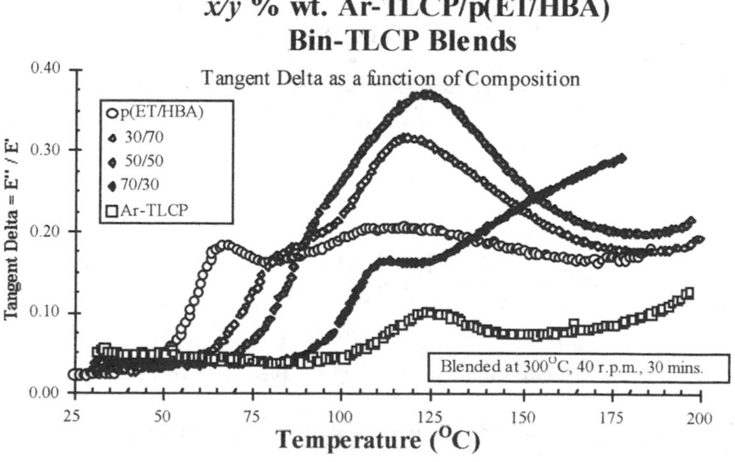

Figure 6. DMTA spectra of the Binary-TLCP blends. See text for compositions.

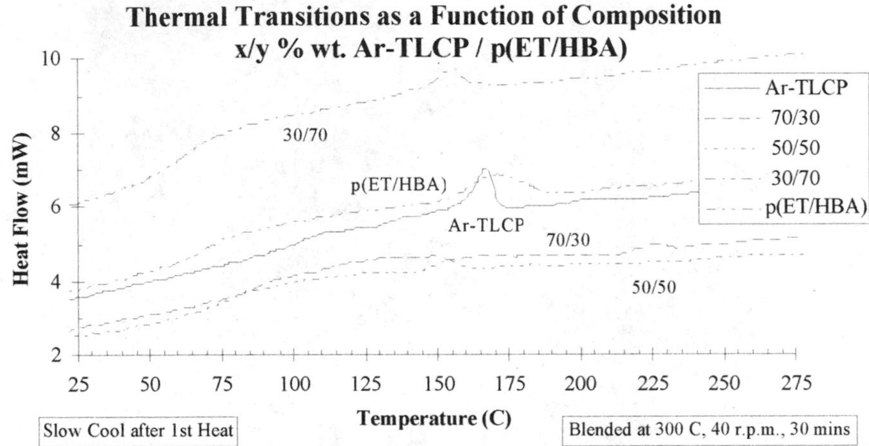

Figure 7. DSC curves of the Binary-TLCP blends during Slow Cool

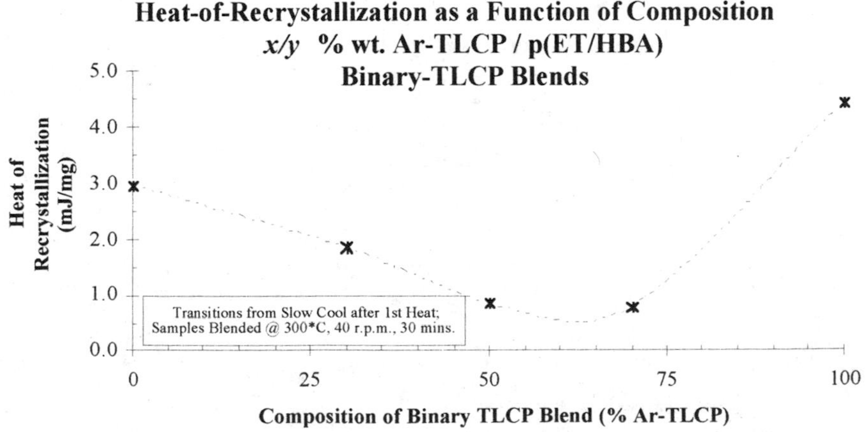

Figure 8. The Binary-TLCP blends' Enthalpy-of-Recrystallization during Slow Cool

PC is poor, with the formation of large aggregated particles. The phase morphology of the system is inverted in the 50:50% wt. PC:Bin-TLCP blends, with PC as the matrix phase and the Bin-TLCP is the dispersed phase. The phase size is decreased to 1-5 μm and evenly dispersed, but with little adhesion at the interface [Figure 9]. When blended as 50:50:0.3% wt. PC:Bin-TLCP:TBOT, the adhesion is substantially increased and the dispersion is further enhanced. In addition to regions of phase sizes on the order of 1-5 μm, there are large regions of submicron dispersion. Strong adhesion of the PC to the dispersed Bin-TLCP can be observed in the SEM micrograph of the fracture surface where the crack path is diverted around the dispersed particles [Figure 10]. DMTA spectra of these PC blends indicate that a thermomechanical change has occurred in the blends, indicative of a chemical or strong physical interaction during the blending [Figure 11].

NMR analysis of the PC extract (polymeric and oligomeric) from the three PC-blends indicates that the thermal treatment has slightly modified the chemical composition of the PC. By comparing total integrated peak areas, terephthalic acid (8.3 ppm) has transesterified into the PC from the PC:Ar-TLCP, PC:Bin-TLCP, and PC:Bin-TLCP:TBOT in amounts of 0.21%, 1.33%, and 1.55%, respectively. In addition, ethylene groups (4.8 ppm) have also transesterified into the PC from the PC:Bin-TLCP:TBOT blend in the amount of 0.23% [Figures 12 a,b]. These proportions represent weight fractions of Ethylene-Terephthalate (ET) groups transesterified into PC from the blends PC:Ar-TLCP, PC:Bin-TLCP, and PC:Bin-TLCP:TBOT in amounts of 0.37%, 0.35%, and 3.10%, respectively. There is also formation of a PC-PET block copolymer exchanged (at 3.7 ppm) in the blends of PC:Bin-TLCP and PC:Bin-TLCP:TBOT as a result of partial transesterification [*25*].

Hence, compatibility of Ar-TLCP's and polycarbonate has been induced and increased by mixing in an aliphatic-containing TLCP to form a randomized multi-block Binary TLCP with minor amounts (ca. 15 wt%) of ethylene linkages. Transesterification of a minimum amount of aliphatic species, initially assumed as block copolymers and then randomized, enhances adhesion between the random-coil PC and the rigid-rod TLCP's. In general, one might assume that a Bin-TLCP blend of a transesterifiable pair of an aromatic and aliphatic-containing TLCP's can be mixed with a thermoplastic capable of undergoing transesterification to generate compatibilization by exchange reactions, if necessary with the help of a catalyst. When compatibilization of the components in a partially miscible blend is accomplished, phase size will be decreased and the dispersion enhanced. Transesterification appears to increase the adhesion between phases.

Further publications are available for: supporting data, details and mechanical properties of these blends; properties of other binary-TLCP blends of either p(HBA/BP/TA/IA/HQ) or p(HBA/HNA) with p(ET/HBA)'s; and, a comparison of block structures of p(ET/HBA)'s and its affect on blending [*26,27*].

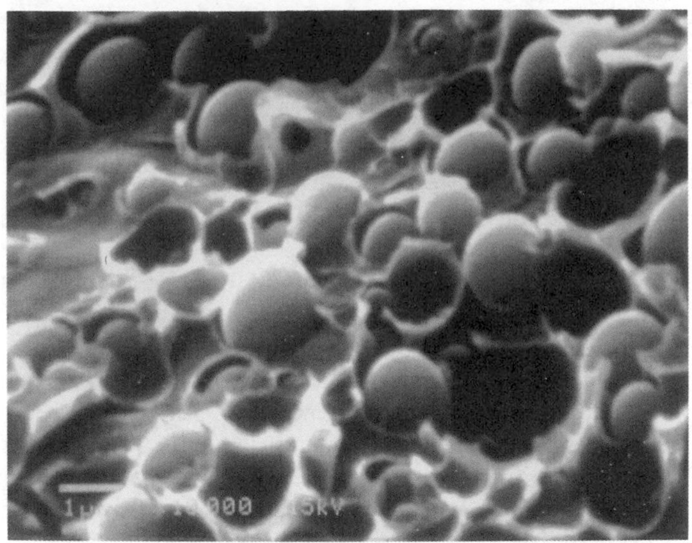

Figure 9. SEM micrograph of 50:50% wt., PC:Bin-TLCP

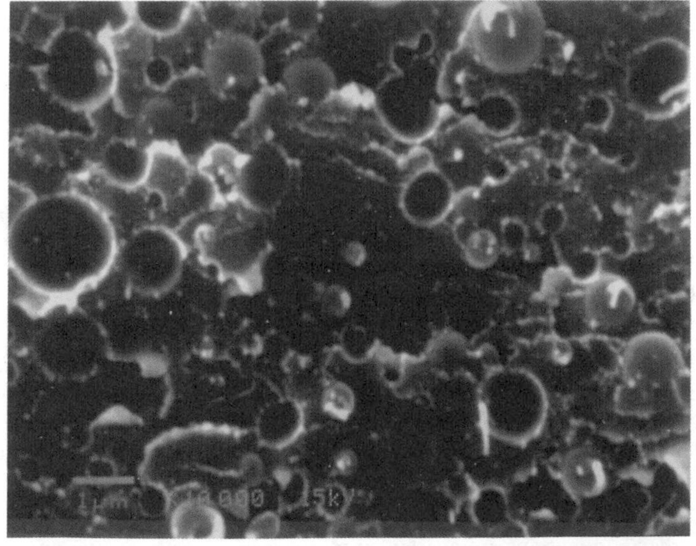

Figure 10. SEM micrograph of 50:50:0.3% wt., PC:Bin-TLCP:TBOT

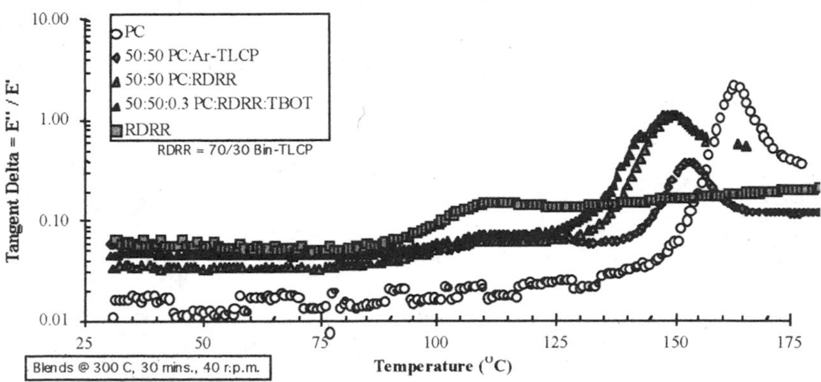

Figure 11. DMTA spectra of 50:50:x % wt., PC:TLCP:TBOT

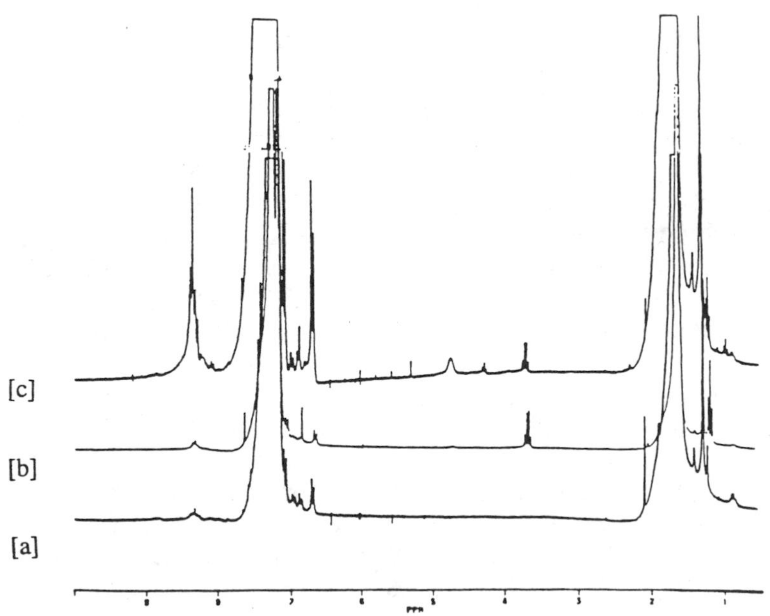

Figure 12 a,b,c. NMR spectra of the extracted portion of PC in [a] 50:50% wt. PC:Ar-TLCP, [b] 50:50% wt. PC:Bin-TLCP, and [c] 50:50:0.3% wt. PC:Bin-TLCP:TBOT. Compare to pure PC [Figure 3].

ACKNOWLEDGMENTS

Many thanks are given to the Tennessee Eastman Corporation for the donation of the p(ET/HBA) copolymers, poly(ethyleneterephthalate-hydroxybenzoic acid), and to the Hoechst-Celanese Corporation for donation of the Vectra™ polymer RD-501 p(HBA/HNA/TA/HQ), poly(hydroxybenzoic acid-r-hydroxynaphthoic acid-r-terephthalic acid-r-hydroquinone).

REFERENCES

1. Paul, D.R., in *Polymer Blends*; D.R. Paul, S. Newman, Eds.; Academic Press: New York, New York 1978, Vol. 2; p.35

2. Porter, R. S. , L-H. Wang, *Polymer*, **1992**, *33*(10), 2019

3. Stachowski, M.J., A.T. DiBenedetto, *SPE ANTEC'94*, **1994**, 2472

4. Dutta, D., H. Fruitwala, A. Kohli, R.A. Weiss, *Poly. Eng. Sci.*, **1990**, *30*(17), 1005

5. Amendola, E., C. Carfagna, P. Netti, L. Nicolais, S. Saiello, *J. Appl. Poly. Sci.*, **1993**, *50*, 83

6. Joseph, E.G, G.L. Wilkes, D.G. Baird, in *Polymeric Liquid Crystals*; A. Blumstein, Ed.; Academic Press: New York, New York, 1982; p.329

7. Lee, W-C. , A.T. DiBenedetto, *Poly. Eng. Sci.*, **1992**, *32*(6), 400

8. Lee, W-C. , A.T. DiBenedetto, *Poly. Eng. Sci.*, **1993**, *33*(3), 156

9. Lee, Wan-Chung, Ph.D. Thesis; *Processing of Thermotropic Liquid Crystalline Polymers and Their Blends*; The University of Connecticut: Storrs, CT, 1992

10. Croteau, J.-F., G.V. Laivins, *J. Appl. Polym. Sci.*, **1990**, *39*, 2377

11. M.J. Stachowski, A.T. DiBenedetto, *ACS Polymeric Materials Science and Engineering*, **1995**, *Spring*, 540

12. Choi, G.D., S.H. Kim, W.H. Jo, M.S. Rhim, *J. Appl. Poly. Sci.*, **1995**, *55*, 561

13. Bladon, P., M. Warner, M.E. Cates, *Macromolecules*, **1993**, *26*, 4499

14. Nobile, M.R., L. Incarnato, G. Marino, D. Acierno, in *Processing and Properties of Liquid Crystalline Polymers and LCP Based Blends*; D. Acierno, F.P. La Mantia, Eds.; Chem-Tec Publishing: Toronto, ON, 1993; p.195

15. DeMeuse, M.T., and M. Jaffe, *Molec. Cryst. Liq. Cryst. Inc. Nonlin. Opt.*, **1988**, *157*, 535

16. DeMeuse, M.T., and M. Jaffe, *Poylm. Prep.*, **1989**, *30*(2), 540

17. DeMeuse, M.T., M. Jaffe, in *Liquid Crystalline Polymers*, R.A. Weiss, C.K. Ober., Eds.; ACS Symposium Series 435; ACS: Washington, DC, 1990, 439

18. DeMeuse, M.T., *SPE ANTEC'93*, **1993**, 1722

19. Lin, Y.G., H.H. Winter, *Poly. Eng. Sci.*, **1992**, *32*(12), 773

20. Ding, R., A.I. Isayev, ., *SPE ANTEC'93*, **1993**, 1176

21 . Lipatov, Y.S., V.V. Tsukruk, O.A. Lokhonya, V.V. Shilov, Y.B Amerik, I.I. Konstantinov, V.S. Grebneva, *Polymer*, **1987**, *28*, 1370

22. Hakemi, H, H.A.A. Rasoul, *Poly. Comm.*, **1990**, *31*, 82

23. Hakemi, H, H.A.A. Rasoul, *U.S. Patents* 4,842,754, **1989** & 4,952,334, **1990**

24. Hayase, S., P. Driscoll, T. Masuda, *Poly. Eng. Sci.*, **1993**, *33*(2) 108
25. Zheng, W-g., Z-h. Wan, Z-n. Qi, *Polym. Int'l*, **1994**, *34*, 301
26. Stachowski, M.J., Ph.D. Thesis; *Compatibilization of Aromatic Liquid Crystal Polymers with Polycarbonate via Transesterification by In-Situ Reactive Blending*; The University of Connecticut: Storrs, CT, 1995
27. Stachowski, M.J., and A.T. DiBenedetto, in preparation

Chapter 6

Development of In Situ Reinforced Polypropylene Fibers for Use in Formable Woven Preforms

C. G. Robertson, J. P. de Souza, and D. G. Baird

Department of Chemical Engineering, Polymer Materials and Interfaces Laboratory, Virginia Polytechnic Institute and State University, Blacksburg, VA 24061–0211

Composite fibers comprised of isotactic polypropylene (PP) reinforced with fibrils of a thermotropic liquid crystalline polymer (TLCP) were generated in a novel fiber spinning process. A patented dual extrusion process was used to allow separate thermal histories to be applied to the two polymers, which have melting temperatures which differ by 119°C, and also to allow continuous streams of the TLCP to be introduced into the PP matrix prior to melt spinning. The in situ reinforced fibers of 50/50 wt.% composition were successfully drawn to draw ratios in excess of 200. The fiber tensile properties leveled off for fibers with draw ratios above 100. The tensile modulus plateau, representing the average modulus for fibers with draw ratios greater than 100, was over 30% greater than the modulus value predicted using the rule of mixtures. The in situ reinforced fibers were woven into fabric preforms. The preforms, pre-wetted with PP, were used to fabricate orthotropic composites. The mechanical performance of the composites was evaluated, and a ten-fold increase in modulus relative to the modulus of neat polypropylene was observed for a composite reinforced with 31.5 wt.% TLCP. Composites reinforced with ~20 wt.% TLCP proved formable with elongation to break values in excess of 15% at 30°C below the melting temperature of the TLCP, and the composites were thermoformed without significant fiber damage.

The addition of a thermotropic liquid crystalline polymer to a thermoplastic matrix is attractive in at least two ways. First, the TLCP may act as a processing aid by reducing the viscosity of the matrix material, so that materials exhibiting extremely high viscosities may be processed with lower energy expenditure (1-3). Second, the TLCP, under adequate processing conditions, deforms into elongated fibrils which often results in the in situ reinforcement of the thermoplastic matrix. Enhancement of the mechanical properties of several polymer matrices upon the addition of TLCPs has been reported by several research groups (4-8).

0097–6156/96/0632–0084$15.00/0

The mechanical properties of TLCP / thermoplastic in situ blends directly correlate to the degree of molecular orientation within the TLCP phase (9,10). Consequently, flow strength, which affects the deformation and orientation of the TLCP phase, will also affect the mechanical properties of TLCP / polymer blends. Mechanical properties of in situ blends processed by means of fiber spinning, where elongational flow prevails, are typically superior to those obtained from processes such as injection molding which predominantly involve shear flow. Fibers with moduli as high as 65 GPa have been obtained by spinning a thermotropic copolyester (9), whereas injection molded samples of the same copolyester have reached moduli of only 12 GPa (3). This illustrates that the reinforcing potential of TLCPs is greatly affected by the processing history.

While fiber spinning is superior to injection molding in the development of TLCP fibril reinforcement, fiber spinning necessitates post-processing to obtain usable composites unlike injection molding. One novel way of utilizing in situ fibers in composites is through the use of woven preforms. The use of such prepregs is advantageous because the fibers which constitute the fabric are pre-wetted with the thermoplastic and are deformable during shaping processes unlike glass and carbon fibers. Fabric preforms also allow the full reinforcing potential of the in situ fibers to be realized by the use of continuous lengths of the fibers within the composite. With these attributes in mind, the focus of this work was to investigate the role of fiber drawing on the tensile property development for in situ reinforced polypropylene fibers, to transfer the PP/TLCP fiber properties to composites with the use of fabric preforms, and to assess the formability of the fabric composites.

Experimental

Materials. The two polymers used in this investigation were Vectra B950 (VB) and Profax-6823 (PP). Vectra B950, a thermotropic liquid crystalline polymer purchased from Hoechst Celanese, is a copoly(ester-amide) with a glass transition temperature of 110°C, a melting temperature of 280°C, and a solid density of 1.41 g/cc (11). Profax-6823 is an isotactic polypropylene produced by Himont Company with a melting temperature of 161°C and a solid density of 0.902 g/cc. This high molecular weight polypropylene has a weight average molecular weight of 600,000 and a polydispersity index of 5 (12).

Fiber Spinning. Polypropylene was in situ reinforced with VB in a fiber spinning operation centered around a dual extrusion melt blending process patented by Baird and Sukhadia (13) and modified later by Sabol (14). Using this process, the VB and PP were separately plasticated in two 25.4 mm Killion model KL-100 single screw extruders as shown in Figure 1. Twelve continuous VB melt streams were injected into the PP matrix melt with the use of a distribution nozzle, and these streams were subsequently split into numerous smaller diameter streams using three 12.7 mm Kenics and four 25.4 mm Koch static mixing elements. The composite melt was then extruded through a 1.83 mm diameter capillary die with a length to diameter ratio of

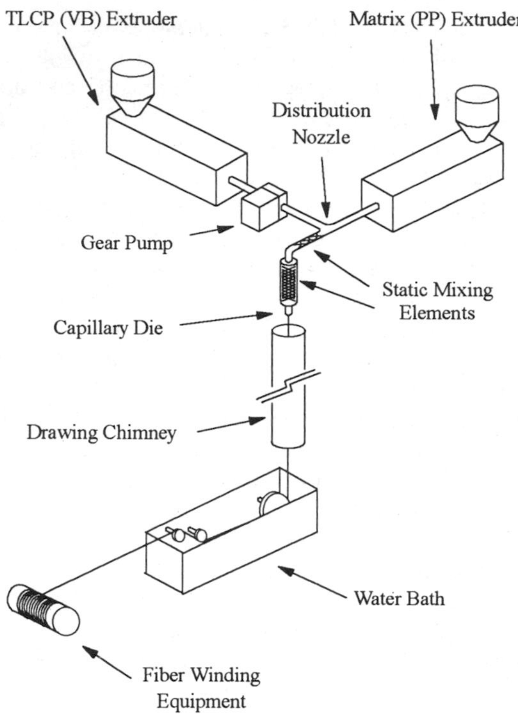

Figure 1: Diagram of fiber spinning process used to generate in situ reinforced fibers.

approximately 1.0. Just prior to their combination at the distribution nozzle, the VB stream was at its maximum temperature of 330°C and the PP melt possessed a temperature of 270°C. The static mixing portion of the process was maintained at a temperature of 290°C, the die exit temperature of the composite extrudate as measured using an external thermocouple. In the two-floor spinning operation, the fiber extrudate was drawn in ambient air through a 3.8 m long chimney used for controlled cooling, quenched in a water trough, and spun onto a spool using variable speed take-up equipment. The linear velocity provided by the fiber winding device was controlled at an increased value relative to the average die exit velocity in order to establish the desired fiber draw-down. For the composite mass flow rate of 20 g/min, the line speed necessary to induce a draw ratio of ~100 was 400 m/min. The final fiber composition of 50/50 wt.% was achieved by control of the total extrudate mass flow rate in conjunction with the use of a calibrated model HD-556 Zenith gear pump to accurately meter the VB melt.

Fabric Preform Weaving. The PP/VB (50/50 wt.%) in situ fibers were woven to create fabric preforms. This weaving process was performed with a manual loom manufactured by Structo Artcraft Loom Company. The fabric was created using a plain weave pattern and bundles of four fibers with an average diameter of 0.177 mm and a corresponding draw ratio of 106.

Composite Fabrication. Orthotropic composites were fabricated using the PP/VB (50/50 wt.%) woven preforms. Four fabric preform layers were placed between layers of polypropylene sheet as shown in Figure 2. In order to produce orthotropic composites, the fabric layers were alternately oriented 90° because the woven preforms possessed different fiber counts in the warp and weft directions. The composites were consolidated at 200°C and 3500 kPa using a model 2696 Carver Laboratory Press with heated plates. The thickness of the PP sheets used in the composite fabrication process was varied in order to achieve the desired final composite composition and to control the distribution of the four fabric layers across the composite cross section.

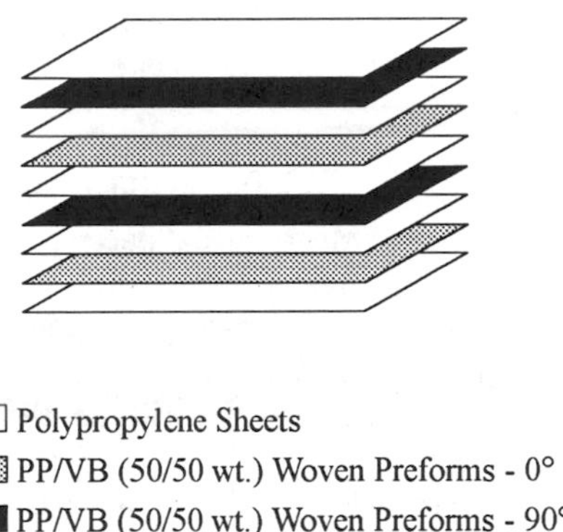

Figure 2: Layer stacking sequence used during orthotropic composite fabrication.

Rheology. All rheological measurements supplementary to this investigation were performed using a Rheometrics Mechanical Spectrometer (RMS 800) with parallel plate fixtures having radii of 12.5 mm. A plate gap of 1.2 mm was used during testing. Additionally, a nitrogen atmosphere was used in the testing chamber for all tests. For the magnitude of the complex viscosity during cooling data presented in this paper, the angular frequency used was 10 rad/s, a 5% strain was employed, and the cooling rate was 7°C per minute.

Morphological Studies. The morphology of the in situ fibers was investigated by scanning electron microscopy using a Cambridge Stereoscan S200 with an accelerating voltage of 25 kV. The fibers were cryogenically fractured perpendicular to the fiber axis and coated with gold using a Bio-Rad sputter coater.

Mechanical Properties. Tensile and flexural mechanical properties were determined using a model 4204 Instron mechanical testing instrument. Tensile properties of fibers were determined using two methods. First, tensile properties of individual fibers were obtained in accordance with ASTM D 3376-75. Also, fibers were fused together in a uniaxial manner at 200°C and 3500 kPa using the aforementioned laboratory press. Plaques of the fused fibers were then cut into tensile bars and tested to obtain the tensile mechanical properties (ASTM D 638-87b). The tensile properties of the orthotropic fabric composites and neat polypropylene were evaluated according to ASTM D 638-87b. A model 2630-25 Instron extensiometer was used during the tensile testing of the composites and plaques of fused fibers in order to accurately determine the strain during tensile loading. Following the guidelines given in ASTM D 790-86, the flexural properties of polypropylene and the fabric composites were assessed with the use of a three point bending apparatus. A pendulum impact tester manufactured by Tinius Olsen Testing Machine Company with sensors and software produced by General Research Corporation was used to determine notched Charpy impact properties (ASTM D 256-87). Properties cited in this paper represent average values for at least five samples.

Dynamic Mechanical Testing. Composite samples were tested in the dynamic torsional mode of a Rheometrics Mechanical Spectrometer (RMS 800) using an angular frequency of 10 rad/s, a strain of 1%, and a nitrogen atmosphere. The dynamic storage and loss moduli of 45 mm long rectangular samples (8 mm x 1.5 mm) were recorded as a function of temperature. The temperature was increased from 50 to 250°C at constant rates and then decreased at the same rates.

Composite Formability Assessment. The formability of the woven preform composites was evaluated by using a temperature chamber constructed by Russells Technical Products in combination with an Instron mechanical tester (model 4204). Composite samples with a thickness of 1.5 mm were cut to a width of 12.7 mm and isothermally stretched at a rate of 50 mm/min to determine the elongation at break as a function of temperature for temperatures ranging from 100 to 250°C.

Fabric composites were thermoformed in a Hydro-Trim model 1620 thermoforming unit using a mold having a 19 mm x 38.1 mm rectangular cross section and a depth of 25.4 mm. The composites were heated radiantly to an average temperature of 250°C prior to being formed between the matched plug and drape components of the mold. Additionally, the thermoformer was used to simultaneously impregnate and form the PP/VB fabric by using a fabric layer between two layers of polypropylene sheet in the aforementioned process.

Results and Discussion

Development of In Situ Reinforced Fibers. Polypropylene was in situ reinforced with VB in a fiber spinning process (Figure 1) to form composite fibers containing PP matrix reinforced with oriented VB fibrils. This patented process involved separate plastication of the two polymers. Separate extrusion allowed different thermal histories to be imposed on the polymers, which have melting temperatures which differ by 119°C, prior to their combination after the distribution nozzle. This was essential in taking advantage of the supercooling behavior of the VB while minimizing the degradation of the PP and maintaining adequate extrudate melt strength. When VB is heated above its melting temperature, solidification will occur at a temperature below its melting temperature of 280°C upon cooling. Increasing the degree to which VB is heated above its melting temperature increases the amount to which the VB supercools below its melting temperature (15) which in turn may allow increased deformation and orientation of the VB during melt spinning. In order to successfully draw the PP/VB (50/50 wt.%) fiber system up to draw ratios over 200, the VB was heated up to 330°C and the average melt temperature of the combined PP/VB stream prior to the die exit was maintained at 290°C. These processing conditions were required to provide the necessary PP/VB extrudate melt strength and provide adequate supercooling of the VB phase to minimize premature solidification during melt drawing.

Consistent with the processing conditions used, Figure 3 illustrates the magnitude of the complex viscosity, $|\eta^*|$, of the two materials when cooled from the respective maximum temperatures realized in the fiber spinning process. This plot clearly illustrates that VB solidifies at a temperature below its melting temperature of 280°C following heating to 330°C. It is also evident in Figure 3 that the viscosity of VB at 290°C after cooling from 330°C is significantly less than the viscosity of PP at 290°C. This implies that the viscosity of the VB was less than that of the PP in the static mixing portion of the dual extrusion process which was a condition favorable for the maintenance of the continuity of the VB streams within the PP matrix. Figure 3 indicates that PP and VB solidify at very different temperatures. This caused the selection of the fiber extrudate exit temperature to be limited to 290°C, a temperature which allowed adequate deformation of the VB phase prior to its solidification during melt drawing. It has been determined from isothermal time sweeps at an angular frequency of 1 rad/s that PP undergoes a 38% reduction in $|\eta^*|$ at 290°C after 20 minutes, indicative of chain scission and degradation. This suggests that the development of in situ reinforcement using only a single extruder would be quite limited due to a processing temperature mismatch for the VB and PP materials.

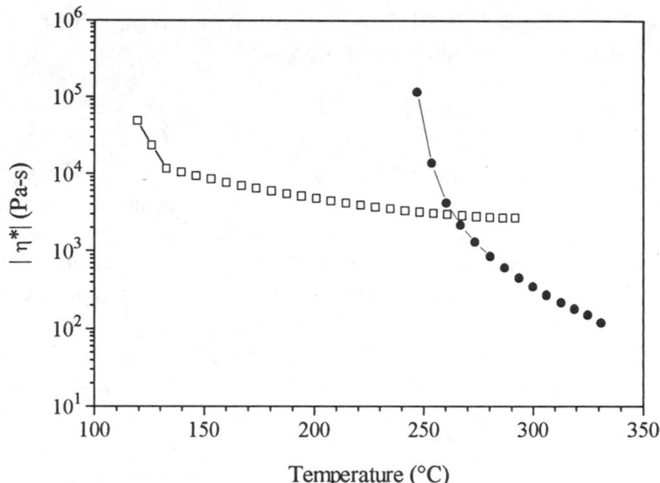

Figure 3: |η*| as a function of temperature for PP cooled from 290°C (—□—) and VB cooled from 330° (—●—). Parallel plate fixtures used. Angular frequency = 10 rad/s. Strain = 5%. Cooling rate = 7°C/min.

The effect of draw ratio on fiber tensile properties was investigated for PP/VB (50/50 wt.%) in situ reinforced fibers. Melt drawing establishes an elongational flow field which increases the aspect ratio of the VB fibrils within the PP matrix and orients the VB phase. Therefore, increasing the draw ratio and corresponding degree of elongation is expected to enhance the fiber tensile properties (16). In situ reinforced fibers were successfully produced up to a draw ratio of 220. Figures 4 and 5 present the tensile properties of the fibers as a function of draw ratio. For this investigation, the draw ratio was determined by dividing the cross-sectional area of the capillary die exit by the cross-sectional area of the fiber. The tensile properties presented were determined for fibers aligned uniaxially and fused together. These properties were confirmed by performing single fiber testing which provided very similar property values but with greater sample standard deviations. Both the tensile modulus and tensile strength of the PP/VB (50/50 wt.%) fibers initially improved with increasing fiber draw ratio but then appeared to level off for draw ratios greater than approximately 100. The diameter of the reinforcing VB fibrils within fibers produced at a draw ratio of ~100 has been determined by scanning electron microscopy to have a minimum value of approximately 1 micrometer and an average value of approximately 4 micrometers.

In addition to experimental fiber modulus data, Figure 4 also illustrates two axial modulus predictions calculated using the rule of mixtures. A previous investigation (14) has indicated that the modulus of neat PP fibers has a

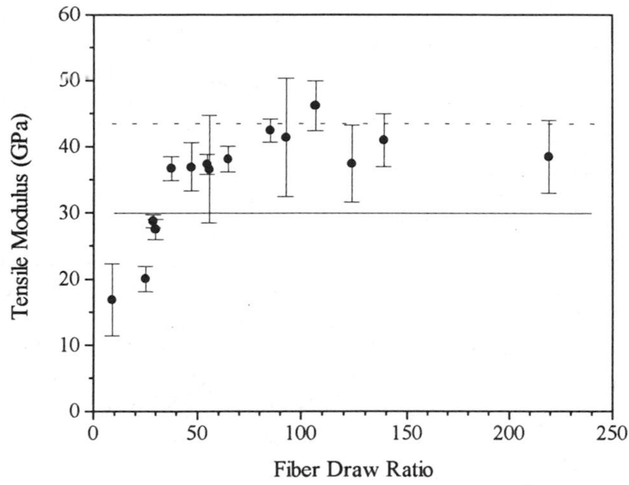

Figure 4: Tensile modulus as a function of draw ratio for uniaxially compression molded PP/VB (50/50 wt.%) fibers. Experimental data (●) illustrated as well as predictions based upon the rule of mixtures using VB modulus values of 75 GPa (—) and 110 GPa (- - -).

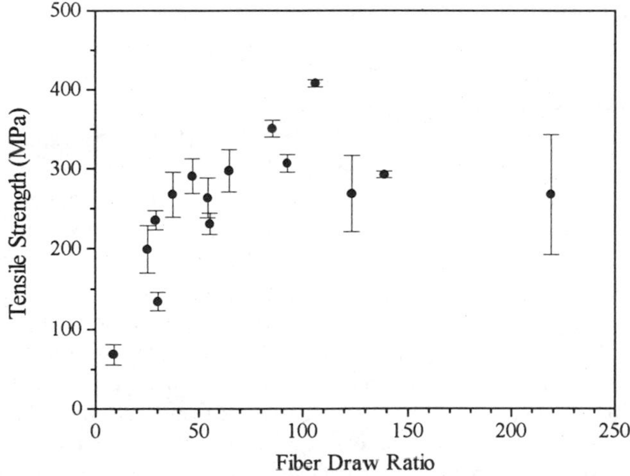

Figure 5: Tensile strength as a function of draw ratio for uniaxially compression molded PP/VB (50/50 wt.%) fibers.

constant value of approximately 1 GPa which is independent of fiber draw ratio using processing conditions very similar to those used for the composite fiber generation in this investigation. Therefore, both modulus predictions utilized a PP modulus value of 1 GPa. In Figure 4, the lower prediction line was calculated using a VB modulus contribution of 75 GPa, which is the maximum attainable axial modulus for VB when spun alone (11,16). The modulus prediction represented by the upper line was calculated using a VB modulus of 110 GPa, the extrapolated modulus for VB assuming complete molecular orientation as reported by Lin and Yee (16). The VB modulus and the fibril aspect ratio are both expected to be functions of fiber draw ratio. The lines representing the rule of mixtures predictions are included to illustrate the expected maximum unidirectional modulus values and are not intended to represent the modulus predictions as a function of draw ratio. The fiber modulus plateau, representing the average modulus for fibers with draw ratios greater than 100, was approximately 30% greater than the modulus prediction using the experimental VB modulus of 75 GPa. This synergistic effect may be due to the insulating effect of the PP matrix which allows the VB fibrils to be further drawn and oriented prior to solidification than is possible when VB is drawn in neat form.

Fabric Preform Composites. The in situ reinforced PP/VB (50/50 wt.%) fibers were woven to create fabric preforms to be used in the generation of composites. Bundles of four fibers with an average diameter of 0.177 mm and a corresponding draw ratio of 106 were woven using a plain weave pattern to create the preforms. The tensile properties of the fibers within the preforms reflect those of PP/VB fibers produced at a draw ratio of 106 as presented in Figures 4 and 5. Due to limitations of the loom used, the resulting weave did exhibit an inconsistency with respect to the warp and weft direction fiber counts. For this reason the four fabric layers used in the fabrication of composites were alternately oriented 90° to form orthotropic composites as previously depicted in Figure 2. The composites were consolidated in a compression molding process at a temperature of 200°C and 3500 kPa. The consolidation temperature of 200°C, 80°C below the melting temperature of VB, was selected in order to minimize the relaxation of the VB reinforcing phase while still maintaining adequate composite consolidation. Because the polypropylene sheet material used in the composite fabrication process was also a component of the in situ reinforced fibers within the preforms, the difficulty in achieving fiber wetting often seen in glass and carbon fiber reinforced systems was not realized to any extent.

Mechanical properties were determined for composites produced with the four preform layers evenly spaced across the final composite cross sections. Mechanical properties and densities are provided in Table I for neat polypropylene and for composites with VB contents of approximately 10, 20, and 30 wt.%. The tensile and flexural properties increased with increasing VB content, and a ten-fold increase in tensile modulus relative to neat polypropylene was noted for a composite with 31.5 wt.% VB reinforcement which possessed a tensile modulus of 10.3 GPa. Polypropylene reinforced with 30 wt.% random fiberglass mat is reported to have a tensile modulus of 4.62 GPa and a tensile strength of 82.8 MPa (17). In order to estimate the tensile modulus of PP randomly reinforced with PP/VB fibers from the modulus of the orthotropic PP/VB composite to provide a better comparison for the modulus of PP reinforced with random fiberglass mat, the tensile modulus value of 10.3 GPa can be multiplied by 2 to convert it to a uniaxial modulus and then multiplied by 3/8 to convert it to a random modulus value (18). Using this approximation, polypropylene randomly reinforced at a loading level of 31.5 wt.% VB with PP/VB fibers should possess a tensile modulus of 7.7 GPa which is noticeably greater than the modulus for the PP reinforced with random fiberglass mat. The PP/VB composite tensile and flexural property improvements with increased VB content were accompanied by only a slight increase in composite density. Significant decreases in the toughness and Charpy impact properties were noted with increased VB wt.% as indicated in Table I. It is possible that improved adhesion between the VB and PP phases would provide greater composite impact performance while retaining or even improving the tensile and flexural properties.

Table I: Composite Properties*

VB Content (wt.%)	Density [1] (g/cc)	Tensile Modulus (GPa)	Tensile Strength [2] (MPa)	Toughness [3] (kJ/m^3)	Flexural Modulus (GPa)	Flexural Strength (MPa)	Notched Charpy Impact (J/m of width)
0 neat PP	0.902	1.01 (0.03)	24.2 (0.4)	1825 (39)	1.28 (0.04)	27.9 (0.7)	347 (156)
9.84 (0.72)	0.949	3.87 (0.80)	45.0 (5.2)	626 (73)	3.22 (0.20)	43.8 (6.3)	–
19.8 (1.7)	0.979	5.60 (1.00)	57.5 (6.2)	471 (87)	6.28 (0.35)	55.1 (3.7)	147 (112)
31.5 (1.0)	1.02	10.3 (2.2)	63.8 (4.5)	445 (91)	10.0 (1.7)	65.6 (7.3)	–

* Composites fabricated with even preform layer distribution
Values in parentheses represent sample standard deviations
[1] Composite density determined from weight and dimensions of composite
[2] Tensile strength determined at yield for neat polypropylene
[3] Toughness determined from the area under the tensile stress-strain curve

In an attempt to maximize flexural properties, PP/VB composites were fabricated with the reinforcing preform layers selectively placed near the composite surfaces where the greatest tension / compression occurs during flexural loading. During the composite fabrication process, PP sheets of varied thickness were used in order to situate two preform layers just inside both composite surfaces. The tensile and flexural mechanical properties for a ~10 VB wt.% composite with such uneven fabric distribution are provided in Table II along with the properties for its counterpart having an even fabric layer distribution. The distribution of the preform layers had no effect on the composite tensile properties as expected. A marked increase in the flexural performance of the surface-weighted distribution relative to the even distribution was clearly evident. This indicates that the composite flexural performance can be improved by placing the composite reinforcing layers in the regions of greatest flexural loading.

Table II: Effect of Fabric Layer Distribution on Composite Mechanical Properties

Distribution of Four PP/VB Fabric Layers Across Composite Cross Section	VB Content (wt.%)	Tensile Modulus (GPa)	Tensile Strength (MPa)	Flexural Modulus (GPa)	Flexural Strength (MPa)
even distribution	9.84 (0.72)	3.87 (0.80)	45.0 (5.2)	3.22 (0.20)	43.8 (6.3)
two fabric layers near both composite surfaces	9.07 (0.36)	3.39 (0.11)	43.6 (2.6)	5.23 (0.22)	56.3 (3.4)

Values in parentheses represent sample standard deviations

Composite Formability. In order to assess the processability of the PP/VB composites, the tensile elongation at break for the fabric composites was determined as a function of temperature. This testing was performed to determine the maximum deformability of the composites with respect to stretching temperature in order to provide insight into the performance of the composites in thermoforming processes. Composite samples were isothermally elongated at an extension rate of 50 mm/min which is similar to the rates observed in typical thermoforming processes. Because the tested composites contained continuous lengths of in situ reinforced fibers in the stretching direction, the deformability of the composite samples reflected the deformability of the PP/VB fibers. For a 19.8 VB wt.% orthotropic composite with even distribution of the four fabric layers across the composite cross section, the effect of temperature on the elongation at break is illustrated in Figure 6. The allowable composite extension increased with temperature up to over 15% elongation at 250°C, a temperature 30°C below the melting temperature of the VB phase. This degree of extension is not possible for composites reinforced with conventional fibers such as fiberglass and carbon.

Dynamic mechanical analysis was used to determine whether the thermal cycle in a thermoforming process would induce composite property loss. The dynamic shear storage and loss moduli were recorded as a function of temperature for a 19.8 VB wt.% orthotropic composite with evenly distributed fabric layers. The moduli measurements were made while increasing the temperature from 50 to 250°C at a fixed rate and decreasing the temperature at the same rate. The dynamic moduli results of this testing performed at heating/cooling rates of 3.6 and 11.1 °C/min are presented in Figure 7. It is clearly evident that heating this composite to 250°C

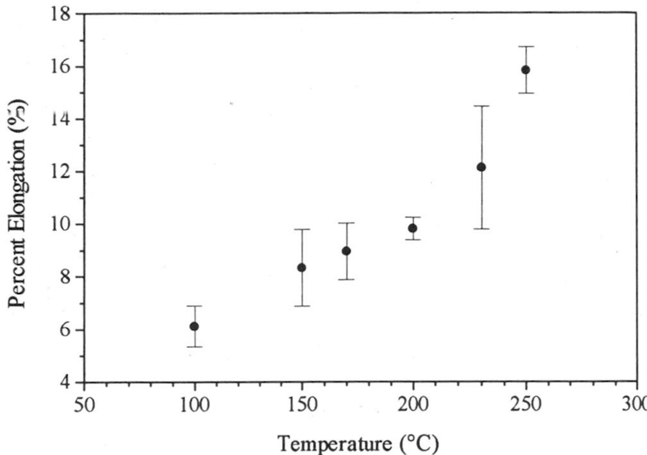

Figure 6: Tensile elongation at break as a function of temperature for 19.8 VB wt.% composite with even fabric layer distribution. Extension rate = 50 mm/min.

resulted in property loss using a 3.6 °C/min rate as can be seen by the final moduli values at 50°C following the downward temperature sweep which were significantly lower than the initial moduli values at 50°C prior to the temperature cycle for this rate. The use of the higher heating/cooling rate of 11.1 °C/min resulted in negligible property loss as can be seen from the lower plot in Figure 7. Because the 11°C/min heating rate represented a total cycle time of 36 minutes, which is far greater than a typical thermoforming cycle time, it is expected that negligible composite property loss should result from a heating cycle to 250°C in a thermoforming process.

The thermoforming of ~20 VB wt.% composites with even fabric distribution proved to be possible without noticeable fiber damage using a 25.4 mm deep tray mold of rectangular (19 mm x 38.1 mm) cross section. In a thermoforming unit, the composites were heated by means of radiation to an average temperature of 250°C. Because the mold required a sample elongation of 66% for a composite sample clamped during the forming process, the composite edges were not constrained during the forming step. This allowed both stretching and draping of the heated composite which reduced the amount of necessary elongation to a degree less than the limit of approximately 15%. Using the thermoforming unit, a combined composite fabrication and forming process was also developed which has economic appeal because it reduces the number of necessary processing steps.

Conclusions

Fibers comprised of polypropylene matrix reinforced with 50 wt.% VB fibrils were processed up to a fiber draw ratio of 220. The reinforcement was developed in situ using a dual extrusion fiber spinning process. The dual extrusion process not only allowed the processing of polymers with widely different melting temperatures but also allowed continuous streams of the TLCP to be introduced into the PP and maintained in the drawn extrudate. The tensile properties of the fibers increased with

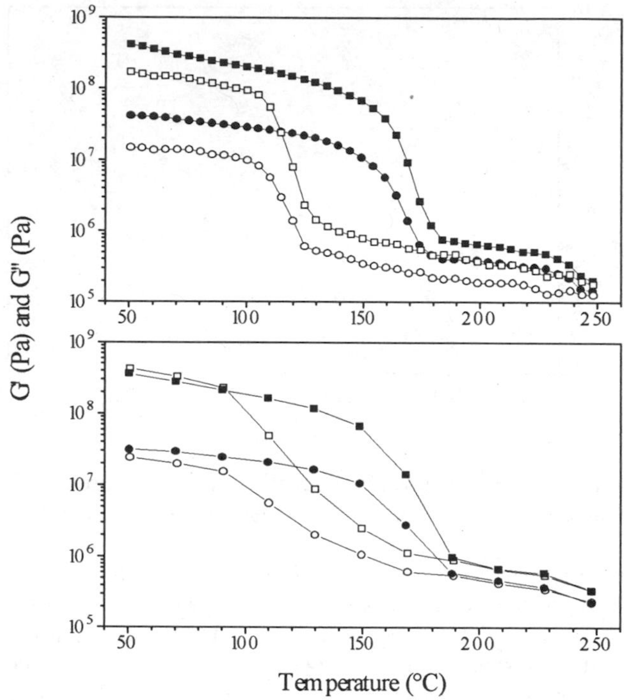

Figure 7: Dynamic shear moduli recovery for a 19.8 VB wt.% composite with even fabric distribution heated to 250°C at heating/cooling rates of 3.6 °C/min (upper plot) and 11.1 °C/min (lower plot). Shown are data for G' (—■—) and G" (—●—) during heating and G' (—□—) and G"(—○—) during cooling. Frequency = 10 rad/s. Strain = 1%.

increasing draw ratio up to a draw ratio of approximately 100 where the properties appeared to level off. The fiber modulus plateau, representing the average modulus for fibers with draw ratios greater than 100, was approximately 30% greater than the modulus prediction using the experimental VB modulus of 75 GPa. The in situ reinforced fibers were woven into fabric preforms. These preforms, pre-wetted with PP, were used to fabricate orthotropic composites. A ten-fold improvement in tensile modulus relative to PP was noted for a 31.5 VB wt.% composite which displayed a tensile modulus over two times greater than that for polypropylene reinforced with 30 wt.% random fiberglass mat. The fabrication of a ~10 VB wt.% composite with the reinforcing preform layers situated in regions of highest compression / tension during flexural loading displayed a flexural modulus and flexural strength of, respectively, 60% and 30% greater than those for a composite with the preform layers evenly distributed. A 19.8 VB wt.% composite proved formable with elongation to break values in excess of 15% at 30°C below the melting temperature of the TLCP. Dynamic mechanical analysis indicated negligible loss of dynamic shear moduli following heating to 250°C at a heating/cooling rate of 11°C/min for a 19.8 VB wt.% composite which suggested that the composites could be formed without thermally induced property loss. The composites were successfully thermoformed without noticeable fiber damage, and a process which combined the composite fabrication and forming steps was developed.

Acknowledgments

Support provided by the Army Research Office, grant number DAAH04-94-G-0282, is gratefully acknowledged.

Literature Cited

1. Handlos, A.A.; Baird, D.G. *J. Macromol. Sci. - Rev. Macromol. Chem. Phys.*, in press.
2. Beery, D.; Kenig, S.; Siegmann, A. *Polym. Eng. Sci.* **1991**, *31*, 459.
3. Baird, D.G.; Bafna, S.S.; de Souza, J.P.; Sun, T. *Polym. Compos.* **1993**, *14*, 214.
4. Metha, A.; Isayev, A.I. *Polym. Eng. Sci.* **1991**, 3, 971.
5. Isayev, A.I.; Modic, M. *Polym. Compos.* **1987**, *8*, 158.
6. Sun, T.; Baird, D.G.; Huang, H.H.; Done, D.S.; Wilkes, G.L. *J. Compos. Mat.* **1991**, *25*, 788.
7. Weiss, R.A.; Huh, W.; Nicolais, L. *Polym. Eng. Sci.* **1987**, *27*, 684.
8. Crevecoeur, G.; Groeninckx, G. *Polym. Eng. Sci.* **1990**, *30*, 532.
9. Chung, T.S. *J. Polym. Sci., Polym. Phys.* **1988**, *26*, 1549.
10. Lin, Q.; Jho, J.; Yee, A.F. *Polym. Eng. Sci.* **1993**, *33*, 789.
11. Hoechst Celanese Product Literature
12. Himont Company Product Literature
13. Baird, D.G.; Sukhadia, A.M. *U.S. Patent 5,225,488*. **1993**.
14. Sabol, E.A. *M. S. Thesis, Virginia Polytechnic Institute and State University*, 1994.
15. Robertson, C.G. *M. S. Thesis, Virginia Polytechnic Institute and State University*, 1995.
16. Lin, Q.; Yee, A.F. *Polym. Compos.* **1994**, *15*, 156.
17. *Modern Plastics Encyclopedia, Vol. 68(11)*; McGraw-Hill: New York, 1992.
18. Mallick, P.K.; *Fiber Reinforced Composites*; Marcel Dekker: New York, 1988.

Chapter 7

Fiber Drawing from Blends of Polypropylene and Liquid-Crystalline Polymers

Y. Qin[1], M. M. Miller[2], D. L. Brydon[1], J. M. G. Cowie[2], R. R. Mather[1], and R. H. Wardman[1]

[1]Scottish College of Textiles, Netherdale, Galashiels TD1 3HF, Scotland
[2]Department of Chemistry, Heriot-Watt University, Edinburgh EH14 4AS, Scotland

This paper discusses the drawing of polypropylene (PP) fibres, blended with liquid crystalline polymer (LCP) at a w/w ratio of 100/10, with the aim of enhancing fibre mechanical performance. After melt extrusion, the blended fibres consist of LCP fibrils surrounded by a PP matrix. These fibrils are, however, split after conventional one-stage drawing, and the tensile properties of the polyblend fibres are poorer than those of the corresponding pure PP fibre. By contrast, two-stage drawing, under carefully optimised conditions, can bring about some enhancement of fibre mechanical performance. In view of the very different structural properties of PP and LCPs, the use of a compatibilising agent has been studied to promote adhesion between the PP and LCP phases in the drawn fibres. It is shown that, whilst a compatibilising agent may indeed promote adhesion across the interface between the two phases, it will also increase LCP fibril fragmentation during the drawing process, with consequent impairment of fibre mechanical performance. A strategy is outlined for overcoming fibril fragmentation during drawing, using compatibilising agents which are themselves liquid crystalline.

The blending of polymers is a technique which is being increasingly applied for improving the mechanical properties of synthetic textile fibres (1,2), for superior properties may be achieved which are not obtainable from any of the individual blend components. By careful control of blend composition and phase morphology, fibre properties can be adjusted to meet

0097–6156/96/0632–0098$15.00/0

specific requirements. Since the development and marketing of new polymers is becoming ever more expensive, the technique of polymer blending is providing an alternative cost-effective approach for producing materials with improved properties (3), and indeed several polyblend fibres have found commercial success.

More recently, the blending of conventional thermoplastics with liquid crystalline polymers (LCPs) has attracted considerable scientific and industrial interest. LCPs generally possess rigid rod-like molecular chains and can exist in oriented polydomains (4). These polydomains can be readily aligned in the direction of flow during fabrication, such that under conventional processing conditions, the LCP can develop a high degree of chain orientation. Indeed, a number of well established commercial high performance fibre types have been developed from aromatic lyotropic polyamide and thermotropic copolyester LCPs (5). These LCP fibres possess tenacities greater than 1.8 N tex^{-1} and initial moduli greater than 40 N tex^{-1}, compared with a fibre tenacity of 0.7 N tex^{-1} and initial modulus of 12 N tex^{-1} for conventional high tenacity polyester (PET) fibre.

With the advance of LCP technology, considerable attention has been directed towards the blending of LCPs with conventional polymers (6-9). Indeed, the presence of a LCP as a minor component in a conventional synthetic fibre offers several major benefits. For example, the low melt or solution viscosity of the LCP facilitates fibre extrusion (7), and the fibre mechanical performance, notably tenacity and initial modulus, can be improved by virtue of the superior mechanical properties of the LCP (8,9). However, incompatibility between the conventional polymer and the LCP will result in poor interfacial adhesion between the two phases, with the result that the reinforcement offered by the LCP may be severely diminished.

Interfacial adhesion may be considerably improved by the addition of a compatibilising agent (10,11), very often in the form of a graft or block copolymer which possesses segments capable of interaction with each blend component. The compatibilising agent acts essentially as a polymeric surfactant, located at the interface between the two phases. A finer, more homogeneous dispersion of the LCP will also result. The improved interfacial adhesion and dispersion generally give rise to increased mechanical performance. For example, the effectiveness of a maleic anhydride grafted polypropylene compatibilising agent in improving the mechanical properties of injection-moulded blends of PP and LCPs has been amply demonstrated by Baird and coworkers (12-14).

In the production of most synthetic fibres, it is also essential to draw (stretch) the fibres after extrusion, in order to increase polymer chain orientation and thus enhance fibre mechanical properties. This paper highlights our research (15-19) on the hot-drawing of melt-extruded fibres consisting of a PP matrix and LCP as a minor component (PP/LCP w/w ratio 100/10).

Experimental

Materials. Three thermotropic LCPs have been used in our studies. Vectra A900, an aromatic copolyester of 1,4-hydroxybenzoic acid and 2,6-hydroxynaphthoic acid, and Vectra B950, a copolymer of 2,6-hydroxynaphthoic acid, 4-aminophenol and terephthalic acid, were supplied from the Hoechst Celanese Corporation. Both LCPs possess a melt temperature of 280°C. Rodrun LC3000, supplied by Unitika, is a copolymer of polyethylene terephthalate (PET) and 1,4-hydroxybenzoic acid, and possesses a melt temperature of 220°C.

Two grades of PP were used: initially, Appryl, supplied from ICI, with a melt flow index (MFI) of 3, and in later work, Statoil 151, from Statoil Limited, with a MFI of 14.

Polybond 1001 (PP-AA), a polypropylene functionalised with 6% w/w acrylic acid, was supplied by BP Chemicals Limited. The functionalised compatibilising agents, FC1, FC2 and FC3, were synthesised from PP-AA. An example is given in Figure 1. Further details are published elsewhere (20).

Melt Extrusion and Hot Drawing. Melt extrusion was carried out using a 25 mm single screw extruder (Extrusion Systems Limited, Bradford, UK) with a 2.5 cm^3 metering pump. For much of the work, a spinneret plate with a single hole of diameter 0.5 mm was used. However, for the initial studies of the extrusion of fibres containing compatibilising agents, a spinneret was used with three holes, each of diameter 0.5 mm.

The temperature profiles adopted for extruding the PP/LCP blends were influenced by the MFI of the PP and the type of LCP. In the earlier work, where PP of MFI=3 was used, the temperature profiles were:

PP/A900	230/285/285/285/280/280°C
PP/B950	230/300/300/300/290/290°C
PP/LC3000	200/270/270/270/270/260°C

Each profile represents the temperatures of the three barrel zones, the metering pump and two heating zones in the die head. Later, when PP of MFI=14 was used, the temperature profile was 200/240/240/240/240/240°C.

Hot drawing of the extruded ('as-spun') fibres was carried out on a small-scale drawing unit consisting of two pairs of advancing rollers and a hot plate. One- or two-stage drawing procedures were used. More detailed accounts of the extrusion and drawing conditions are published elsewhere (17,18).

Tensile Testing. Tensile properties were measured on a Nene tensile tester at 20°C and 65% relative humidity. The as-spun fibres were tested with a gauge length of 20 mm and an extension rate of 2 mm min^{-1}. The drawn fibres were tested with a gauge length of 20 mm and an extension rate of 20 mm min^{-1}. Tests were carried out five to ten times for each sample, with a standard deviation generally less than 5%.

Hot-stage Microscopy. Microscopy of samples was carried out using either a Leitz or Olympus BH2 polarising microscope. Samples were prepared by placing a small length of fibre between two glass slides and heating to 180–185°C, whereupon the PP melted and the LCP morphology could be observed.

Fibre Crystallinity. Fibre crystallinity was estimated by differential thermal analysis, using a Mettler FP90 instrument. Samples were heated over the temperature range 40–200°C, at a heating rate of 20°C min^{-1}. Details of the procedure have been described elsewhere (18).

Fibre Extrusion and Drawing

As-spun Fibres. The as-spun PP/LCP blended fibres all consisted of two separate phases. Vectra A900 and B950 showed well developed fibrils of 2-5 um diameter with apparently smooth surfaces. Rodrun LC3000 consisted of fibrils of similar cross-section, but of a ribbon-like appearance with far less clearly defined surfaces (17). As the PET component of Rodrun LC3000 contains $-CH_2-CH_2-$ units, Rodrun LC3000 will exhibit some compatibilising action in the polyblend. The PP matrix possesses a structure of high extensibility, in which the polymer chains have little orientation.

The improvement in the initial modulus of as-spun PP fibres which is provided by the LCPs is illustrated in Table I. This observation accords with those of Baird and coworkers (12,13), who noted improvements in the moduli of injection-moulded samples of PP where LCP was incorporated. The tensile strengths are, however, still low, so that fibre drawing remains essential, to develop an oriented molecular structure for the PP matrix.

One-stage Drawing. Table I also shows the effect of conventional one-stage drawing at 150°C to maximum draw ratio. It is clear that the pure PP fibres possess the highest draw ratio and the best tensile properties. The results suggest that the lower draw ratios obtained with the polyblend fibres are the result of resistance to fibre drawing by the LCP fibrils. It is noteworthy that the draw ratio for blended fibres containing Rodrun LC3000, possessing short aliphatic hydrocarbon segments, is closest to that of the pure PP fibres, and also appreciably greater than the draw ratios for the other two blended fibres.

Table I. Drawing Conditions and Fibre Properties of PP and PP/LCP Blended Fibres (PP/LCP w/w ratio 100/10)

	PP	PP/A900	PP/B950	PP/LC3000
<u>As-spun fibres</u>				
Initial modulus, N tex^{-1}	1.08	1.18	1.80	1.21
Increase over PP, %		9.2	66.6	12.0
<u>One-stage drawn fibres</u> (drawn at 150°C)				
Maximum draw ratio	12.3	10.5	9.7	11.9
Tenacity, N tex^{-1}	0.931	0.812	0.610	0.866
Initial modulus, N tex^{-1}	8.74	7.84	6.53	7.41
<u>Two-stage drawn fibres</u>				
Draw ratio at 120°C	6.0	5.7	5.9	6.2
Draw ratio at 165°C	2.5	2.6	2.5	2.4
Overall draw ratio	15.1	14.7	14.6	15.0
Tenacity, N tex^{-1}	0.986	0.974	0.808	1.04
Initial modulus, N tex^{-1}	13.5	14.0	12.8	10.5

SOURCE: Adapted from ref. 17.

Hot-stage photomicrographs have revealed that on one-stage drawing the Vectra A900 and B950 fibrils were split into small fragments, but the fragments still appeared to maintain a smooth surface structure, indicating low interfacial adhesion with the PP matrix (17). In the PP/LC3000 blend, the LCP phase changed from a ribbon-like structure to a rough-surfaced sheet-like structure (17), suggesting greater interfacial adhesion.

Two-stage Drawing. In view of the reduced mechanical performance of the one-stage drawn polyblend fibres, a two-stage drawing process was devised. Optimum drawing conditions were carefully established for the PP/A900 blend. The maximum draw ratio and temperature for the first stage were established as 6 and 120°C, respectively. For the second stage, the optimum conditions were a temperature of 165°C to maximum draw ratio obtainable (17).

It is clear from Table I that in all cases the overall draw ratio was considerably increased by two-stage drawing. For pure PP fibres, there was a large increase in initial modulus and a small increase in fibre tenacity in comparison with one-stage drawing. For the polyblend fibres, however, there were significant increases in both fibre tenacity and initial modulus. Indeed, the PP/A900 fibre blend showed a 4% increase in initial modulus and the PP/LC3000 fibre blend a 5.4% increase in tenacity, respectively, over pure PP fibres.

Using hot-stage microscopy, it has been shown that the two-stage drawing process reduces the extent of splitting of the Vectra A900 and B950 fibrils (15,16). In two-stage drawn PP/LC3000 fibres, the LCP phase exists in a sheet-like structure (as in the one-stage drawn polyblend fibres), but the sheets seem to form a network across the drawn fibres (17).

In Figure 2, fibre tenacities and initial moduli are plotted as a function of the overall draw ratio in the one- and two-stage drawing processes. For both mechanical properties, there is generally an apparent linear dependence on draw ratio, although on this basis the tenacity values of the PP/B950 blended fibres and the modulus values of the PP/LC3000 blended fibres are lower than expected. The reduced tenacity in the PP/B950 fibres may result from the presence of amide linkages in the LCP main chain. The reduced modulus in the PP/LC3000 fibres may reflect the greater flexibility of the copolyester chain in the LCP as a result of its constituent $-CH_2-CH_2-$ units.

Use of Compatibilising Agents

For the LCP to impart to the drawn PP fibres the desired improvement in mechanical performance, good adhesion between the two phases is required (21). The interfacial tension across the phase boundaries must be low. Baird and coworkers (12) have noted that poor adhesion between the PP and LCP phases indicates incompatibility between the polymers, giving rise to reduced tensile strength and only a modest increase in initial modulus. Indeed, the two phases have highly incompatible structures: the PP matrix possesses a flexible aliphatic hydrocarbon structure, whereas the LCP fibrils possess a more rigid, aromatic polar structure. As Baird and coworkers have done for injection-moulded PP/LCP polyblends (12-14), we have investigated the use of compatibilising agents as adhesion promoters between the two phases, and this aspect of the work has concentrated on PP/LC3000 blended fibres, using a PP/LCP/compatiblising agent weight ratio of 100/10/2.5. Much of our attention has been centred on a commercially available product, Polybond 1001 (Figure 1). From Polybond 1001, hereafter referred to as PP-AA, a wide variety of graft copolymers has been synthesised (20), and Figure 1 provides an example, FC1.

Since PP-AA is reported to decompose at temperatures significantly higher than 230°C, a lower extrusion temperature, 240°C, was used than before. Moreover, Rodrun LC3000 can be processed at 240°C owing to its melt temperature of 220°C. PP of MFI 14 was more

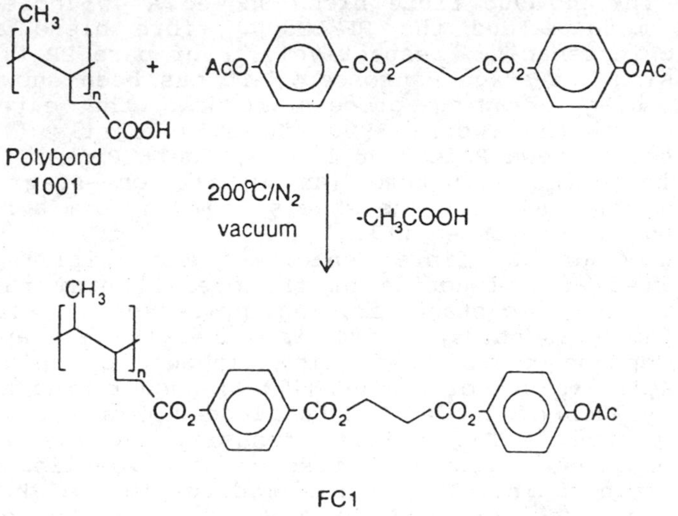

CH3
n
COOH
Polybond
1001
+
AcO
CO2
CO2
OAc
200°C/N2
vacuum
-CH3COOH
CH3
n
CO2
CO2
CO2
OAc
FC1

Figure 1. Synthesis of FC1 from PP-AA. (Adapted from reference 20.)

Tenacity (N tex⁻¹)
Initial Modulus (N tex⁻¹)
Draw Ratio
1·0
0·9
0·8
0·7
0·6
0·5
16
14
12
10
8
6
10
11
12
13
14
15
1
1
2
2

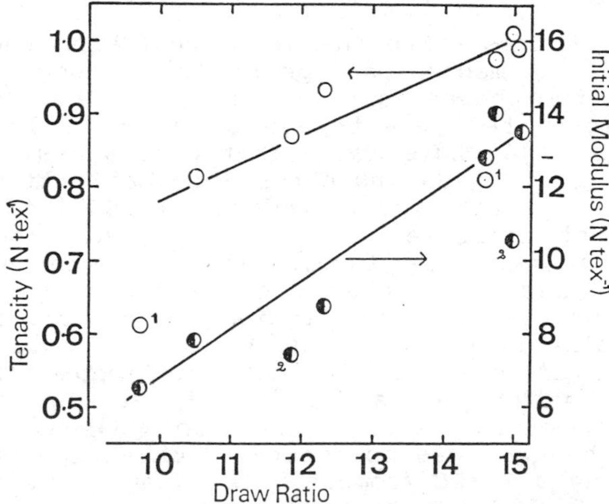

Figure 2. Variation of fibre tenacity, ○, and initial modulus, ◐ , with draw ratio. Points labelled 1 refer to tenacity values for PP/B950. Points labelled 2 refer to initial moduli for PP/LC3000.

suitable for spinning at 240°C than the PP of MFI 3 used previously at the higher melt extrusion temperatures.

Effect of Compatibilising Agent FC1. The effect of incorporating the functionalised compatibilising agent, FC1, on LCP morphology within the as-spun blended fibres has been studied (18). The blended fibre containing FC1 possesses a more widely dispersed LCP fibril structure, indicative of reduced interfacial tension between the PP matrix and LCP phase. Polyblend fibres compatibilised with PP-AA contain a much more oriented structure, and the LCP fibrils have a distinctly higher aspect ratio.

Table II lists some tensile data obtained for one-stage and two-stage drawn fibre samples containing pure PP, PP/LC3000 and PP/LC3000/FC1, respectively. The compatibilised fibre exhibits the highest tenacity amongst the one-stage drawn samples but the lowest tenacity amongst the two-stage drawn samples. It is noteworthy too that after the second drawing stage the tenacity of the compatibilised fibre falls, in contrast to the behaviour of all the other polyblend fibres we have studied. This fall in tenacity is a consequence of the increased adhesion between the LCP and PP phases promoted by the compatibilising agent, and increased adhesion would be expected to have a detrimental effect on fibres drawn by a two-stage process.

Table II. Effect of Compatibilising Agent, FC1, on the Tenacities of One-stage and Two-stage Drawn PP/LC3000 Blended Fibres (PP/LCP w/w ratio 100/10)

	PP	PP/LC3000	PP/LC3000/FC1
One-stage drawn fibres			
Draw ratio	11.6	11.5	11.7
Tenacity, N tex^{-1}	0.643	0.584	0.713
Two-stage drawn fibes			
Draw ratio	18.1	15.0	14.5
Tenacity, N tex^{-1}	0.776	0.672	0.662

SOURCE: Adapted from ref. 18.

Microscopic evidence supports this explanation (18). Little difference in LCP structure has been observed amongst the one-stage drawn samples, but the LCP fibrils in the two-stage drawn PP/LC3000/FC1 fibres are considerably smaller than in the corresponding PP/LC3000 fibres. The greater fragmentation of the LCP fibrils in the compatibilised fibres can be seen as a direct consequence of the increased adhesion between the PP and LCP phases.

Effect of PP-AA as a Compatibilising Agent. Table III highlights the tensile properties of two-stage drawn PP/LC3000 fibres using PP-AA itself as a compatibilising agent. There appears to be little difference in the tenacities of the fibre samples compared. Moreover, we have found no significant difference in the LCP fibrillar structure of PP/LC3000/PP-AA in comparison with the fibrillar structure of the other two samples (18). These observations suggest that the compatibilising action of PP-AA differs markedly from that of FC1. Work by Xanthos et al. (22) on blends of PP-AA with PET established that the presence of the acrylic acid groups caused a four-fold reduction in interfacial tension compared with standard PP. This reduction was attributed to enhanced specific physical interactions between the polar components of the blend. Porter and Wang have (23) discussed the effect of transesterification reactions and their effect on compatibility in polymer blends, including those containing LCPs. Such reactions could occur between the acrylic acid moieties present in PP-AA and the ester linkages in the LCP during melt processing. The compatibilisation effect appears, therefore, to arise from a specific interaction between the polar acrylic acid groups and the LCP itself, which reduces the interfacial tension in the blend, as indicated by the pronounced difference in LCP fibril structure in the as-spun fibre sample. This specific interaction does not appear to have such a detrimental effect on fibre properties during two-stage drawing: the increase in interfacial adhesion between the PP and LCP phases is less than the increase promoted by FC1.

Table III. Effect of PP-AA on Two-stage Drawn PP/LC3000 Blended Fibres (PP/LCP w/w ratio 100/10)

Sample	Fibre thickness (tex)	Draw ratio	Extension (%)	Tenacity (N tex^{-1})
PP	2.56	24.1	14.9	0.856
PP/LC3000	5.81	12.9	15.9	0.838
PP/LC3000/PP-AA	3.37	12.4	21.2	0.873

SOURCE: Adapted from ref. 18.

Fibre Crystallinities. It is evident from the data listed in Table IV that PP-AA causes a pronounced increase in the crystallinity of the polyblend fibres. 20% and 10% increases are observed over all the other samples for the as-spun and two-stage drawn fibres, respectively (18). This result can also be explained in terms of blend interactions arising from the acrylic acid groups present in PP-AA. No equivalent increase in crystallinity is promoted by the compatibilising agent, FC1, an observation providing further evidence for the different compatibilising effects of PP-AA and FC1.

Table IV. Crystallinity Data for Fibre Samples

Sample	Crystallinity, %	
	As-spun	Two-stage drawn
PP	40.1	61.8
PP/LC3000	42.6	64.3
PP/LC3000/FC1	43.4	61.8
PP/LC3000/PP-AA	61.7	70.7

SOURCE: Adapted from ref. 18.

Liquid Crystalline Graft Compatibilising Agents. The above results demonstrate that a graft copolymer compatibilising agent functionalised on PP-AA can indeed effect a significant increase in adhesion between the LCP fibrils and PP matrix in polyblend fibres. However, this strong interfacial adhesion also promotes fragmentation of the fibrils during the drawing process. A strategy is, therefore, required which would considerably reduce adhesion during the actual drawing process but would promote good adhesion in the final drawn fibres. To this end, we are incorporating compatibilising agents synthesised from PP-AA, which themselves possess liquid crystalline properties (19). Two examples, FC2 and FC3, are shown in Figure 3. FC2 possesses a nematic phase over a small temperature range well below the temperatures used for the drawing process. FC3, on the other hand, exhibits nematic properties over a wide temperature range, which covers the fibre drawing (and extrusion) temperatures. Table V shows the properties of

Table V. Tensile Properties of Two-stage Drawn PP/LC3000 Monofilament Fibres (PP/LC3000 ratio w/w 100/10)

	Draw ratio	Tenacity (N tex^{-1})	Initial modulus (N tex^{-1})
PP	10.4	0.935	8.53
PP/LC3000	10.5	0.986	5.56
PP/LC3000/FC2	9.8	0.860	4.31
PP/LC3000/FC3	12.1	0.995	6.27

SOURCE: Adapted from ref. 19.

some fibres drawn by a two-stage process after extrusion through a monofilament spinneret. It is noteworthy that

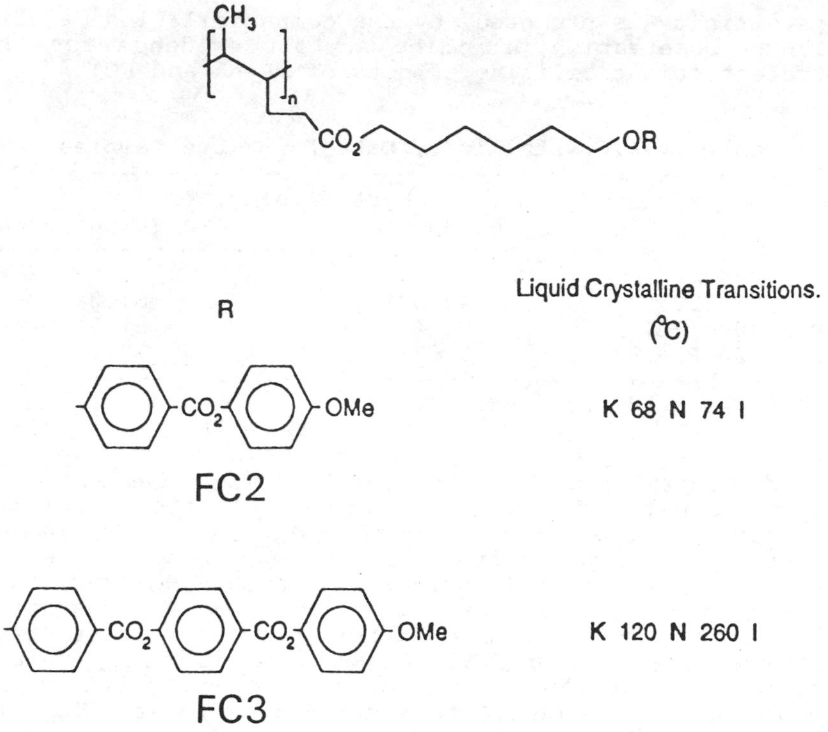

R	Liquid Crystalline Transitions. (°C)
FC2	K 68 N 74 I
FC3	K 120 N 260 I

Figure 3. Liquid crystalline functionalised compatibilising agents. (Reproduced with permission from reference 19. Copyright 1996 Elsevier.)

there is a small, but definite increase in fibre tenacity in the sample containing FC3 over standard PP/LC3000 fibre, and a significant increase in initial modulus. By contrast, the fibre tenacity and modulus of the sample containing FC2 are lower than those of PP/LC3000. It is suggested that, since FC3 is in a nematic state during the drawing process, it lubricates the stretching of the PP chains over the LCP fibrils. In the drawn fibre, however, it promotes adhesion between LCP and PP. FC2, however, is not in a nematic state during fibre drawing and, therefore, does not confer a lubricating effect. There is a consequent loss in tenacity and initial modulus in the drawn fibres.

Acknowledgements

The authors are grateful to the former U.K. Science and Engineering Research Council for financial assistance (Grant Ref. Nos. GR/F58776 and GR/H28417), and to Bonar Textiles Limited for financial assistance, technical discussions and the provision of polymer samples.

Literature Cited

1. Paul, D.R. and Newman, S., (Eds.) '**Polymer Blends**', Academic Press, New York, 1978.
2. Hersh, S.P., in '**High Technology Fibres**, Part A' (Eds. M.Lewin and J. Preston), Marcel Dekker Inc., New York, 1985.
3. MacDonald, W.A., in '**High Value Polymers**' (Ed. A.H. Fawcett), The Royal Society of Chemistry, Cambridge, 1991, pp428–454.
4. Beers, D.E. and Ramirez, J.E., **J. Text. Inst.**, 1990,**81**,561.
5. Yang, H.H., '**Aromatic High-Strength Fibres**', John Wiley and Sons, New York, 1989.
6. Isayev, A.I. and Modic, M., **Polym. Compos.**, 1987,**8**,158.
7. Siegmann, A., Dagan A. and Kenig, S., **Polymer**, 1985,**26**,1325.
8. Kiss, G., **Polym. Eng. Sci.**, 1987,**27**,410.
9. Chung, T.S., **Plast. Eng.**, 1987,**43**,39.
10. Gaylord, N.G., **J. Macromol. Sci. Chem (A)**, 1986,**26**,1211.
11. Fayt, R., Jerome, T. and Teyssie, P., **Makromol. Chem.**, 1986,**187**,837.
12. Datta, A., Chen, H.H. and Baird, D.G., **Polymer**, 1993,**34**,759.
13. Datta, A. and Baird, D.G., **Polymer**, 1995,**36**,505.
14. O'Donnell, H.J. and Baird, D.G., **Polymer**, 1995,**36**,3113.
15. Qin, Y., Brydon, D.L., Mather, R.R. and Wardman, R.H., **Polymer**, 1993,**34**,1196.
16. Qin, Y., Brydon, D.L., Mather, R.R. and Wardman, R.H., **Polymer**, 1993,**34**,1202.
17. Qin, Y., Brydon, D.L., Mather, R.R. and Wardman, R.H., **Polymer**, 1993,**34**,3597.
18. Miller, M.M., Cowie, J.M.G., Tait, J.G., Brydon, D.L. and Mather, R.R., **Polymer**, 1995,**36**,3107.
19. Miller, M.M., Brydon, D.L., Cowie, J.M.G. and Mather, R.R., submitted to **Polymer Communications**.
20. Miller, M.M., Brydon, D.L., Cowie, J.M.G. and Mather, R.R., **Macromol. Rapid Commun.**, 1994,**15**,857.
21. Dutta, D., Fruitwala, H., Kohli, A. and Weiss, R.A., **Polym. Eng. Sci.**, 1990,**30**,1005.
22. Xanthos, M., Young, M.W. and Biesenberger, J.A., **Polym. Eng. Sci.**, 1990,**30**,355.
23. Porter, R.S. and Wang, L.H., **Polymer**, 1992,**33**,2019.

Chapter 8

Fibers of Blends with Liquid-Crystalline Polymers: Spinnability and Mechanical Properties

F. P. La Mantia[1], A. Roggero[2], U. Pedretti[2], and P. L. Magagnini[3]

[1]Dipartimento di Ingegneria Chimica dei Processi e dei Materiali, Università di Palermo, Viale delle Scienze, 90128 Palermo, Italy
[2]Eniricerche S.p.A., Via Maritano 26, 20097 San Donato Milanese, Italy
[3]Dipartimento di Ingegneria Chimica, Chimica Industriale e Scienza dei Materiali, Università di Pisa, Via Diotisalvi 2, 56126 Pisa, Italy

Several blends of flexible polymers (PET, PC, Nylon-6, PP) with wholly aromatic or semi-aromatic liquid crystalline polymers (LCPs) were spun under different conditions and their melt viscosities, their spinnability (evaluated through the melt strength, MS, and the breaking stretching ratio, BSR), and their mechanical properties were studied. The measured MS and BSR values of some of the blends were found to be higher than expected on the basis of the viscosity reduction brought about by the LCP minor phase, and were interpreted as being the result of good phase compatibility in the molten state. The fiber mechanical properties were appreciably enhanced over those of neat polymers only when good spinnability (granting LCP droplets fibrillation, and orientation of the matrix polymer) was accompanied by sufficient interphase adhesion in the solid state.

It is well known that the addition of relatively small amounts of a thermotropic liquid crystalline polymer (LCP) into flexible thermoplastic polymers may mean a considerable energy saving, due to a strong reduction of melt viscosity (*1*). Moreover, a reinforcing effect can also be expected, especially if flow conditions favoring LCP fibrillation are used for processing. In particular, if a LCP/polymer blend is melt spun, the elongational flow at the die inlet, as well as that taking place as a result of the draw-down force applied to the molten filament at the spinneret exit, may favor such fibrillation. Thus, melt spun fibers displaying enhanced tensile properties can be expectedly obtained as a result of the *in situ* addition of LCP into the matrix resin, provided that the compatibility of the two phases is sufficient. However, the accurate design of LCP/polymer blend fibers should be made considering that, whereas the mechanical properties of as-spun filaments of neat thermoplastic polymers may often be improved by furthering molecular orientation through cold drawing, this is generally not possible for LCP/polymer fibers due to the high rigidity of the LCP fibrils. Therefore, it is

0097–6156/96/0632–0110$15.00/0

necessary that spinning conditions granting, not only the fibrillation of the LCP minor phase, but also a sufficient molecular orientation of the matrix resin are employed, or the advantage of LCP reinforcement may be even outweighed by the lower matrix orientation. This means that high draw-down ratios, i.e., the ratios of windup to extrusion speeds, must be used while spinning LCP/polymer blends.

The melt strength (MS) and the breaking stretching ratio (BSR) are the two most important characteristics defining the spinnability of a polymer. A polymer having high MS and BSR will expectedly sustain very high draw ratios during melt spinning. For a given polymer, MS and BSR increase on increasing the molecular weight and the melt viscosity. Thus, the drawability of a molten thermoplastic might be expected to worsen as a result of the LCP addition, because this is known to bring about a strong viscosity reduction.

In this work, the rheological behavior of several LCP/polymer blends has been studied. Their spinnability has also been investigated by measuring the MS and BSR, as well as the mechanical properties of as-spun fibers. The results are discussed with reference to the temperature dependence of the viscosities of the two phases, which clearly influences the blends behavior during the non-isothermal fiber drawing stage.

Experimental

Four thermoplastic polymers: polypropylene (PP), polycarbonate (PC), poly(ethylene terephthalate) (PET), and nylon-6 (NY), and four LCPs: Vectra A-900 (VA), Vectra B-950 (VB), SBH 1:1:2 (SBH), and SBHN 1:1:3:5 (SBHN) were used. The source and some of the characteristics of the polymers, as well as the compositions of the LCPs, are shown in Table 1.

Table 1
Characteristics of the polymers used for the blends preparation

Sample	M_w	[η], dl/g	Name	Manufacturer
PP	680.000		D60P	Himont
PC	36.000		Sinvet 301	Enichem Polimeri
PET		0.62		Enichem Polimeri
NY	62.000		ADS40	SNIA
VA			Vectra A-900	Hoechst Celanese
[*4-hydroxybenzoic acid (H) (73%), 2-hydroxy-6-naphthoic acid (N) (27%)*]				
VB			Vectra B-950	Hoechst Celanese
[*N (60%), terephthalic acid (20%), 4-aminophenol (20%)*]				
SBH			SBH 1:1:2	Eniricerche SpA
[*sebacic acid (S) (25%), 4,4'-dihydroxybiphenyl (B) (25%), H (50%)*]				
SBHN			SBHN 1:1:3:5	Eniricerche SpA
[*S (10%), B (10%), H (30%), N (50%)*]				

The LCP/polymer blends were prepared with a laboratory single screw extruder (D=19 mm, L/D=25; Brabender, Germany), equipped with a die assembly for ribbon extrusion. Blends with 0, 5, 10, and 20% LCP (w/w) were prepared. The die temperatures used for the different blends were as follows: PP/SBH and NY/SBH, 250°C; PET/SBH, PET/SBHN and PET/VA, 270°C; PC/SBH and NY/VB, 290°C; PC/VB, 300°C.

The shear viscosity measurements were made with a capillary (D = 1 mm, L/D = 40) viscometer (Rheoscope 1000 by CEAST). The temperatures were as follows: PP/SBH, NY/SBH, 250°C; PC/SBH, PET/SBHN, PET/VA, 270°C; PC/VB, NY/VB, 290°C. Elongational viscosity was estimated using Cogswell's analysis (*2*):

$$\lambda = \frac{9}{32} \cdot (n+1)^2 \cdot \frac{\Delta P^2}{\dot{\gamma}^2 \eta}$$

where η is the shear viscosity, $\dot{\gamma}$ is the apparent shear rate, n is the power law index, and ΔP is the pressure drop at the die inlet evaluated through the Bagley plot. Capillaries with L/D=5, 10 and 20 were used. The average extension rate was evaluated by means of the relation:

$$\dot{\varepsilon} = \frac{4}{3(n+1)} \cdot \eta \cdot \frac{\dot{\gamma}^2}{\Delta P}$$

Fibers were spun using the tensile attachment of the same equipment, with a shear rate of 24 s^{-1}. The spinning temperatures were as follows: PP/SBH and NY/SBH, 250°C; PET/SBH, PET/SBHN and PET/VA, 270°C; PC/SBH and NY/VB, 290°C; PC/VB, 300°C. During the spinning tests, the MS and the BSR were measured by drawing the extruded monofilaments with a linearly increasing extension rate (acceleration: 1 rpm/s). The fiber mechanical properties were measured with an Instron 1122, at an elongational rate of 0.66 min^{-1}.

Results and Discussion

The shear viscosities of some of the investigated LCP/polymer blends, measured at a shear rate of 24 s^{-1}, are shown in Figure 1 as a function of the LCP content. For all the blends, a considerable decrease of melt viscosity was observed as a result of the LCP addition. This is in agreement with literature data (*1*). A minimum is seen for some of the viscosity/composition curves (cf. NY/VB and PET/VA). This was shown to occur when the LCP's viscosity is higher than, or similar to, that of the matrix (*3, 4*).

Because the shear viscosities of thermoplastic polymers are reduced by the addition of small amounts of LCPs, one would expect their MS and, therefore, their spinnability to decrease accordingly. This is not always so. In fact, literature data show that many LCP/polymer blends can easily be spun with high draw-down ratios. On the other hand, the decrease of the shear viscosity of a polymer, due to the LCP addition, is not accompanied, in most cases, by a concomitant decrease of the elongational viscosity, as expected for materials obeying the Trouton rule ($\lambda = 3\eta_0$). As an example, the elongational viscosities of NY, VB, and 90/10 NY/VB are shown in Figure 2 as a function of the extension rate. It may be seen that, whereas the elongational viscosity, λ, of neat NY is almost constant, and not far from the Trouton value, that of the LCP is very much higher, and 50-100 times larger than $3\eta_0$. This behavior, which is in agreement with that observed by others (*5, 6*), has been attributed to the considerable

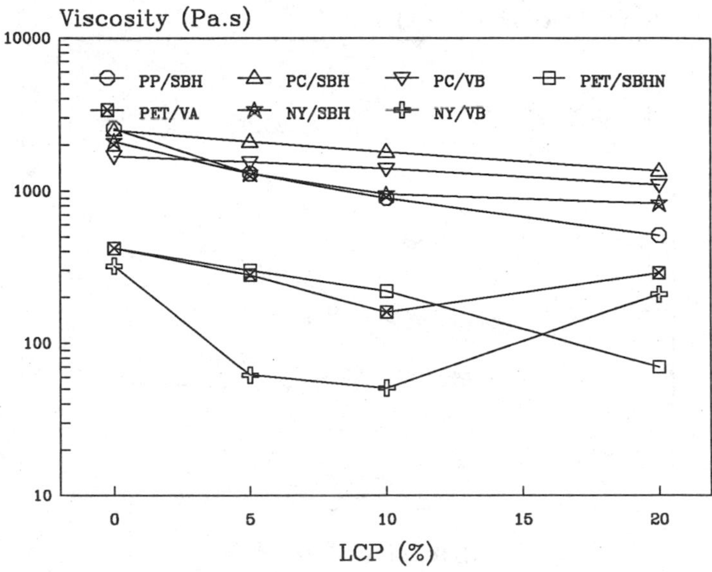

Figure 1
Viscosity/composition curves measured at a shear rate of 24 s^{-1}, and at the following temperatures: 250°C (PP/SBH, NY/SBH); 270°C (PC/SBH, PET/SBHN, PET/VA); 290°C (PC/VB, NY/VB).

amount of energy required for deforming and orienting the LCP domains in the convergent flow at the die entrance. Thus, it is not surprising that, contrary to shear viscosity, the elongational viscosity of the 90/10 NY/VB blend is considerably higher than that of the pure matrix.

In order to get more direct information on the spinnability of LCP/polymer blends, measurements of MS and BSR, carried out under non-isothermal conditions on the spinning line, have been made. As it is shown in Figure 3, a decrease of MS with increasing the LCP content is actually observed only for some of the blends, namely: PC/VB, PC/SBH, PET/SBH, PET/SBHN and NY/SBH. For the PP/SBH, PET/VA and NY/VB blends, on the contrary, an increase of MS is obtained. These results should be interpreted by the consideration of the different factors influencing the tensile strength of the molten filament, i.e., the viscosity, the viscosity changes taking place along the spinning line while the extruded filament rapidly cools down in the draw region (*7*), etc. Actually, under non-isothermal conditions, the viscosity of some blends may come out to be higher than that of the neat thermoplastic matrices, in some temperature intervals, if the viscosities of the two components have different activation energies.

As for the BSR values of the LCP/polymer blends, i.e., the ratios of the windup speeds to the extrusion speeds corresponding to filament breaking, the values measured at a shear rate of 24 s^{-1} for the investigated blends are collected in Table 2. It may be observed that the BSR of all the investigated blends decreases upon increasing the LCP content, as it might be expected on the basis of the biphasic nature of the blends. However, the extent of BSR reduction, which depends mainly on the interphase adhesion in the molten state, changes markedly with the chemical structure of the two

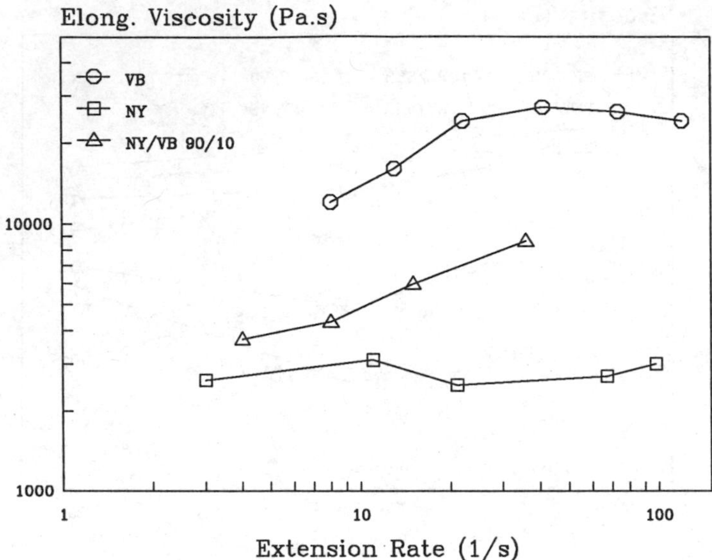

Figure 2
Elongational viscosity of VB, NY, and NY/VB 90/10.

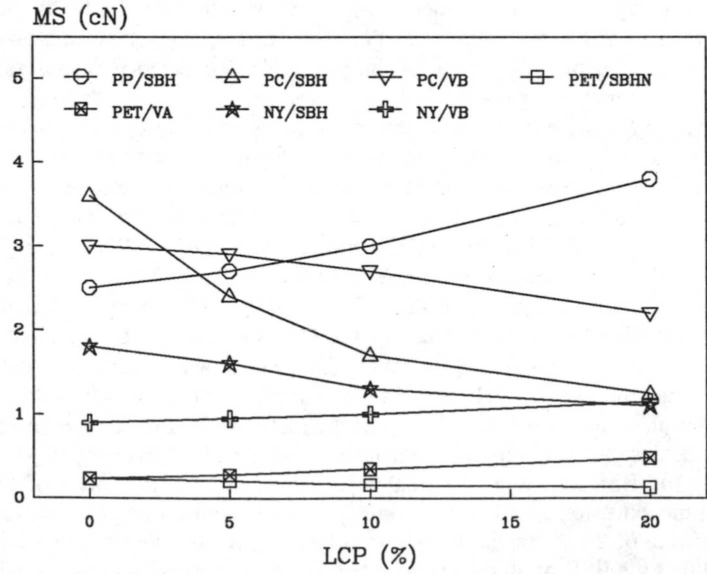

Figure 3
Melt strength of some LCP/polymer blends as a function of the LCP content.

components of the blends. For PET, for example, the decrease of BSR is very small when it is blended with semi-aromatic LCPs, such as SBH and SBHN, whereas a more significant drop of stretchability is found when it is blended with a wholly aromatic LCP, such as VA.

On the contrary, both PC and NY experience almost the same BSR reduction when added with aromatic or semi-aromatic LCPs. The spinning behavior of the PP/SBH blend is peculiar in that an increase of the LCP content leads to a marked MS increase, accompanied by a fairly small BSR reduction. This means that, in the molten state, the adhesion between the PP and the SBH incompatible phases is not bad. From these experimental data, it may be concluded that the spinnability of PP is not severely worsened by the addition of a semi-aromatic LCP, such as SBH.

Table 2
Breaking stretching ratios (BSR) of the LCP/polymer blends, measured at a shear rate of 24 s^{-1}

LCP (%)	PP/ SBH	PC/ SBH	PC/ VB	PET/ SBH	PET/ SBHN	PET/ VA	NY/ SBH	NY/ VB
0	650	730	1400	710	710	710	1400	2000
5	610	520	1100	710	700	620	950	1000
10	500	400	650	700	660	400	670	850
20	450	200	400	660	620	300	540	750

The elastic modulus of the fibers, prepared with different windup speeds, was also measured in order to get an idea of the reinforcing ability of the different LCPs, toward the different thermoplastics. The tensile modulus increases with the draw-down ratio.

In Figure 4, the ratios of the maximum moduli measured for the blends and the neat thermoplastics are plotted versus the LCP content, for the different blends. For the correct interpretation of the data shown in Figure 4, it must be kept in mind that the values of the tensile moduli whose ratios are plotted in the figure are those measured on fibers prepared with the highest windup speeds that could be used for either the blends and the neat polymers. Thus, the ratio of the maximum tensile moduli accounts also for the orientation of the matrix, due to the filament attenuation induced by drawing.

The strongest modulus improvement is found for NY, especially when a wholly aromatic copolyesteramide (VB) is added into it (E ratio = 2.75, for an LCP content of 20%). However, even with a semi-aromatic copolyester (SBH), NY experiences fairly good reinforcement. The behavior of the PC based blends is very similar: the comparatively lower values of the E ratios of the PC blends, with respect to the NY blends, can be explained considering that the tensile modulus of neat PC fibers, prepared under comparable conditions, is much higher than that of neat NY fibers (6.0 GPa vs. 0.8 Gpa).

As for the PET based blends, apparently opposite results have been obtained. In fact, whereas the addition of a semi-aromatic LCP, such as SBH or SBHN, leads to a remarkable modulus enhancement, the effect of the addition of an intrinsically much stiffer wholly aromatic LCP, such as VA, is surprisingly negligible. This is even more unexpected because, for compression molded specimens, the modulus improvement was found to be higher for PET/VA blends than for, e.g., PET/SBH blends (*8*).

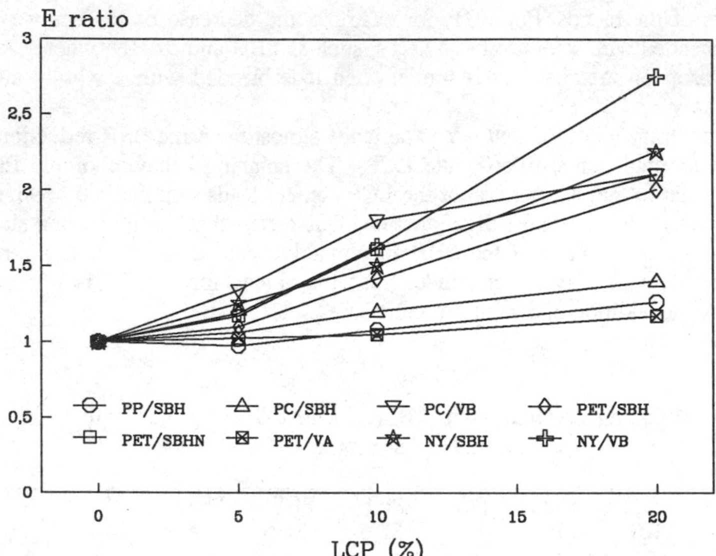

Figure 4
Ratio of the maximum tensile moduli measured for blend and neat thermoplastic fibers, as a function of the LCP content.

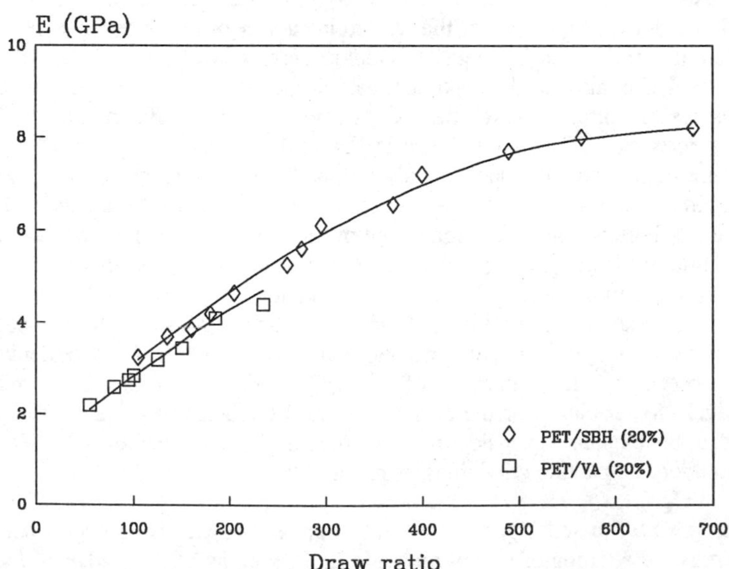

Figure 5
Tensile moduli of 80/20 PET/SBH and PET/VA fibers, as a function of the draw-down ratio.

However, as shown in Table 2, the stretchability of the PET/VA blends is approximately the half of those of the PET/SBH and PET/SBHN blends. It must be emphasized that, in order to avoid PET degradation, the PET/VA blends were spun at a temperature (270°C) lower than that of the crystal-nematic transition of VA. Therefore, it is possible that the LCP melting was uncompleted under the spinning conditions, thus preventing the use of the high draw-down ratios needed in view of a pronounced orientation of the PET macromolecules. The plots of the tensile moduli of the PET/VA and PET/SBH blends with 20% LCP, against draw-down ratio, shown in Figure 5, demonstrate that the PET/VA fibers could not be spun with draw ratios higher than ca. 250, and that their modulus remained much lower than the hypothetical asymptotic value. On the other hand, an increase of the spinning temperature worsened the MS of the blends besides leading to initial PET degradation.

The modulus improvement of the PP fibers resulting from the addition of SBH is much lower than it might be anticipated in view of the good spinnability of the PP/SBH blends. The apparent conflict between the very good MS and BRS values of these blends and their disappointing mechanical properties can perhaps be explained assuming that the interphase adhesion between PP and SBH, which has been shown to be fairly good in the molten state, is strongly depressed in the solid state, as a result of the matrix crystallization.

Conclusions

Despite the fact that the LCPs are known to reduce the shear viscosity of molten flexible polymers, the MS and the BSR of the LCP/polymer blends may still be high enough to grant good spinnability. High MS and BSR values may be taken as an indication of good phase compatibility in the molten state. On the other hand, the LCP phase may actually play a considerable reinforcing effect only if good spinnability (granting the fibrillation of the LCP particles and the orientation of the matrix macromolecules) is accompanied by good interphase adhesion is good in the solid state.

Acknowledgments

This work was financially supported by C.N.R., by the Italian Ministry of University and Scientific and Technological Research (MURST), and by Eniricerche S.p.A.

Literature Cited

1. *Thermotropic Liquid Crystal Polymer Blends*; La Mantia, F. P., Ed.; Technomic Publ. Co. Inc., Lancaster, U.S.A., 1993, Chapters 4, 5, and references therein
2. Cogswell, F. N. *Polym. Eng. Sci.*, **1972**, *12*, 64.
3. La Mantia, F. P.; Valenza, A. *Makromol. Chem., Macromol. Symp.*, **1992**, *56*, 151.
4. La Mantia, F. P.; Valenza, A. *Polym. Networks & Blends*, **1993**, *3*, 125.
5. Beery, D.; Kenig, S.; Siegmann, A. *Polym. Eng. Sci.*, **1992**, *32*, 14.
6. Turek, D .E.; Simon, G. P.; Smejkal, F.; Grosso, M.; Incarnato, L.; Acierno, D. *Polymer Commun.*, **1993**, *34*, 204.
7. La Mantia, F. P.; Valenza, A.; Scargiali, F. *Polym. Eng. Sci.*, **1994**, *34*, 799.
8. La Mantia, F. P.; Pedretti, U.; Città, V.; Geraci, C. *Polym. Networks Blends*, **1994**, *4*, 151.

Chapter 9

Processing and Physical Properties of Ternary In Situ Composites

Byoung Chul Kim, Seung Sang Hwang, Soon Man Hong, and Yongsok Seo

Division of Polymers, Korea Institute of Science and Technology, P.O. Box 131, Cheongryang, Seoul, Korea 136–791

The ternary blends containing a thermotropic liquid crystalline polymer (LCP) as a reinforcing component were investigated in terms of processing characteristics, mechanical properties, and phase morphology. The ternary LCP blends were advantageous over the binary LCP blends if the third isotropic component could impart a compatibilizing effect to the blend systems. A proper choice of the third component reduced cleavage at the interface in the blends and it was helpful in solving problems associated with phase separation between LCP and matrix phases. In addition, the third component played an auxiliary role in matching the viscosity ratio of LCP to matrix phases during melt blending, and allowed a greater degree of elongational deformation of the LCP domains.

It has been well recognized that melt blending of a thermotropic liquid crystalline polymer (LCP) and an isotropic polymer produces a composite in which fibrous LCP domains dispersed within the blend act as a reinforcement (*1,2*). The so-called *in-situ* composite possesses several advantages in comparison with the inorganic reinforced thermoplastic composites. Firstly, LCP lowers the blend viscosity in the actual fabrication temperature range *(3-5)*. Hence, the enhanced processability endows moldability for fine and complex shaped products.

0097–6156/96/0632–0118$16.00/0

Secondly, the wetting problems of the dispersed fibrous LCP domains to the isotropic matrix phase are less serious. Thirdly, in some cases LCP imparts a nucleating effect to the crystallization of the matrix phase *(6,7)*. Fourthly, the dimensional stability of the fabricated products may be greatly improved because of the low thermal expansion coefficient of LCP. Finally, the products based on the *in-situ* composite are reusable, which is helpful to solve the ecological and environmental problems.

Up to now, most of the efforts to develop a high-performance *in-situ* composite have been focused on the binary LCP blends, and a number of papers have been disclosed during the last decade. The binary *in-situ* composites, however, frequently lead to unsatisfactory mechanical properties *(8,9)*. It is largely because of poor dispersion and poor elongational deformation of the LCP domains, along with interface instabilities between LCP and matrix phases resulting from phase separation.

Recently, several researchers *(8,10-12)* have reported that compatibilized LCP blends exhibit much enhanced mechanical properties, and they have attributed it to the improved interface adhesion between LCP and matrix phases. In this study, an isotropic polymer was incorporated to the binary LCP blends as a third component to solve the problems caused by interface instabilities and poor deformation of LCP domains. The third component was selected from the commercially available block copolymers on the experimental basis, or prepared by synthesizing block copolymers by molecular design.

EXPERIMENTAL

Materials. As a matrix polymer, polyetherimide (PEI) Ultem 1000 of General Electric and polyphenylenesulfide (PPS) Ryton GR-02 of Phillips Petroleum were used. The reinforcing component was thermotropic liquid crystalline copolyester (LCP) Vectra B950 of Hoechst-Celanese. The chemical structures of PEI and LCP are presented in Figure 1.

As a third component, polysulfone (PSF) Udel P-1700 of Amoco was tested for PPS/LCP blend, and three kinds of newly designed imide polymers, polyetherimide (PEsI), PET/PEsI block copolymer (BHETI), and PBT/PEsI block copolymer (BHBTI), were tested for PEI/LCP blend. The three compatibilizing agents, PEsI, BHETI and

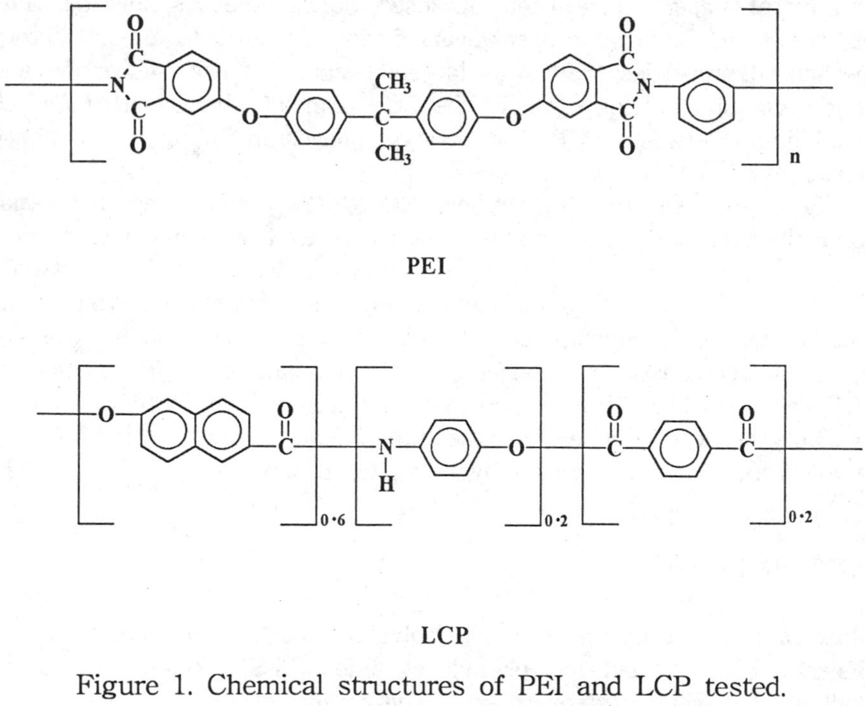

PEI

LCP

Figure 1. Chemical structures of PEI and LCP tested.

BHBTI, were prepared *via* multistep reactions. PEsI was prepared according to the reaction scheme shown in Figure 2. The reaction schemes for preparing the two imide block copolymers, BHETI and BHBTI, are given in Figure 3. The physical properties of BHETI and BHBTI are given in Table 1.

Preparation of Blends and Specimens. The formulated components were melt blended in a Brabender twin-screw extruder in the nitrogen atmosphere for 3 to 5 minutes. In the case of PPS blends, the blending temperature ranged from 290 to 300°C, and PEI blends were compounded in the temperature range of 320 to 330°C according to the blending ratio. The spinning experiments were carried out using a capillary die attatched to the twin-screw extruder, whose radius and length were 2 and 20mm, respectively. The spinning temperature of PPS blends was 290°C and that of PEI blends was 320°C.

Measurement of Physical Properties. The dynamic rheological properties were measured by RDS-7700 with the concentric parallel plates. The thermal properties were measured with a Du Pont Thermal Analysis 2000 equipped with a 910 DSC. The fractured surface was observed by Hitatchi scanning electron microscope S-510. The tensile properties were measured by Instron tensile tester.

RESULTS AND DISCUSSION

1. PEI/LCP/PEsI Blend Systems

It is well recognized in the binary polymer blends that the two component is partially compatible and the binary blend forms a biphase if the glass transition temperatures (T_g) of the two components approach each other. In the 75/25 PEI/LCP blend, T_g of PEI is decreased from 218 (T_g of pure PEI) to 214°C *(5)*. In the 70/30 LCP/PEsI blend, T_g of LCP is increased from 124 (T_g of pure LCP) to 152°C and T_g of PEsI is decreased from 225 (T_g of pure PEsI) to 190°C. These DSC results verify that PEsI is partially compatible with both LCP and PEI.

PEI is an amorphous polymer of high melt viscosity. As frequently observed in other LCP blends, LCP lowers the viscosity of PEI in the processing temperature range, suggesting that LCP acts as plasticizer

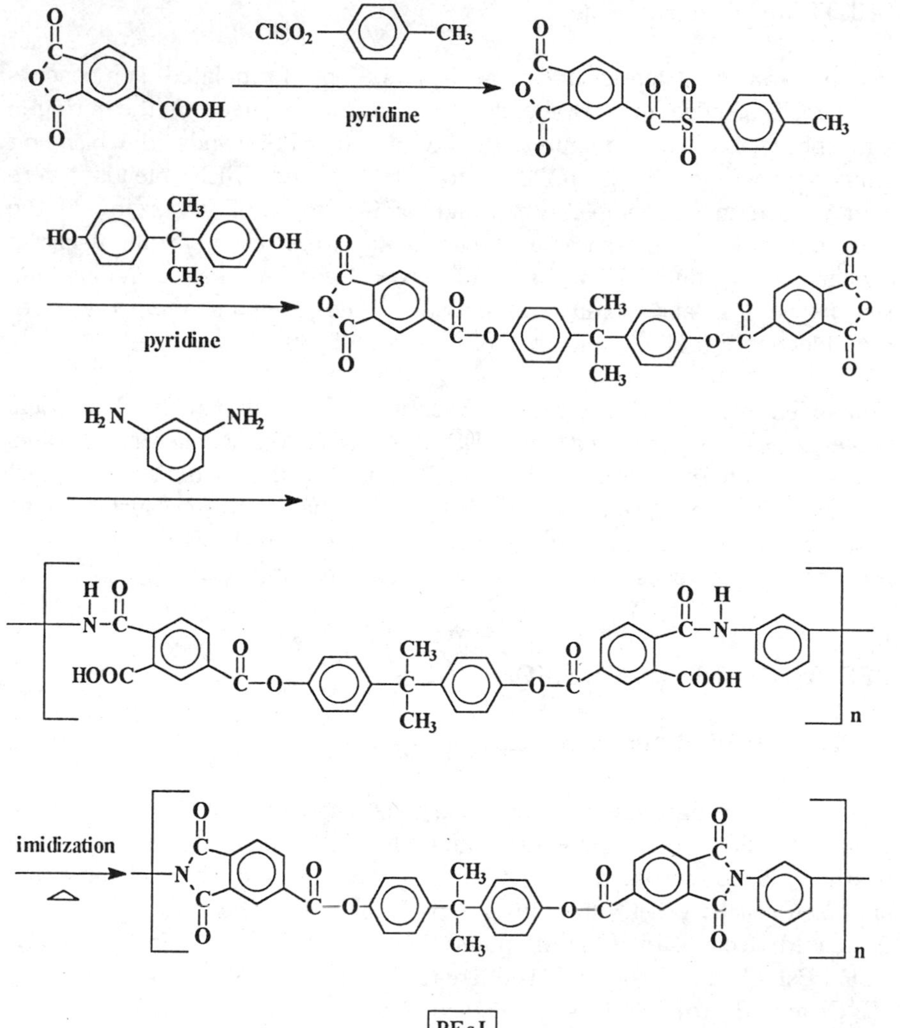

PEsI

Figure 2. Schematic representation of preparative method of PEsI.

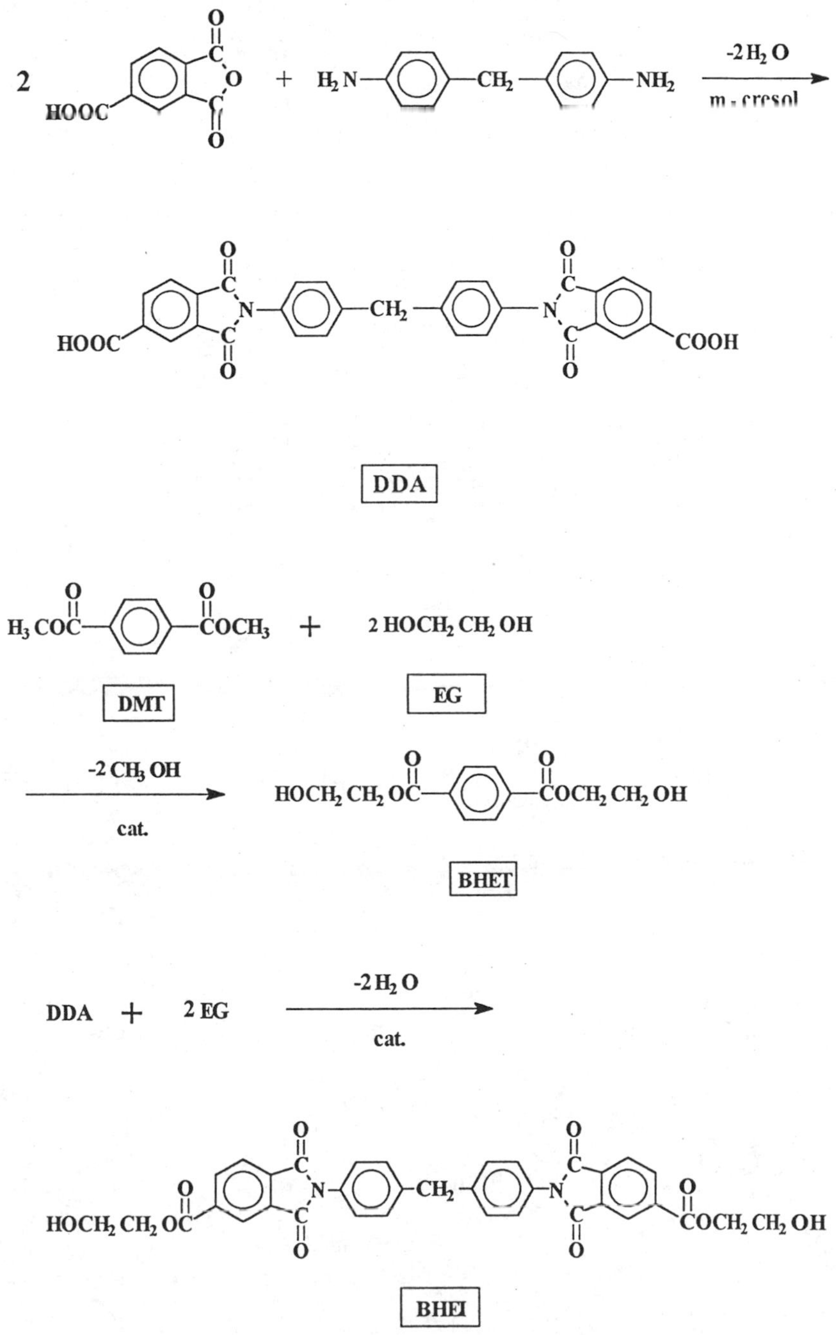

2 HOOC–(trimellitic anhydride) + H_2N–⟨⟩–CH_2–⟨⟩–NH_2 $\xrightarrow[\text{m-cresol}]{-2H_2O}$

HOOC–(imide)–N–⟨⟩–CH_2–⟨⟩–N–(imide)–COOH

DDA

$H_3COC(=O)$–⟨⟩–$C(=O)OCH_3$ + 2 $HOCH_2CH_2OH$

DMT EG

$\xrightarrow[\text{cat.}]{-2\,CH_3OH}$ $HOCH_2CH_2OC(=O)$–⟨⟩–$C(=O)OCH_2CH_2OH$

BHET

DDA + 2 EG $\xrightarrow[\text{cat.}]{-2H_2O}$

$HOCH_2CH_2OC(=O)$–(imide)–N–⟨⟩–CH_2–⟨⟩–N–(imide)–$C(=O)OCH_2CH_2OH$

BHEI

Figure 3. Schematic representation of preparative methods of BHETI and BHBTI. Continued on next page.

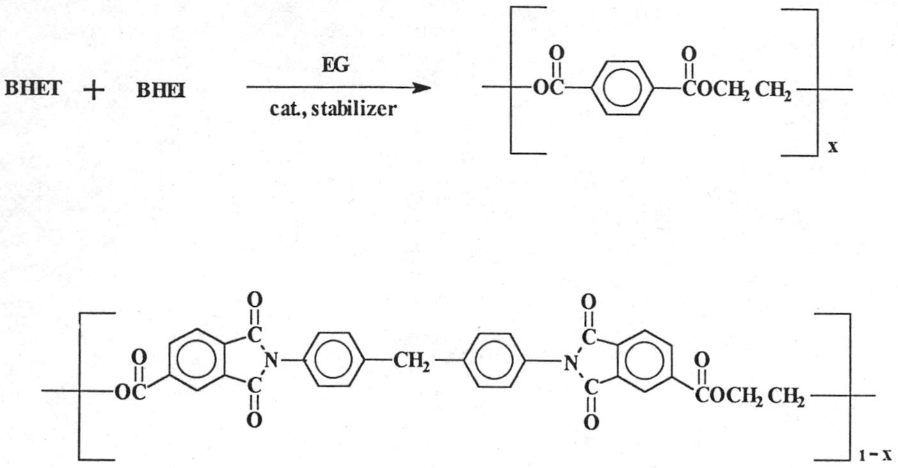

BHET + BHEI —EG / cat., stabilizer→

BHETI

BHBT + BHBI —BD / cat., stabilizer→

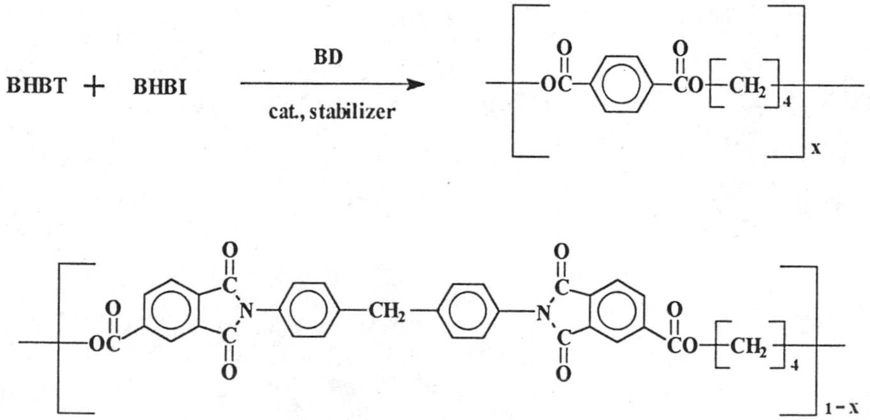

BHBTI

Figure 3. Continued.

or lubricant *(5)*. In addition, the loading level of LCP appears to be a desicive factor to the extent of fibrillation of the LCP phase in the binary blends of PEI and LCP. The blend morphology of the blends reveals that the dispersed LCP domains produce a maximum aspect ratio at the LCP content of *ca.* 25wt% *(5)*. So the effect of introducing PEsI on the mechanical properties is investigated for the 75/25 PEI/LCP blend.

Figure 4 exhibits viscosity curves of 75/25 PEI/LCP blends at several PEsI contents at 340°C. An addition of PEsI slightly increases the viscosity of the binary blend.

The effect of introducing PEsI on the tensile strength of 75/25 PEI/LCP blend fibers is shown in Figure 5. The draw ratio of the fibers is 12. As shown in the figure an introduction of PEsI improves the tensile properties of the binary blend. The most notable improvement is obtained at the loading level of PEsI of 1.5phr.

Figure 6 shows SEM photographs of the fractured surface of 75/25 PEI/LCP blend fibers at several PEsI contents. The fibers have been fractured after freezing in the liquid nitrogen. The photomicrographs provide a morphological evidence for the improvement in the tensile properties by incorporating PEsI. In the binary blends without PEsI, the domain size of the dispersed LCP phase is large and lots of holes are observed, from which LCP fibrils have been pulled out. This is probably because interface adhesion between PEI and LCP is poor. However, the LCP domain size is greatly reduced and the dispersion of LCP domains is more uniform at the loading levels of PEsI of 0.75 and 1.5phr. In addition, the fibrous LCP domains seem to be coupled with the matrix phase in the ternary blends of PEI, LCP and PEsI, which is comparable with the binary blend morphology in which the LCP domains are simply embedded in the PEI matrix. This indicates that adding PEsI makes the interface adhesion better. The enhanced adhesion between PEI and LCP phases is attributable to intermolecular interactions and transesterification reaction through PEsI.

Interestingly it has been found that the mixing order affects the resultant blend morphology of the PEI/PEsI/LCP blends as well. The weight-average diameter of the dispersed LCP domains (D) in the ternary blends at the identical blend composition is smaller when PEI and PEsI are premixed (D=*ca.* 5.56 μm) than when LCP and PEsI are premixed (D=*ca.* 6.03 μm). In the case of binary blend without PEsI, D is measured to be *ca.* 7.13μm. This experimental result indicates that PEsI is more compatible with LCP than with PEI. In other words, PEsI

Table 1. Physical Properties of Polyesterimides

	M_n [*1]	IV [*2]	T_g(℃)	T_m(℃)
BHETI	24,600	0.59	91.6	225.7
BHBTI	29,400	0.71	50.2	208.7

[*1] Number-average molecular weight measured by high temperature GPC using orthochlorophenol as eluent at 100℃.
[*2] Inherent viscosity measured in the 0.5g/dl solution in orthochlorophenol at 30℃.

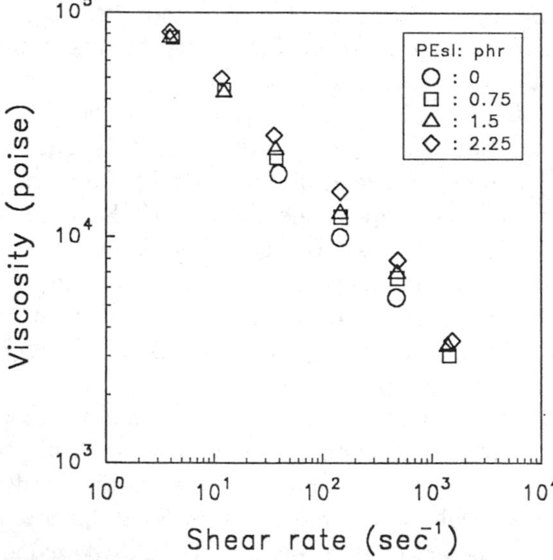

Figure 4. Viscosity curves of 75/25 PEI/LCP blend at several PEsI contents at 340°C.

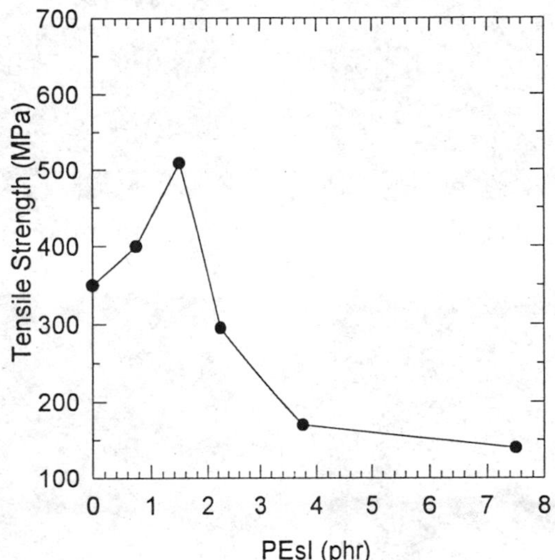

Figure 5. Effect of PEsI on tensile strength of 75/25 PEI/LCP blend fibers (draw ratio; 12).

(a) PEsI: 0 phr (b) PEsI: 0.75 phr

(c) PEsI: 1.5 phr (d) PEsI: 2.25 phr

Figure 6. SEM photographs of cryogenically fractured surfaces of 75/25 PEI/LCP blend at several PEsI contents; scales in (a), (b) and (d) are the same as given in (c).

would move to the interface of LCP and PEI when the premixure of PEsI and PEI is mixed with LCP because PEsI is more compatible with LCP. Hence smaller LCP domains result. Whereas lots of PEsI would still remain within the LCP phase even after the subsequent mixing with LCP when PEsI and LCP are premixed. Hence larger LCP domains result.

Figure 7 exhibits the effect of introducing PEsI on the elongation at break of 75/25 PEI/LCP blend fibers. The draw ratio of the fibers is 12. This result is a little surprising because it is commonplace in polymeric materials that the tensile properties are generally improved at the cost of elongation at break. This result is reproducible, but the reason can not be so clearly elucidated. Figure 8 presents the effect of introducing PEsI on the impact strength of 75/25 PEI/LCP blend fibers. The impact strength was measured by impacting the non-notched fibrous extrudate to the vertical direction of the fiber. In the impact measurements, fibers of draw ratio 1 are used. The measured impact value has been normalized to the impact strength of the absolutely incompatible blend. The normalization factor is 8KPa. An incorporation of 1.5phr of PEsI gives the greatest impact strength as predicted by elongation at break and tensile strength.

On the other hand, if the loading level of PEsI exceeds a critical value, the size of LCP domains is increased again and a poor fibrillation and dispersion of the LCP domains result as observed at the loading level of PEsI 2.25phr in Figure 6(d). Thus, the mechanical properties of the *in-situ* composites are deteriorated at the excessive loading level of PEsI. This is ascribable to the fact that the LCP domains are coagulated before they are deformed if the PEsI level is too high, just like in the emulsion systems stabilized with a surface active agent *(13)*. Consequently, the optimum range of loading levels of PEsI for the 75/25 PEI/LCP blend has turned out to be 1.5phr.

2. PEI/LCP/BHETI Blend Systems.

The binary blends of PEI and BHETI give two distinct T_g's. However, unlike the absolutely incompatible polymer blends they change with blend composition. For example, T_g of BHETI is increased by 70°C and that of PEI is decreased by 40°C at the blend composition of PEI/BHETI 75/25 when compared with that of pure BHETI and pure PEI. In the ternary blends of PEI, LCP and BHETI, T_g of PEI is

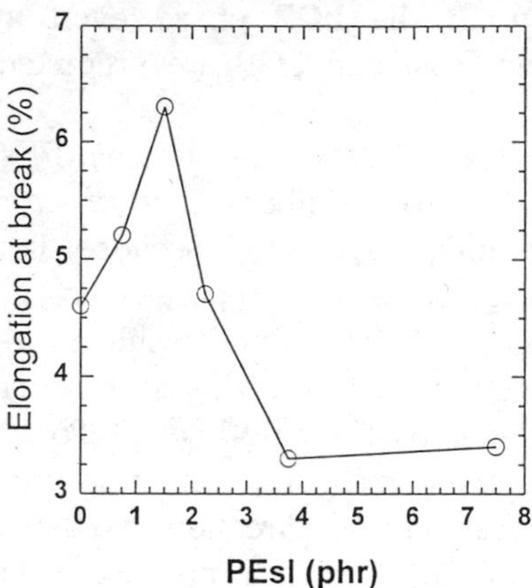

Figure 7. Effect of PEsI on elongation at break of 75/25 PEI/LCP blend fibers (draw ratio; 12).

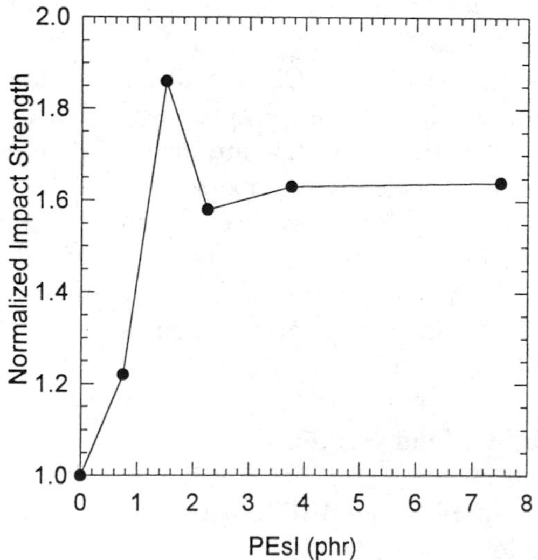

Figure 8. Effect of PEsI on impact strength of 75/25 PEI/LCP blend fibers (draw ratio; 1).

decreased with increasing the BHETI content as shown in Figure 9. These facts stand for a partial compatibility between PEI and BHETI.

The viscosity curves of 75/25 PEI/LCP blends at several BHETI contents at 290°C are shown in Figure 10. The viscosity of the blends is decreased as the loading level of BHETI is increased. This is a little surprising if the high molecular weight of BHETI is considered. However, reduction of the blend viscosity at the processing temperature range is expectable if one takes the low T_g of BHETI into consideration. The decreased viscosity is advantageous in molding of fine and delicate products.

The effect of introducing BHETI on the tensile properties of 75/25 PEI/LCP blend fibers is shown in Figure 11. The draw ratio of the fibers is 8. The tensile stress of the blend fibers gives rise to a maximum value at the BHETI level of *ca.* 2.0phr. The photomicrographs of tensile fractured surfaces of 75/25 PEI/LCP blends at several BHETI contents in Figure 12 clearly verify this result. That is, the domain size of the dispersed LCP is notably decreased at the loading level of BHETI of *ca.* 2.0phr.

3. PEI/LCP/BHBTI Blend Systems

The binary blends of PEI and BHBTI exhibit thermal behavior similar to the PEI/BHETI blends. The T_g of BHBTI is increased by 30°C in the 50/50 blend. In the ternary blends of PEI, LCP and BHBTI, T_g of PEI is decreased with increasing the BHBTI content as shown in Figure 13, representing a partial compatibility between PEI and BHBTI.

The viscosity curves of 75/25 PEI/LCP blends at several BHBTI contents at 290°C are shown in Figure 14. The viscosity of the blends is decreased as the loading level of BHBTI is increased. This may also result from the low T_g of BHBTI.

The effect of introducing BHETI on the tensile properties of 75/25 PEI/LCP blend fibers is shown in Figure 15. The draw ratio of the fibers is 8. The tensile stress of the blends gives rise to a maximum value at the BHETI level of *ca.* 1.0phr. The photomicrographs of tensile fractured surfaces of 75/25 PEI/LCP blends at several BHETI contents in Figure 16 support this result. That is, the domain size of the dispersed LCP is notably decreased at the loading level of BHETI of *ca.* 1.0phr.

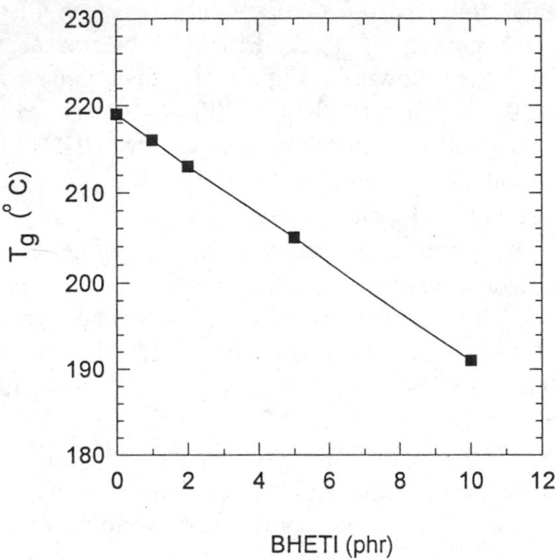

Figure 9. Variation of T_g of PEI with BHETI content in PEI/LCP/BHETI blends.

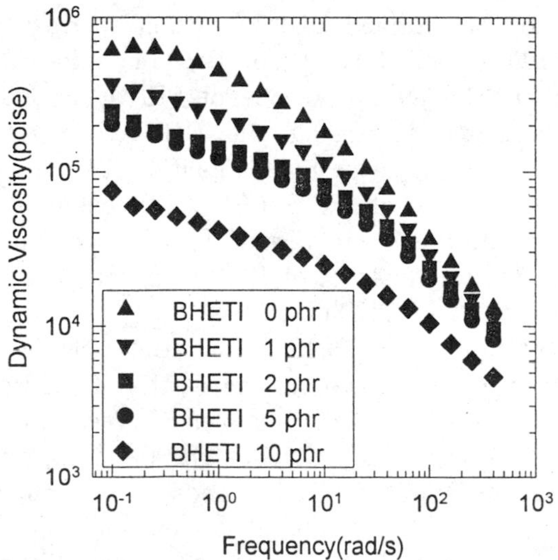

Figure 10. Viscosity curves of 75/25 PEI/LCP blend at several BHETI contents at 290°C.

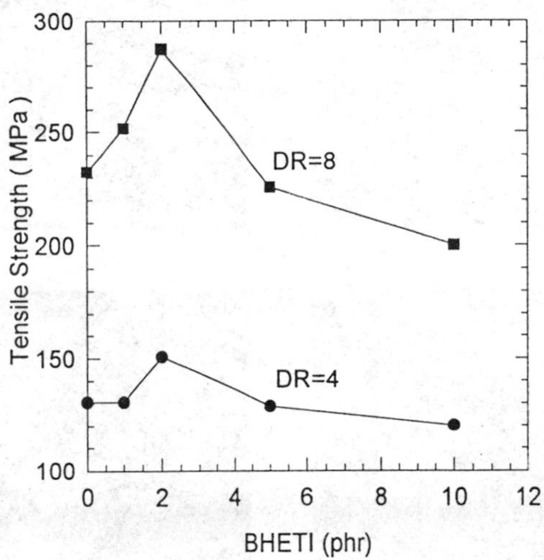

Figure 11. Effect of BHETI on tensile strength of 75/25 PEI/LCP blend fibers (draw ratio; 8).

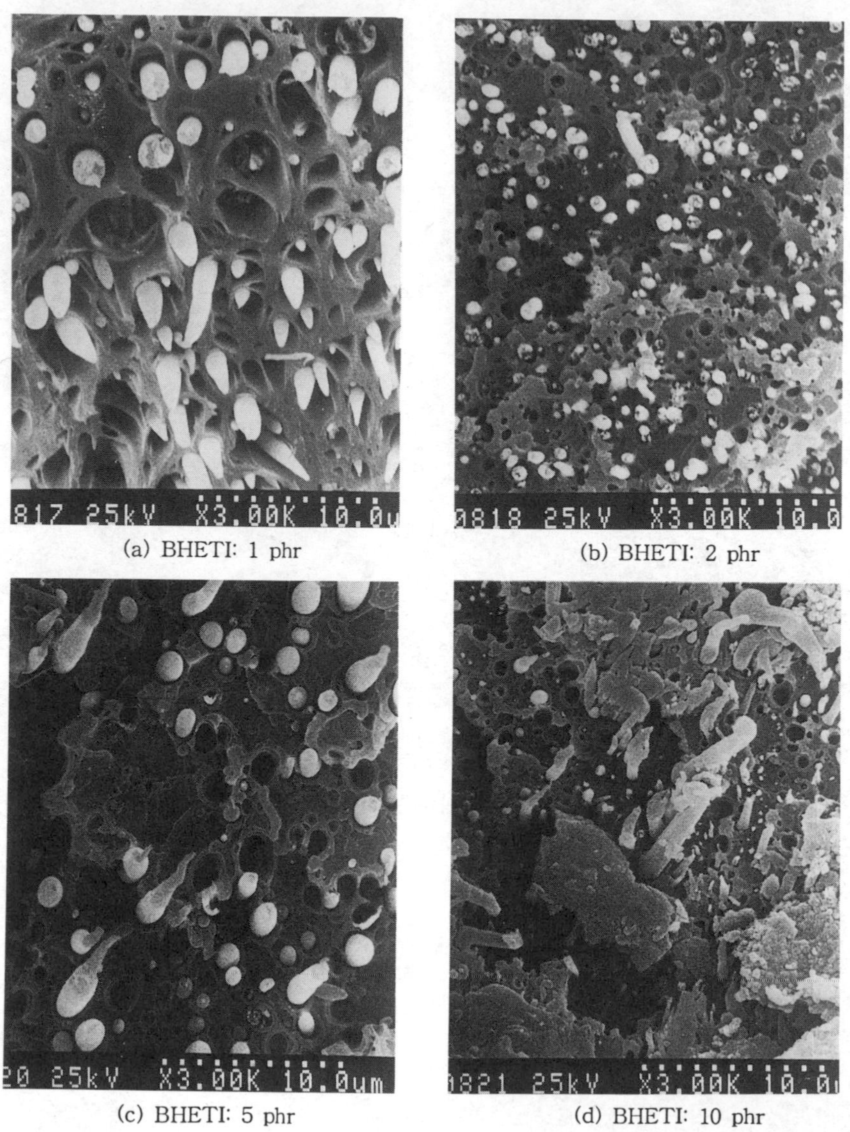

(a) BHETI: 1 phr

(b) BHETI: 2 phr

(c) BHETI: 5 phr

(d) BHETI: 10 phr

Figure 12. SEM photographs of tensile fractured surfaces of 75/25 PEI/LCP blend at several BHETI contents.

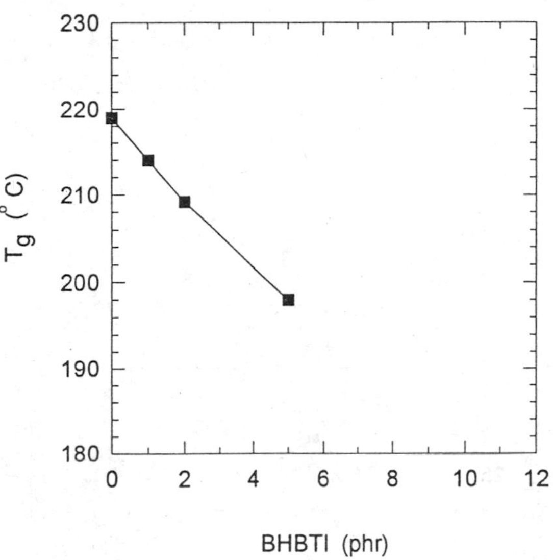

Figure 13. Variation of T_g of PEI with BHBTI content in PEI/LCP/BHBTI blend.

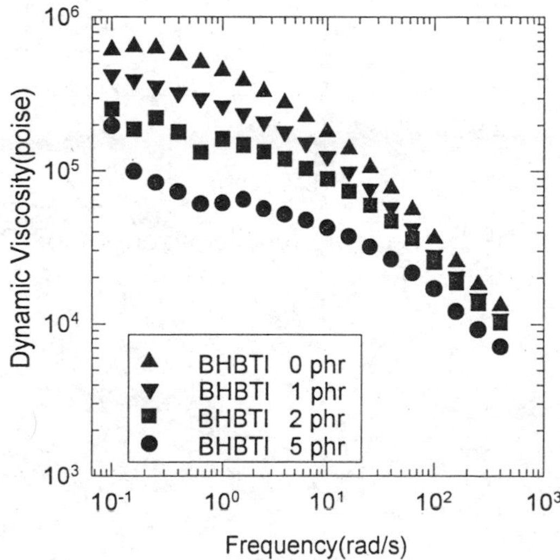

Figure 14. Viscosity curves of 75/25 PEI/LCP blends at several BHBTI contents at 290°C.

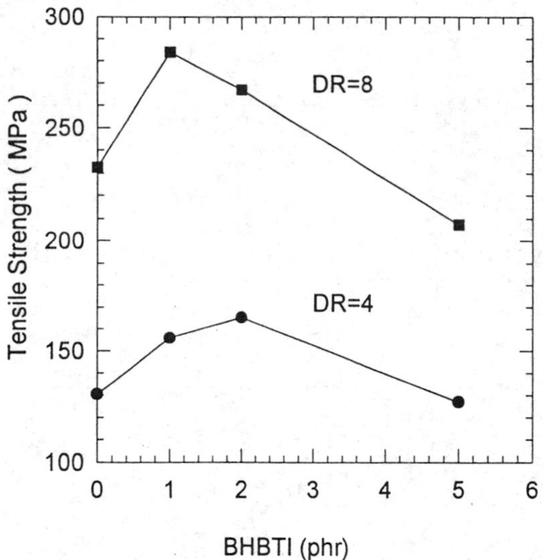

Figure 15. Effect of BHBTI on tensile strength of 75/25 PEI/LCP blend fibers (draw ratio; 8).

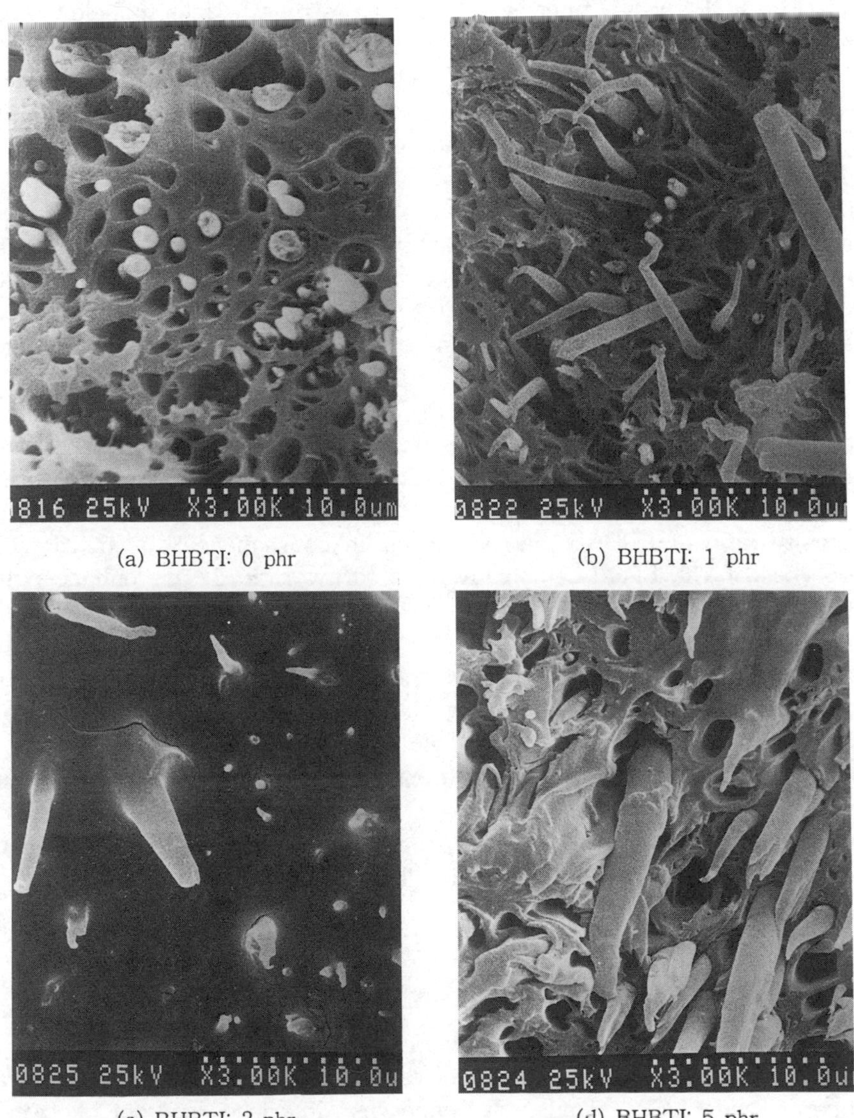

(a) BHBTI: 0 phr

(b) BHBTI: 1 phr

(c) BHBTI: 2 phr

(d) BHBTI: 5 phr

Figure 16. SEM photographs of tensile fractured surfaces of 75/25 PEI/LCP blend at several BHBTI contents.

4. PPS/LCP/PSF Blend Systems

The PPS/LCP blend fiber gives tensile strength falling below that predicted by the additivity rule, exhibiting a minimum value in the vicinity of the blending ratio 50/50. This has been ascribed to cleavage, cavitation, and interfacial debonding through phase separation between PPS and LCP and poor dispersion and elongation of LCP domains *(14,15)*. The microstructure of the PPS/LCP blends also reveals that phase separation occurs regardless of the blend composition. Further, the LCP component is dispersed as droplet domains in the PPS phase up to the LCP content of 50wt% (refer to Kim, B.C.; Hong, S.M.; Hwang, S.S.; Kim, K.U. *Polym. Eng. Sci.*, in press). This is obliged to the fact that the viscosity of the matrix PPS is too low to deform and break the dispersed LCP domains *(16)*.

An introduction of PSF into the PPS/LCP blends may selectively increases the viscosity of the PPS phase in the processing temperature range. Hence, the elongational deformation of the dispersed LCP domains would be promoted on account of adequate matching of viscosity ratio of LCP to matrix phases. In consequence, the mechanical properties of the PPS/LCP blend fibers (draw ratio 3) are enhanced by introducing PPS as shown in Figure 17 (refer to Kim, B.C.; Hong, S.M.; Hwang, S.S.; Kim, K.U. *Polym. Eng. Sci.*, in press). The morphological illustration for the enhanced mechanical properties by incorporating PSF is given in Figure 18. The flexible amorphous PSF reduces free volume at the interface of the crystalline PPS and LCP polymers and seems to facilitate the fibril formation and the molecular orientation of LCP under shearing. Hence, the spherical or ellipsoidal LCP domains observed in the PPS/LCP blend are deformed into rod-like or thread-like fibrils in the PPS/LCP/PSF blend.

CONCLUSION

Introducing a suitable third polymeric component into the binary LCP blends has proved to be effective in improving the mechanical properties of LCP blends. It enhances interface contact between reinforcing LCP and matrix isotropic polymer phases as well as dispersion and elongational deformation of LCP domains. Further, the ternary LCP blends provide freedom for property design of product

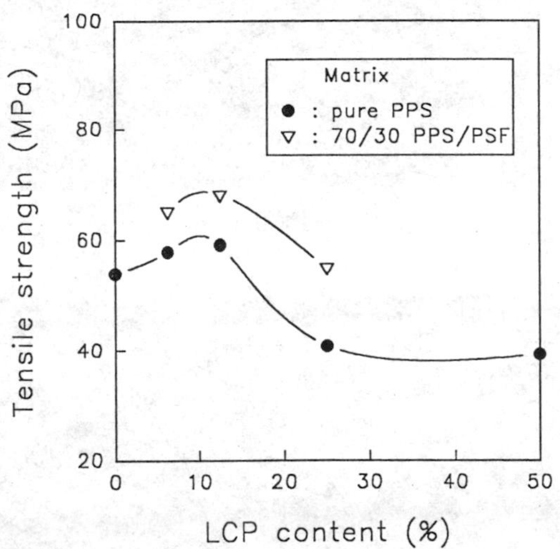

Figure 17. Effect of PSF on tensile strength of PPS/LCP blend fibers (draw ratio; 3).

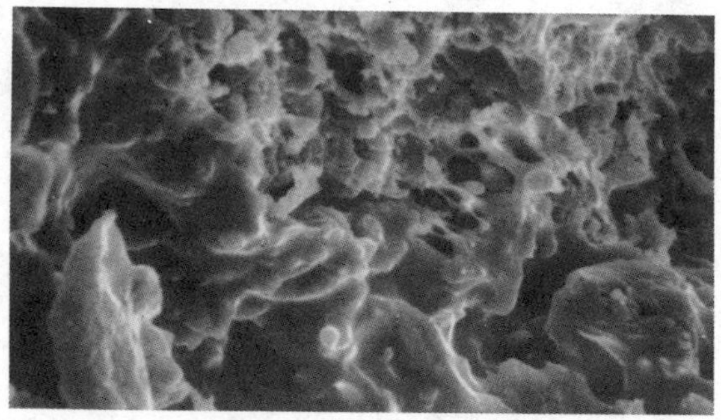

(a) PSF: 0 phr

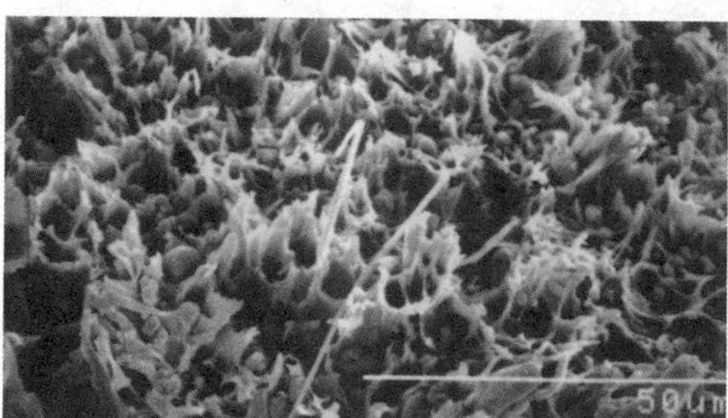

(b) PSF: 30 phr

Figure 18. SEM photographs of tensile fractured surfaces of PPS/LCP blends with and without PSF; the scale of (a) is the same as given in (b).

when compared with the binary blend systems. Thus the ternary LCP blend concept may be effectively utilized in manufacturing high performance *in-situ* composites.

LITERATURE CITED

1. Kardos, J.L.; McDonnel, W.L.; Raisoni, J. *J. Macromol. Sci. Phys.*, **1972**, B6, 397.
2. Kiss, G. *Polym. Eng. Sci.*, **1987**, 27, 410.
3. Hong, S.M.; Kim, B.C.; Kim, K.U.; Chung, I.J. *Polym. J.*, **1991**, 23, 1347.
4. Hong, S.M.; Kim, B.C.; Hwang, S.S; Kim, K.U. *Polym. Eng. Sci.*, **1993**, 33, 630.
5. Lee, S.; Hong, S.M.; Seo, S.; Park, T.S; Hwang, S.S; Kim, K.U.; Lee, J.W. *Polymer*, **1994**, 35, 519.
6. Sukhadia, A.M.; Done, D.; Baird, D.G. *Polym. Eng. Sci.*, **1990**, 30, 519.
7. Hong, S.M.; Kim, B.C.; Kim, K.U.; Chung, I.J. *Polym. J.*, **1992**, 24, 727.
8. Lee, W.C.; Dibenedetto, A.T. *Polymer*, **1993**, 34, 684.
9. Skovby, M.H.B.; Kops, J.; Weiss, R.A. *Polym. Eng. Sci.*, **1991**, 31, 954.
10. Datta, A.; Chen, H.H.; Baird, D.G. *Polymer*, **1993**, 34, 759.
11. Kobayashi, T.; Sato, M.; Takeno, N; Mukaida, K-I. *Eur. Polym. J.*, **1993**, 29, 1625.
12. Datta, A.; Baird, D.G. *Polymer*, **1995**, 36, 505.
13. Piirma, I.; *Polymeric Surfactants*, Marcel Dekker, New York, 1992; pp 184-196.
14. Cheung, M.F.; Golovoy, A.; Plummer, H.K.; Oene, H.V. *Polymer*, **1990**, 31, 2299.
15. Minkova, L.I.; Paci, M.; Pracella, M.; Magagnini, P. *Polym. Eng. Sci.*, **1992**, 32, 57.
16. Chin, H.B.; Han, C.D. *J. Rheol.*, **1979**, 23, 557.

Chapter 10

Blends of Poly(amide imides) and Liquid-Crystalline Polymers

Avraam I. Isayev and T. R. Varma

Institute of Polymer Engineering, University of Akron, Akron, OH 44325–0301

Poly(amide imides) (PAI) are high performance polymers which are capable of replacing metals. Though they are thermoplastics, they need highly specialized machinery for processing in order to achieve maximum property development. The difficulty in their processing is mainly due to high viscosity and melt reactivity. Attempts have been made to process the PAI precursor material using conventional thermoplastic processing methods by blending it with a liquid crystal polymer (LCP) which is in itself a high performance polymer with easy processability and self-reinforcing capabilities. The LCP is also capable of reducing the melt viscosity when blended with other thermoplastics. The PAI precursors were melt blended with the LCP using two methods: a corotating twin screw extruder, and a single screw extruder attached to a static mixer. The blends were injection molded and characterized, both before and after heat treatment. Studies were done to determine the thermal, flow and mechanical properties of the blends and the pure components.

Blending two or more characteristically different polymers resulting in a new material with improved properties is a very old practice. As an alternative to synthesizing new polymers, blending can be used to modify the thermal, mechanical, and processing characteristics of various existing polymers. High performance engineering thermoplastics, a class of polymers with high load bearing characteristics, resistant to chemical attack and having high continuous use temperatures have been modified by blending to suit specific application needs.

Polyimides are engineering thermoplastics which are used when high quality parts and performance are required (1). The main drawback of these materials,

0097–6156/96/0632–0142$19.75/0

which overshadows the excellent properties of the fabricated parts, is their intractability. The intractability makes it very difficult, if not impossible, to melt process this class of materials and hence limits the mass production of polyimide parts. Therefore, attempts were made to develop a more tractable material without serious loss in properties. By using the synthesis route, polyimides have been modified by incorporating flexible functional groups, such as ether, ester or amide, into the main chain. This has led to the development of modified polyimides such as poly(bismalein imides), poly(ester imides), poly(ether imides) and poly(amide imides).

Poly(amide imide) precursors have excellent mechanical properties after imidization, at temperatures up to 260 °C. It was first synthesized in the 60's as a material for high temperature wire enamel (2). This high performance thermoplastic was developed from research done by James Stephens of Amoco Chemicals (see Ref. 3). It became available as an injection molding resin in 1976. These materials were first applied to make burn-in electrical pin connectors. At present the usage has spread into the aerospace, transportation, chemical processing and electronics industries.

The highly intractable chemical structure which imparts the outstanding mechanical properties also makes the PAI's very difficult to process (4, 5). In the fully imidized form PAI is not processable hence a poly(amic acid) (PAA) precursor is the usual form in which they are supplied and fabricated. The precursors themselves have very high viscosities in the melt state and hence the flow characteristics tend to be very poor. Semicrystalline and amorphous polyamides (6) and aromatic sulfone polymers such as poly(phenylene sulfide), poly(ether sulfone) and polysulfone (7) have been blended with the precursor to PAI, to obtain better flow characteristics.

Thermotropic liquid crystal polymers (LCP's) are a special class of engineering thermoplastics which form a highly ordered structure in molten states (8). Their rigid rod-like molecular conformation and the stiffness of the backbone chains impart a high degree of orientation during melt processing and forms fibrous structures in the final product. They are very easy to process and possess outstanding mechanical properties and high chemical resistance. However, they show a high degree of anisotropy in properties (9) which may be overcome by blending with flexible polymers (10). LCP's have also shown that they can impart fiber reinforcement when melt blended with other engineering thermoplastics (10-14). Brief reviews on studies of blends of isotropic polymers and thermotropic LCP's are given by Brostow (15), by Isayev and Limtasiri (16) Dutta et al. (17), and Handlos and Baird (18). Since the LCP forms the fibers in the melt state, there is considerably lower wear and tear on the processing machinery in comparison to conventional reinforcing glass fibers. There is also a viscosity reduction reported in the above studies which imparts better flow properties to the blends. Present authors are aware of only two studies reported in open literature (19, 20) where PAI blended with LCP. In these papers some results on thermal stability, processability and shear stability of these blends have been reported.

The present study seeks to improve the flow properties of the PAA precursor in the melt state by melt-blending it with an LCP. It is hoped that this results in a blend which can be melt processable using conventional equipment with thermoplastic screws as opposed to highly specialized equipment with thermoset screws that are currently being used. The present work also seeks to understand the effect of the LCP on the imidization process in order to improve the mechanical properties of the PAI by utilizing the in-situ fibre forming capability, during processing, of the liquid crystal polymer with the intrinsic strength of the imidized PAA matrix.

Materials and Methods of Investigation

The PAI's used were Torlon 4000Tf-40 (PAI-1) and Torlon 4203L (PAI-2) supplied by Amoco Performance Products. The thermotropic liquid crystal polymer (LCP) used was Vectra A950, which is a random copolyester delivered from HBA-HNA (Hoechst Celanese). The PAI-1 was in fine powder form with no additives while the PAI-2 was in pellet containing 0.5% PTFE and 3% TiO_2.

PAI-2 was melt-blended with the LCP using a six-element static mixture (Koch Engineering) attached to the exit of a 1" single screw extruder (Killion Inc.) at the screw speed of 30 rpm and temperature of 350 °C. The PAI-1 and PAI-2 were melt-blended with the LCP using a modular, corotating, fully intermeshing twin-screw extruder with mixing elements (ZSK-30, Werner & Pfleiderer Corp.) at the screw speed of 200 rpm, the barrel temperature of 330 °C. The static mixer and the twin-screw blends are referred below as STM and TS blends, respectively.

Molding was carried out using an injection molding machine (Boy15S) at barrel and mold temperature 330 °C and 220 °C, respectively, and at maximum injection speed. The molding conditions are listed in Table 1. End-gated dumbbell shaped test bars of two sizes were molded: mini-tensile bars (MTB's) (0.155m x 0.013m x 0.0033m), and standard tensile bars (MTB's) (0.065m x 0.0031m x 0.0052m). Only TS blends were molded since the STM blends showed discoloration and voids in the extrudate indicating degradation. From the twin-screw mixed blends involving PAI-2, only the higher concentration (90 and 75% PAI-2) were molded since the other blend compositions (50, 25 and 10% PAI-2) showed distinct phase separation and delamination due to poor mixing. This was visible to the naked eye by the green color of PAI and the cream color of LCP. All blends involving PAI-1 were molded. Some of the molded samples of each blend were heat treated (cured) according to the recommended procedure by Amoco. The curing cycle was as follows: 166 °C, 216 °C, and 243 °C for 1 day each, and 260 °C for 2 days in a vacuum oven at -28 in. Mercury. PAI-1 powder was not injection molded since it cannot be molded using conventional machines. Molding of PAI-2 was also tried unsuccesfully.

An Instron capillary rheometer (Model 3211) was used to measure the rheological properties. A capillary die of diameter D=0.00127m and a length to diameter ratio, L/D, of 28.7 was used. Measurement was done at 330 °C and 350 °C for all PAI-1/LCP blends, but only for the 10% and 25% LCP concentration blends with PAI-2 since at all other concentrations a poor mixing was noticeable.

A Dupont DSC 9900 Thermal Analysis System was used to analyze granulated extrudates of the blends and pure resins, both before and after heat treatment. The heating rate was 20 °C/min.

A Monsanto Tensiometer (Model T-10) was used with an extensiometer to measure the tensile properties (ASTM D648) of the MTB's, both before and after heat treatment, at a crosshead speed of 5mm/min.

To measure the dynamic mechanical properties, a Dynamic Mechanical and Thermal Analyzer (DMTA, Polymer Laboratories) interfaced with a HP300PC was used. A single cantilever, and a heating rate of 4 °C/min. was used at the frequency of 1 Hz. and strain of 4%. An Impact Tester (Testing Machines, Inc.) was used to conduct the Izod impact test (ASTM D235C) on notched STB's of both cured and uncured samples. A Scanning Electron Microscope (SEM), Model ISI-SX-40m was used in the morphological study of the surface of MTB's broken in liquid nitrogen. Gold-palladium plasma coating technique was utilized to make the surfaces conductive. The skin and core morphologies of both cured and uncured samples were studied.

Rheology

Fig. 1 gives a comparison of flow curves of pure PAI-1 and PAI-2 and their blends with LCP. From that figure it follows that PAI-1 melt show higher viscosity than PAI-2 melt. That could be due to the presence of PTFE in PAI-2 which is used as processing aid for PAI. The viscosity of the 90/10 PAI/LCP blend is significantly lower than that of pure PAI. Possibly, the lower viscosity LCP melt in the blend migrates to region of higher shear rate leading to drop in viscosity of the blend. The drop is more significant in the case of the TS blends than the STM blends as indicated by Fig. 2 for PAI-2/LCP blends. The lower viscosity LCP acting as an external lubricating agent in combination with the lower residence time in the twin-screw extruder could cause the 90/10 PAI-2/LCP blend viscosity to be lower than that of the corresponding STM blend. The long residence time in the static mixer-single screw extruder system causes an increase in the viscosity by chain extension in the reactive PAI. In the case of the 75/25 PAI-2/LCP blend, the higher concentration of the LCP (lower concentration of PAI) causes the viscosity increase due chain extension to be lower.

Various workers have proposed theoretical models to predict the flow behavior of polymer blends. Einstein studied the shear flow behavior of a suspension of rigid spheres in Newtonian fluids. Taylor (21, 22) extended this concept to include dispersions of one liquid in another liquid based on their shear viscosities and also accounted for circulation in the droplets. According to his model, for a component 2 which is dispersed in a component 1, the blend viscosity η_b, is given by the following equation:

$$\eta_b = \eta_1\left[1+\left\{\left\{(\eta_1+2.5\eta_2)/(\eta_1+\eta_2)\right\}\phi_2\right\}\right] \tag{1}$$

The main drawback of Taylor's equation is that it only predicts an increase in viscosity with addition of component 2 even if the viscosity of component 2 is lower

Table 1 - Injection Molding Conditions for PAI/LCP Blends

Barrel Temperature	=	330 °C
Mold Temperature	=	220 °C
Max. Shot Size	=	1.24 oz
Back Presure	=	0 Pa
Nozzle	=	100%
Mold Cooling Time	=	15 sec
Injection Pressure	=	13.75 MPa
Injection Speed	=	Max.
Screw Speed	=	250 RPM
Clamp Pressure	=	220 kN
Injection Molding Screw Characteristics		
Compression Ratio	=	3.2
Length/Diameter Ratio	=	17
Diameter	=	24 mm
Back Flow Valve	=	Yes

* Note * All barrel zones were kept at the same temperatures.

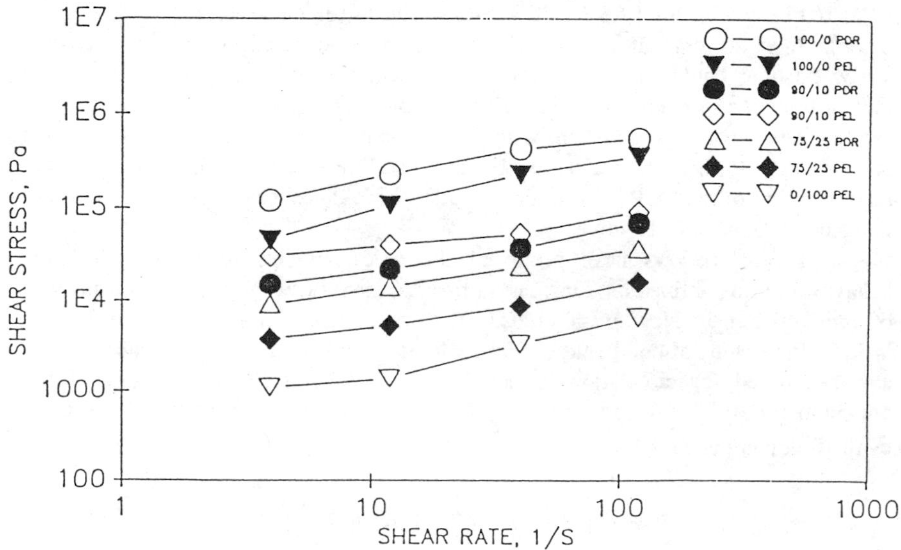

Figure 1. Shear stress as a function of shear rate at 350 °C: Comparison of twin-screw mixed PAI-2/LCP (PEL) and PAI-1/LCP (PDR).

than that of the matrix polymer, component 1. This equation assumes that no chemical interaction occurs between the dispersed and continuous phases, the dispersed phase forms uniform spherical droplets, and that the droplets retain their spherical shape in the shear flow field. This equation does not include the effect of droplet size on the viscosity. According to Chuang and Han (23), if no chemical interaction occurs between the phases, the experimentally observed viscosities would generally be lower than the theoretically predicted values.

Figure 3 compares Taylor's prediction to the observed shear viscosities at two shear rates: 3.93 and 118.1 s^{-1}. If PAI is taken as component 2 most of the predicted values, though lower, closely follow the experimental values. At PAI concentration of 90% the under-prediction is substantial. The predictions are better for higher shear rates at all concentrations. Blend viscosities show an increase in value with increasing PAI concentration which indicates a lack of interaction between the two phases. According to Han (24), studies on HDPE - EVA blends have shown that when deformation of the droplets is absent, the mixture gives viscosities greater than the viscosity of the less viscous of the two polymers. Since no maximum is observed in the case of the PAI/LCP system studied here, it is safe to presume that no interaction between the phases have taken place. Han also theorized that the viscosity of the droplet phase is much larger than that of the suspending medium. He also claimed that an inflection takes place when there is a phase inversion. A minimum of the viscosity occurs when elongated droplets are present giving rise to threadlike fibrils. In the PAI/LCP system, this is not observed.

Figure 4 shows the predicted viscosity values when LCP is taken as component 2. As can be seen, the values are substantially over-predicted and at variance with the observed values.

Heitmiller et al. (25) derived an expression for the viscosity of a mixture based on an inverse volume-weighted rule and assuming concentric layers of component 1 in component 2. Assuming a large number of layers the viscosity of the mixture is given as:

$$1/\eta_b = \phi_1/\eta_1 + \phi_2/\eta_2 \tag{2}$$

According to Heitmiller's equation, the blend viscosity varies monotonically with the volume fraction ϕ_1 and ϕ_2 $(\phi_1 + \phi_2 = 1)$. As in the case of Taylor's prediction, Heitmiller's prediction also under-predicts the viscosity values, but, the predicted values compare better at higher, rather than lower, shear rates. This is shown in Figure 5 which gives the comparison of the curves of the experimental viscosity values and the predicted viscosity values using the Heitmiller equation.

Hashin (26) extended the equations used successfully for the prediction of the upper and lower bounds for elastic modulus of a composite material and obtained a viscosity envelope for polymer blends.

In case of mixture of Newtonian fluids the viscosity equation is:

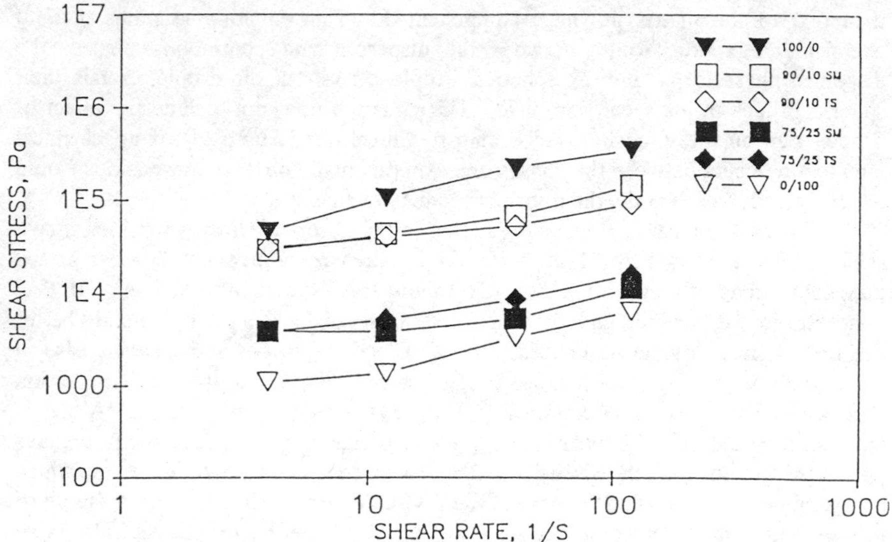

Figure 2. Shear stress as a function of shear rate at 350 °C: Comparison of twin-screw and static mixer mixed PAI-2/LCP blends.

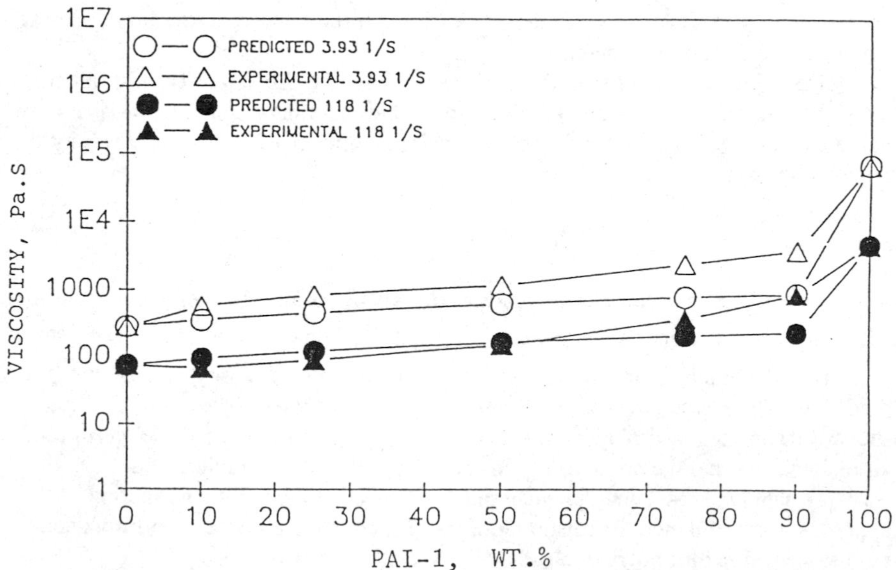

Figure 3. Viscosity as a function of PAI-1 concentration at 330 °C at two shear rates: Comparison of observed viscosity and viscosity predicted by Taylor's equation $\left(\eta_1 = \mathrm{LCP}, \eta_2 = \mathrm{PAI\text{-}1}\right)$.

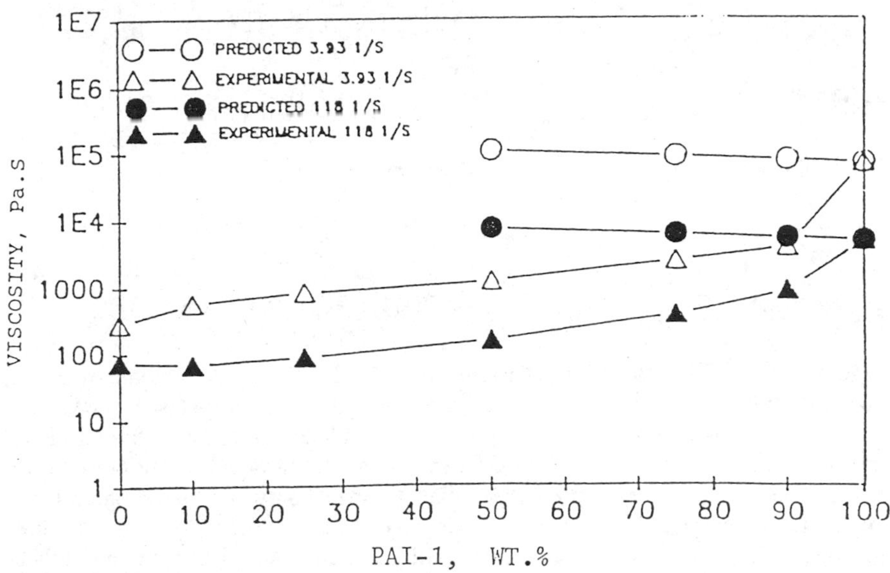

Figure 4. Viscosity as a function of PAI-1 concentration at 330 °C at two shear rates: Comparison of observed viscosity and viscosity predicted by Taylor's equation $\left(\eta_1 = \text{PAI-1}, \eta_2 = \text{LCP}\right)$.

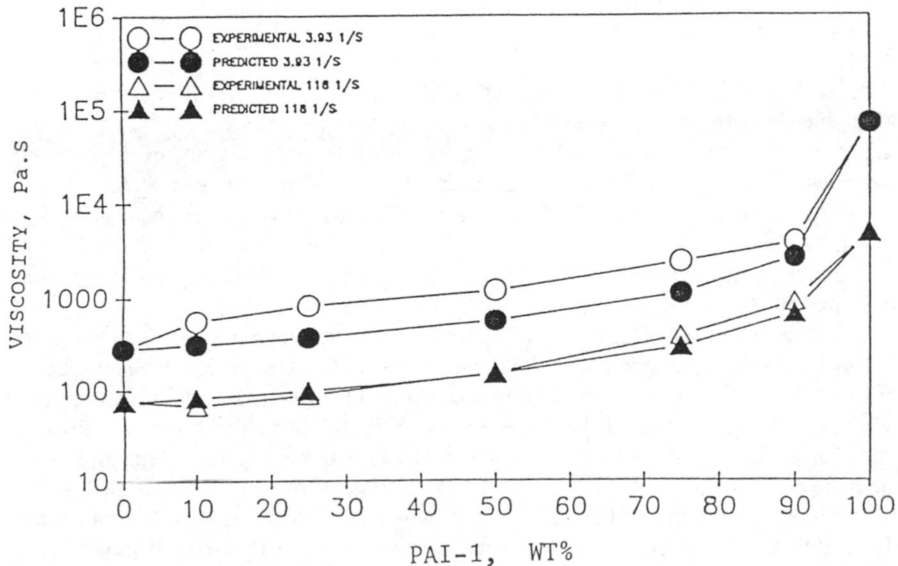

Figure 5. Viscosity as a function of PAI-1 concentration at 330 °C at two shear rates: Comparison of observed viscosity and viscosity predicted by Heitmiller's equation for twin-screw mixed blends.

Upper Bound: $$\eta_b = \eta_2 + \left[\phi_1 / \{(1/[\eta_1 - \eta_2]) + (2/5)(\phi_2 / \eta_2)\}\right]$$

Lower Bound: $$\eta_b = \eta_1 + \left[\phi_2 / \{(1/[\eta_2 - \eta_1]) + (2/5)(\phi_1 / \eta_1)\}\right] \quad (3)$$

For non-Newtonian fluids the above equation becomes

Upper Bound: $$\eta_b = \eta_2 + \left[\phi_1 / \{(1/[\eta_1 - \eta_2]) + (1/2)(\phi_2 / \eta_2)\}\right]$$

Lower Bound: $$\eta_b = \eta_1 + \left[\phi_2 / \{(1/[\eta_2 - \eta_1]) + (1/2)(\phi_1 / \eta_1)\}\right] \quad (4)$$

Figure 6 shows the experimental results of the viscosity as a function of PAI-1 concentration along with the predicted results according to the Hashin model. In case of the PAI-1 blend system, all the values lie close to the lower bound of the envelope with the higher shear rate viscosities falling slightly below but close to the lower bound of the envelope. According to the theory, the latter indicates a complete segregation with the fluid having the lower viscosity forming the outer annulus. This theory is also reinforced by the observations of Everage and others (27, 28) who reported encapsulation of the higher viscosity component by the lower viscosity component. Hence all indications are that the lower viscosity LCP encapsulates the higher viscosity PAI. It should be noted that the cure inhibition and the reduced moisture absorption seen in our DMTA and DSC studies also indicate the same phenomena.

Temperature Transitions

The Differential Scanning Calorimeter (DSC) scans of the PAI-2 and PAI-1 and their blends with LCP are presented in Figs. 7a and 7b, respectively. The TS and STM blends both uncured and cured are compared. From the scans of the uncured samples, no significant difference is observed between the TS and STM samples (Fig. 7a). The peak seen at 320 °C is that of PTFE which is a component of PAI-2. Uncured PAI-2 shows a T_g of 260 °C whereas cured PAI-2 shows a T_g of 290 °C and melting peak around 310 °C. Brooks (7) based on a DSC scan has also reported a T_g of 260 °C for uncured PAI-2.

The cured blends show no T_g transitions but melting peaks can be seen instead. There is no significant difference between the DSC scans of heat treated 90/10 PAI-2/LCP samples which were blended in the twin-screw or the static mixer. However, the DSC scan of heat treated 75/25 PAI-2/LCP blend shows a distinct difference. In case of the twin-screw mixed blend, two melting peaks, one at 270 °C and the other around 280 °C, can be seen. In case of the 75/25 PAI-2/LCP static mixer mixed blend, the DSC scan is similar to that for pure cured PAI-2. Since there is no difference between the DSC scans of uncured 75/25 PAI-2/LCP mixes using twin-screw and static mixer, the only explanation seems to be that the twin-screw mixed components separate out during heat treatment. The DSC scan of heat treated LCP shows some changes. In particular, the heat treated LCP shows a melting peak about 330 °C which is larger in area and higher in temperature than

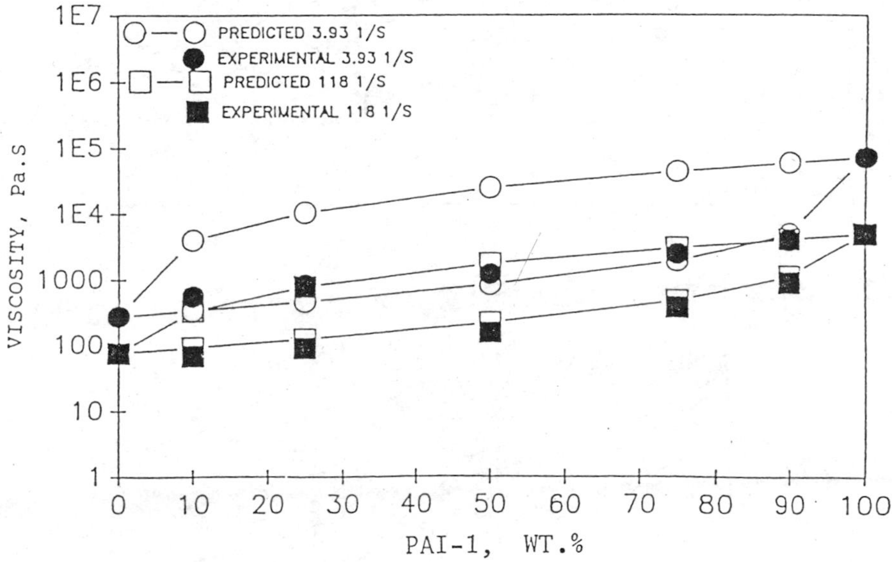

Figure 6. Viscosity as a function of PAI-1 concentration at 330 °C at two shear rates: Comparison of observed viscosity and viscosity predicted by equations of Hashin et al. for twin screw mixed blends.

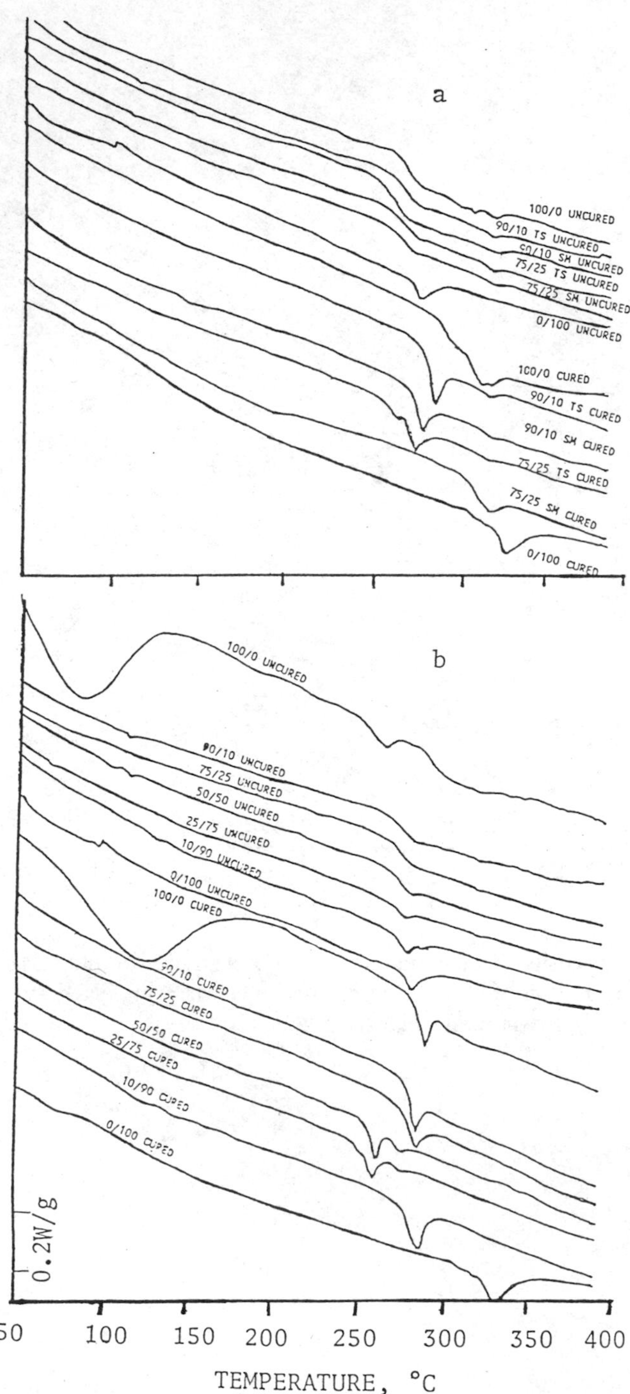

Figure 7. DSC curves of uncured and cured PAI-2/LCP (a) and PAI-1/LCP (b) blends.

that of the non-heat treated LCP. It should also be noted that LCP often exhibits dual endothermic peaks depending on heat treatment history (16).

Figure 7b shows DSC scans of PAI-1/LCP blends. Pure PAI-1 has a very high moisture absorption rate compared to PAI-2 as can be seen by the peak around 100 °C. The cured sample of PAI-1 shows volatiles being released up to 160 °C. The volatiles peak is larger in case of the cured PAI-1. This can be attributed to the fact that the volatiles being more on the surface of the uncured PAI-1 and not as strongly held as in case of the cured PAI-1. The T_g of PAI-1 is higher (280 °C) when compared to PAI-2 (260 °C). This could be due to the neat PAI-1 which has more surface area due to its powder form, curing during the DSC temperature scan. The neat PAI-2 is in pellet form and hence has lesser surface area than the pure PAI-1. The latter retards curing effects during the DSC scan. Addition of only 10% LCP decreases the moisture absorption of the PAI-1 blend as observed by the disappearance of the peak around 100 °C which is attributed to moisture.

There is a gradual change from a T_g transition in the case of pure uncured PAI-1 to a melting peak in the case of the pure non-heat treated LCP. 10/90 PAI-1/LCP uncured blend shows two peaks, indicating possible phase separation. On curing, the T_g transitions are transformed into melting peaks and shifted to a higher temperature. According to Jellinek (29), this could be due to a molecular aggregation within amorphous regions after heat treatment. 25/75 and 50/50 PAI-1/LCP cured blends show two melting peaks indicating possible phase separation. Since the temperature at which the melting peaks observed in case of each heat treated blend composition falls below that of the heat treated pure component, the presence of the LCP possibly inhibits the imidization of the blend. Since the improvement in mechanical properties of PAI is directly proportional to the degree of imidization, this inhibition of cure can cause deterioration of properties when an incompatible component is added. The deterioration of properties can be seen from the results of the mechanical tests such as tensile and impact tests presented below.

DSC studies of uncured and cured PAI-1 confirm that the endothermic peak around 100 °C, seen in Figure 7b, is surface moisture and the entrapped n-methyl pyrolidone (NMP, boiling point: 115 °C). The surface moisture is due to the amide moiety in the main chain of the copolymer. The wide peak extending up to 160 °C can only be the trapped solvent which, according to various studies, could be as much as 25 weight % (1, 29). To proof this solvent effect, the uncured and cured samples were heated up to 150 °C in a closed aluminum pan, allowed to cool to room temperature and then reheated to 400 °C at 20 °C/min. During the initial heating, both scans show peaks around 100 °C which disappear on cooling and subsequent heating from room temperature to 400 °C.

Melting peak of the LCP in the blends is not seen until an LCP concentration of about 50% has been reached. This result agrees well with studies done on PC-LCP (10) and PEI-LCP systems (13). Two reasons have been proposed for this behavior: (1) at very low LCP concentrations, the LCP peak may be too small to become evident in a DSC scan, and (2) at higher LCP concentrations, the amorphous PAI may start to interfere with the LCP crystal structure. Since the T_g of PAI and T_m peak of LCP are very close to each other, compatibility cannot be accurately substantiated by DSC scans alone; but the indications from the 50/50,

25/75 PAI/LCP cured blends and the 10/90 PAI/LCP uncured blend are that the components are incompatible in the entire range. The fact that there is a reduction in the surface moisture absorption as seen from the DSC plots and a viscosity reduction, due to the addition of just 10% LCP points to the possibility of a lubricating effect by the LCP. Since the lower viscosity components possibly tend to migrate to the region undergoing the highest shear (24), it is conceivable that the lower viscosity LCP has formed a lubricating film over the blend, hence also possibly inhibiting the imidization.

On diffusion studies done on PET/LCP blends using methylene chloride as a diffusing solvent, Joseph et al. (31) found that as the LCP content in the blend increased, the rate of diffusion of methylene chloride decreased i.e. the LCP acted as a barrier for diffusing solvents. The present LCP has 73% HBA in its HBA/HNA chemical structure. This could explain the inhibition of cure in the blends. Studies (30) have shown that diffusion of condensation products of the imidization reaction is very important to obtain good mechanical properties of the PAI. The reaction stops if the byproduct is not removed.

MECHANICAL PROPERTIES

Tensile Properties

Mechanical properties such as tensile strength and modulus of the blends depend on the degree of interfacial bond strength between the two polymeric phases and morphology. Generally, the formation of fibrils of one polymer within the other with high aspect ratio increases the tensile properties of the blend. In particular, reinforcement of the fibers reportedly (32) occurs above some critical aspect ratio below which no reinforcement occurs: A fibrous filler transfers the applied mechanical load from the matrix to the reinforcement and in fibers with aspect ratios lower than a critical value, a complete load transfer may not be achieved.

PAI-2/LCP Blends

Stress-strain curves for various blends of cured and uncured blends of PAI-2/LCP indicated fracture behavior of brittle material. It was found that the toughness of the pure LCP increases on curing, which is also corroborated by the DMTA (increased storage modulus) and the impact strength results. The core section of the SEM photographs show a grainy texture for non-heat treated LCP but a platelet formation after heat treatment which also explains the improvement in toughness of the samples after the heat treatment.

Figure 8 presents a comparison of tensile strength (a) secant modulus (b) of injection molded MTB's for the twin-screw mixed PAI-2/LCP blend samples. Except for pure LCP, all the blends and the pure PAI-2 show improvement in ultimate strength after the heat treatment. Also, elongation at break is found to increase for both pure components and the blends. Heat treatment improves the modulus of the LCP and, to a lesser extent, the 75/25 PAI-2/LCP blend. The modulus of the 90/10 blend does not show any improvement after the heat

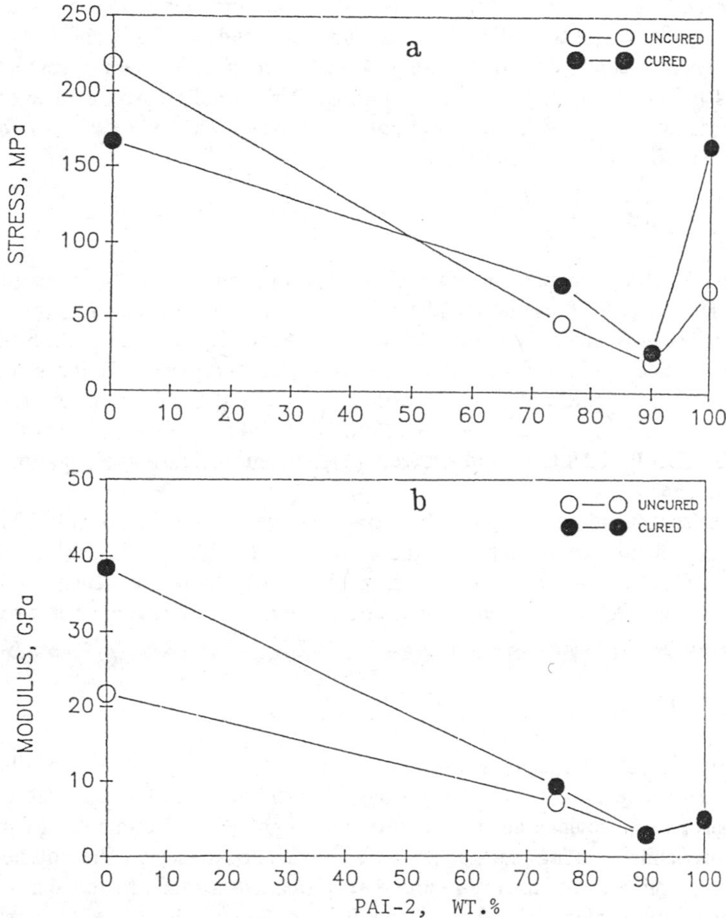

Figure 8. Break stress (a) and secant modulus (b) as a function of PAI-2 concentration for cured and uncured MTB samples.

treatment. This could be attributed to the inhibition of cure by the LCP combined with the degradation of the PAI due to high shear. Improvement in modulus of the LCP due to heat treatment is possibly caused by solid state polymerization as indicated in (8) for heat treated LCP fibers.

There is little improvement in modulus upon addition of LCP to PAI-2. Some deterioration of properties is seen at the 90% PAI-2 concentration. The increase in the mechanical properties of cured PAI-2 is due to chain extension on imidization. Comparing the properties of cured and uncured blends of 75% and 90% PAI-2 concentration, it is observed that there is some but no significant improvement in the modulus and other tensile properties on curing. This could be attributed to the LCP component interfering with the curing of the blends. The values shown for pure PAI-2 are reported book values (5).

PAI-1/LCP Blends

Figure 9 presents a comparison of tensile strength (a) and modulus (b) for the PAI-1/LCP blend system. It can be seen from Figure 9a that the tensile strength increases on heat treatment only in the cases of 50/50, 75/25 and 100/0 PAI-1/LCP blends. In all other cases there is a decrease in value on heat treatment. At the same time, results showed a significant increase in the elongation at break after heat treatment only in case of the pure components and the 10/90 PAI-1/LCP blend. On the other hand the 25/75 PAI-1/LCP blend showed a significant decrease in elongation due to heat treatment.

The secant modulus (Figure 9b) shows that, except for the 10 and 50% PAI-1 composition blends, curing improves the modulus of the blends. In case of the 10% PAI-1 composition, the LCP content is probably too high and hence due to the mutual incompatibility of the individual components, no improvement in properties occurs.

Impact Properties

The measurement of impact resistance of materials gives a value for the total fracture energy necessary for crack propagation and ultimate failure. The notched Izod impact test provides an insight into the toughness and durability of material when stress-concentrating features such as notches are present. The notched Izod impact test evaluates the energy required to propagate a crack to the shatter point but it does not evaluate the energy required to initiate the crack. Long fibers dramatically increase the impact resistance due to more stress transfer capability whereas fillers with spherical shapes generally decrease the impact resistance of thermoplastics.

Figures 10 and 11 show of the Izod impact test results of cured and uncured blends of PAI-2 and PAI-1, respectively. From the figures it can be seen that there is a decrease in toughness on blending due to the mutual incompatibility of components. Generally, with exception of the 90/10 PAI-1/LCP blends, there is an increase in toughness after curing. This was also seen from the observations in the tensile and DMTA results. The lack of improvement in the impact properties in the

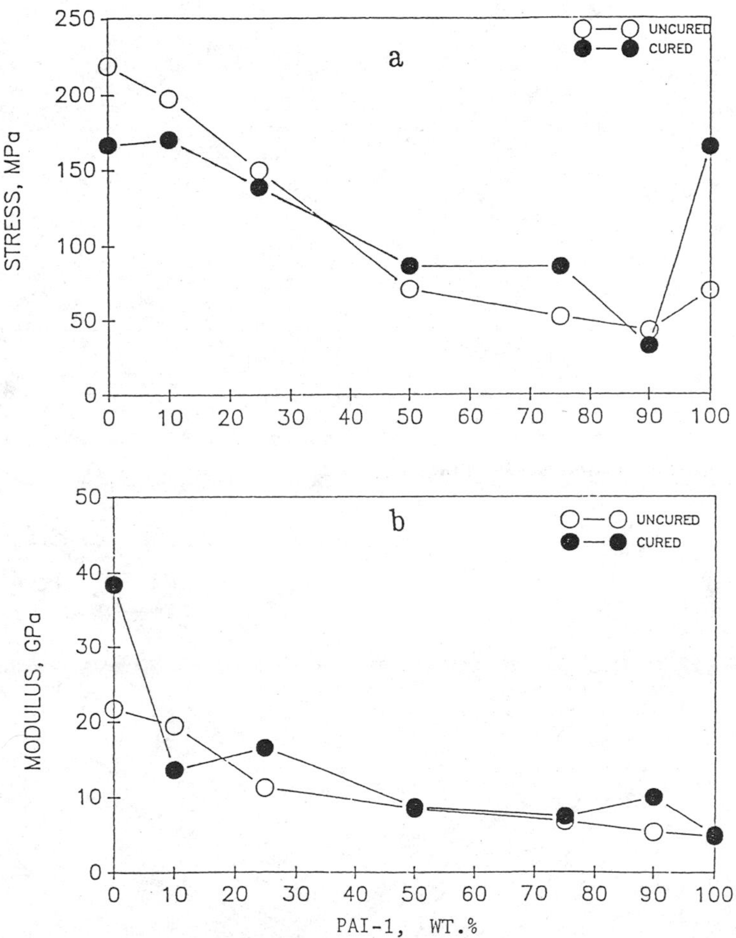

Figure 9. Break stress (a) and secant modulus (b) as a function of PAI-1 concentration for cured and uncured MTB samples.

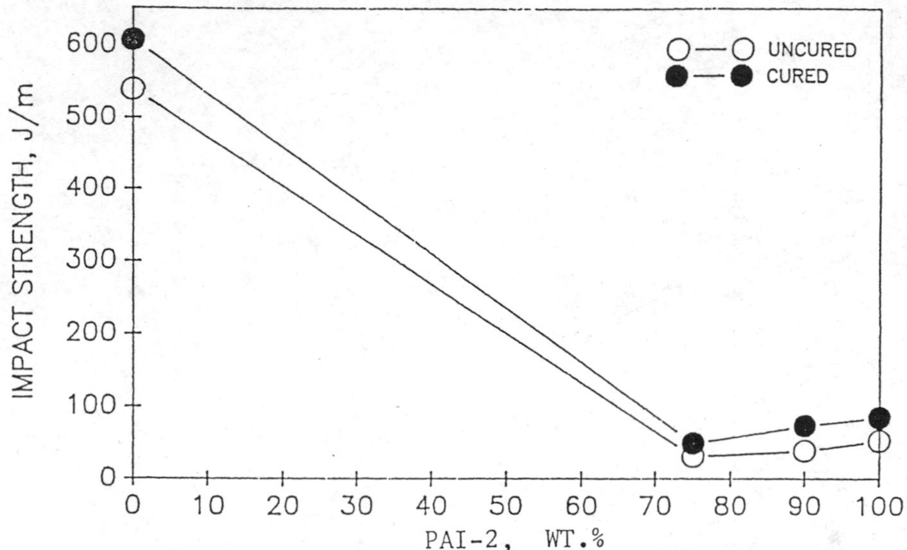

Figure 10. Notched Izod impact strength as a function of PAI-2 concentration for cured and uncured STB samples.

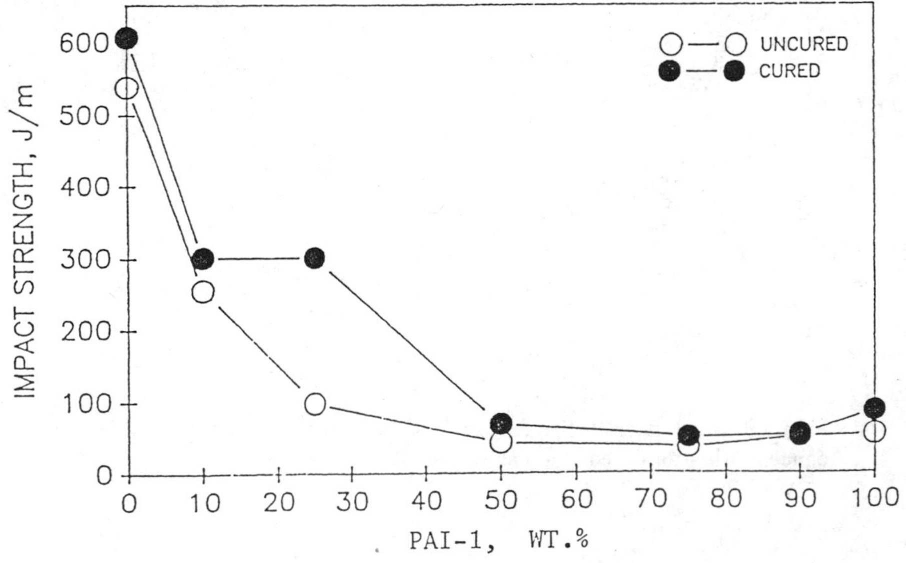

Figure 11. Notched Izod impact strength as a function of PAI-1 concentration for cured and uncured STB samples.

case of cured 90/10 PAI-1/LCP blend indicates that addition of LCP inhibits the imidization of PAI-1 which is required if property improvement is to take place. The degradation of the large PAI component, due to the shear induced by the screw, is also a factor which has to be taken into consideration when explaining the deterioration of the final properties of the 90/10 PAI-1/LCP blend.

Morphological Studies

SEM photographs of MTB samples broken in liquid nitrogen were obtained. Photographs are of representative samples of skin and core sections of each cured and uncured blend composition. Skin and core sections of pure unheat treated LCP show fiber formation. However, the core has a grainy appearance with finer fibers as indicated in Figure 12a, whereas the skin shows lesser and coarser fibers. Skin and core sections of heat treated LCP indicate platelet morphology with little evidence of fibrillation as shown in Figure 12b for the core section. Evidently the fibrous structures seen in the unheat treated samples are lost due to release of orientation and solid state polymerization upon heat treatment. This could explain the improvement in toughness on heat treatment on the LCP.

Figure 13 shows the skin and core sections of uncured (a, b) and cured (c, d) 10/90 PAI-1/LCP TS samples, respectively. Fiber formation can be observed in both the sections of each cured and uncured sample. In both cases, the core section shows higher aspect ratio fibers than the skin. Figure 13 shows phase separation with a few low aspect ratio fibers in the skin (a) and a large concentration of fibers of about the same aspect ratio in the core (b).

Figure 13 also shows phase separation with the LCP as the continuous phase and a spherical PAI-1 phase of about 10 micron in diameter in the core (d). The skin (c) section does not show any PAI-1 phase. Hence, encapsulation of the PAI-1 by the LCP can be inferred. There is less fibrillation seen in the cured samples than in the uncured samples.

Figure 14 shows photographs of uncured (a, b) and cured (c, d) 25/75 PAI-1/LCP samples, respectively. The 25/75 PAI-1/LCP blend shows fiber formation in the uncured sample (Figure 14) in both the skin (a) and core (b). Distinct phases are absent which indicates better dispersion. The core also shows finer fibers than the skin. The fiber structures are destroyed on heat treatment (Figure 14c, d). The core section shows phase separation with LCP being the continuous and PAI-1 the discrete phase. The cured samples also show voids due to the release of volatiles. No agglomeration of the PAI-1 phase seems to have occurred in this case.

Figure15 shows photographs of uncured (a, b) and cured (c, d) 50/50 PAI-1/LCP samples, respectively. The uncured samples (Figure 15) show very fine fibers in the core (b) and coarse fibers in the skin (a). The fibers are less evident in the cured samples (Figure 15c, d). Figure 15a shows an LCP rich skin with large unevenly oriented fibers. The core section shows phase separation with fibrillation (Figure 15b). This fibrillation is seen to have been destroyed on curing (Figure 15c, d). The core seems to be PAI-1 rich (Figure 15d), as can be seen from the few LCP fibers present, but the skin is LCP rich as seen from the coarse LCP fibers present (Figure 15c).

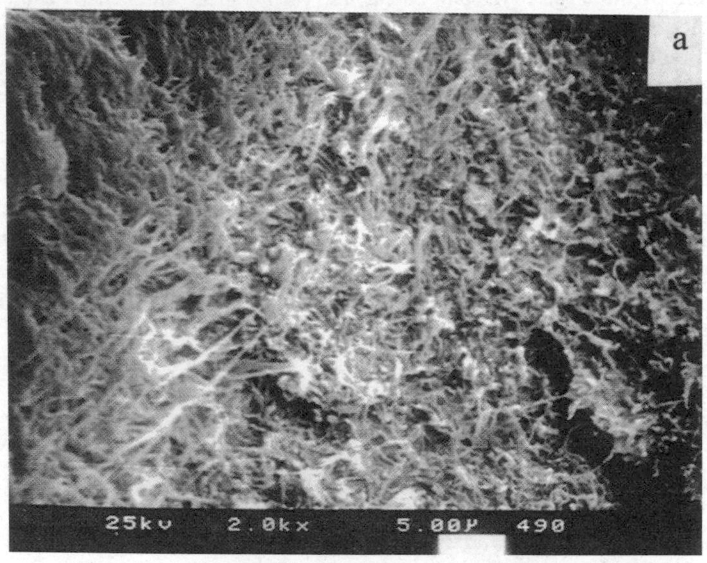

Figure 12. SEM photographs of core of pure unheat (a) and heat (b) treated LCP MTB samples.

Figure 16 shows photographs of uncured (a, b) and cured (c, d) 75/25 PAI-1/LCP blends, respectively. Uncured samples (Figure 16a, b) show fiber formation with the core (b) fibers being much finer than the skin (a) fibers. Cured samples (Figure 16c, d) show no evidence of fiber formation but voids. Phase separation can be observed in the core (d). In case of the uncured sample (Figure 16a, b), ample evidence of phase separation can be observed in the PAI-1 rich core, which shows fine fibrillation. The LCP rich skin (a) has coarser fibers than the core (b) section. The fibers do not seem to be oriented. On curing, the fibrillation disappears (Figure 16c, d) and discontinuous specks of LCP in a continuous PAI-1 phase can be observed in the core (d).

Figure 17 shows photographs of uncured (a, b) and cured (c, d) 75/25 PAI-2/LCP blends, respectively. In case of uncured 75/25 PAI-2/LCP samples (Figure 17a, b), the LCP rich skin (a) also shows platelet formation and the PAI-2 rich core (b) shows uneven discontinuous specks of LCP. Interestingly, on curing (Figure 17c, d) the platelets are transformed into fibrils in the skin (c) and the LCP specks into fibrils in the core (d). No orientation of the fibrils is observed. In both uncured and cured samples, phase separation is evident in the core.

Figure 18 shows photographs of uncured (a, b) and cured (c, d) 90/10 PAI-1/LCP blends, respectively. An interesting observation is the presence of fibers in the core (b) of the uncured 90/10 PAI-1/LCP sample. The skin (a) does not show any fibrillation. The observed fibers in the core (b) are of very low aspect ratios and hence could not possibly reinforce the blend. This fibrillation is destroyed on heat treatment (Figure 18c, d). Phase separation can be observed on heat treatment by the presence of uneven specks of LCP in a continuous PAI-1 phase.

Figure 19 shows photographs of uncured (a, b) and cured (c, d) 90/10 PAI-2/LCP blends, respectively. The LCP rich skin of the uncured blend (Figure 20a) shows platelet formation and the PAI-2 rich core shows uneven specks of LCP (Figure 19b). Interestingly, on curing (Figure 19c,d) the platelets are transformed into fibrils in the skin (c) and the LCP specks into fibrils in the core (d). No orientation of the fibrils is observed. In both uncured and cured samples, phase separation is evident in the core.

Figure 20 shows photographs of uncured (a, b) and cured (c, d) pure PAI-2 samples, respectively. The skin (Figure 21a) section of the uncured sample shows some platelet formation which is not evident in the core (Figure 20b). Void formation due to the release of volatiles is evident in the cured sample (Figure 20c, d). The skin of the uncured sample shows platelet formation.

The SEM photographs suggest that the curing creates voids and destroys the fibers which are formed during injection molding. There is ample evidence of poor mixing, which contributes to the decrease in physical properties of the blend when compared to the pure components. The skin is rich in LCP which provides lubrication and hence brings down the apparent viscosity of the blends compared to that of pure PAI. The lack of property improvement is also due to the inhibition of the imidization reaction by the LCP which has been shown to have barrier properties. Since diffusion is the main factor in the degree of imidization of PAI, the blends have a lesser degree of imidization than the pure PAI with the consequent lack of property improvement. The incompatibility and the poor mixing are also

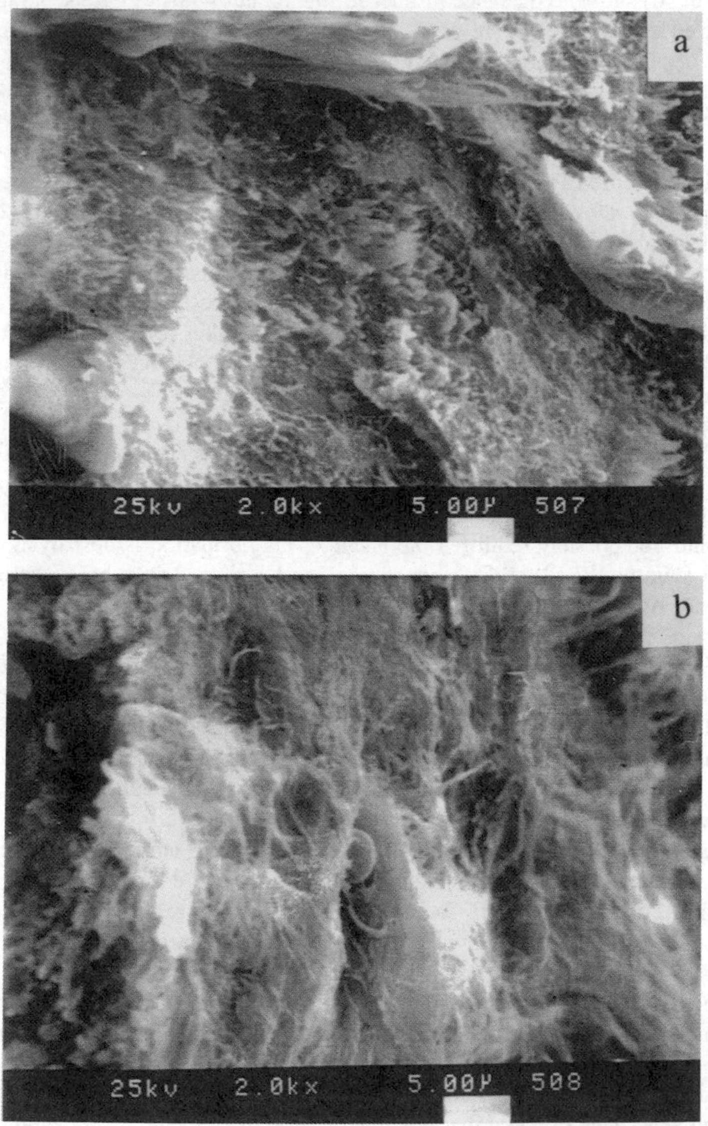

Figure 13. SEM photographs of uncured (a,b) and cured (c,d) 10/90 PAI-1/LCP MTB samples: (a,c) skin, (b,d) core.

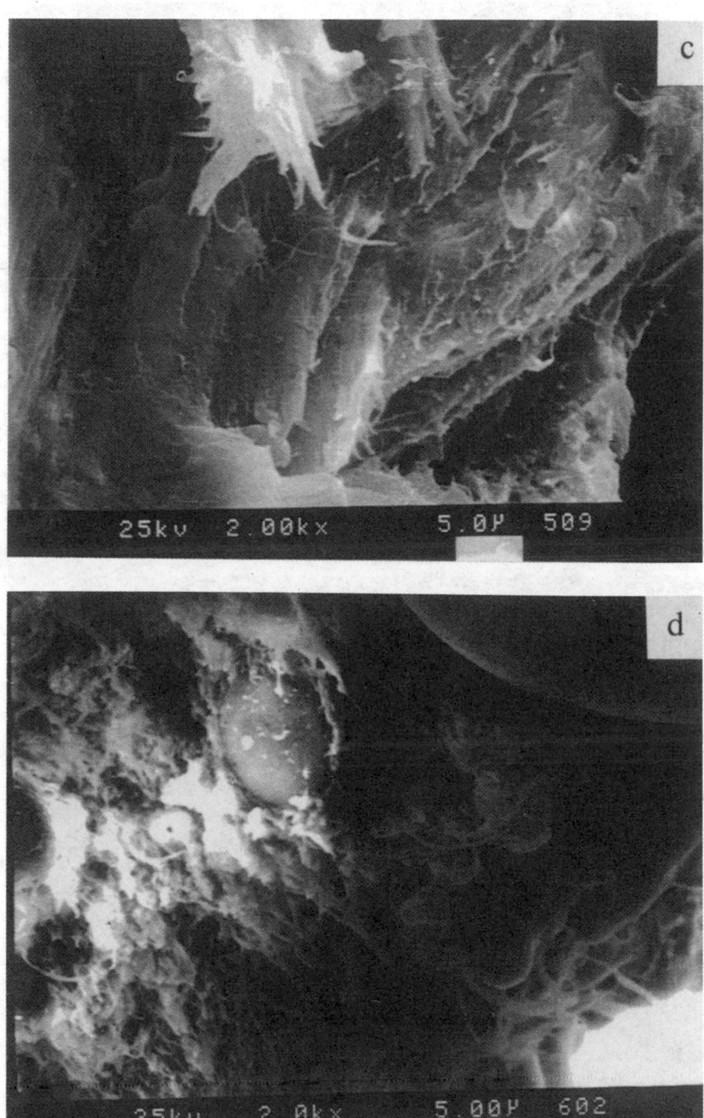

Figure 13. Continued.

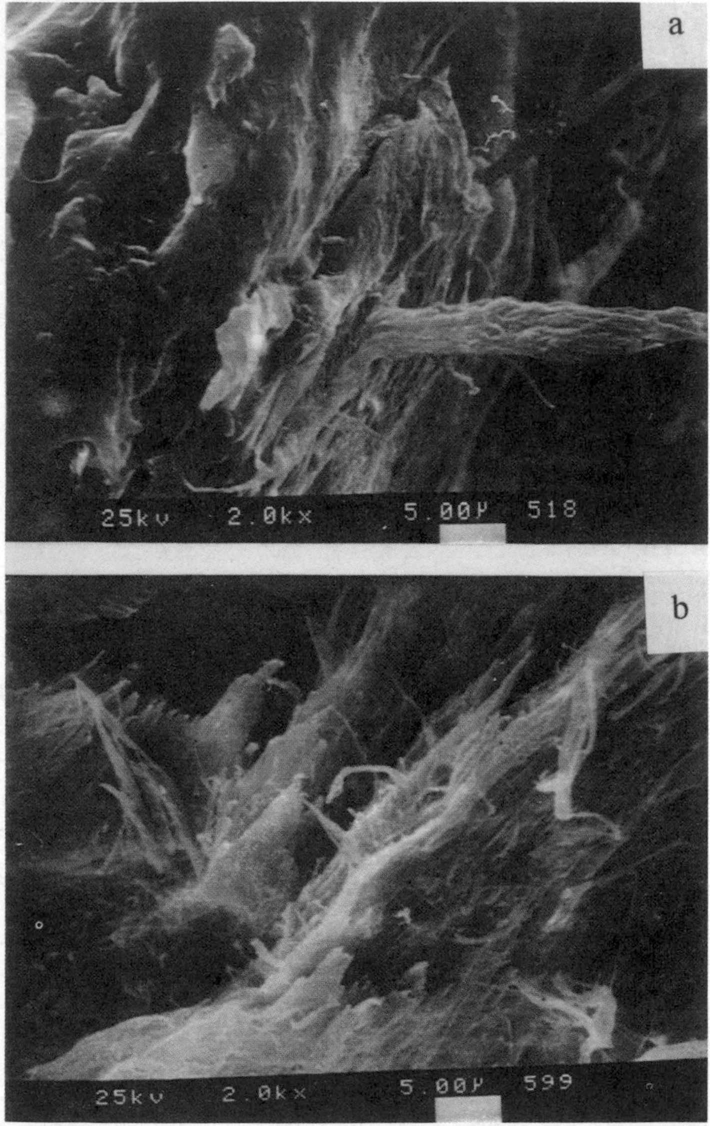

Figure 14. SEM photographs of uncured (a,b) and cured (c,d) 25/75 PAI-1/LCP MTB samples: (a,c) skin, (b,d) core.

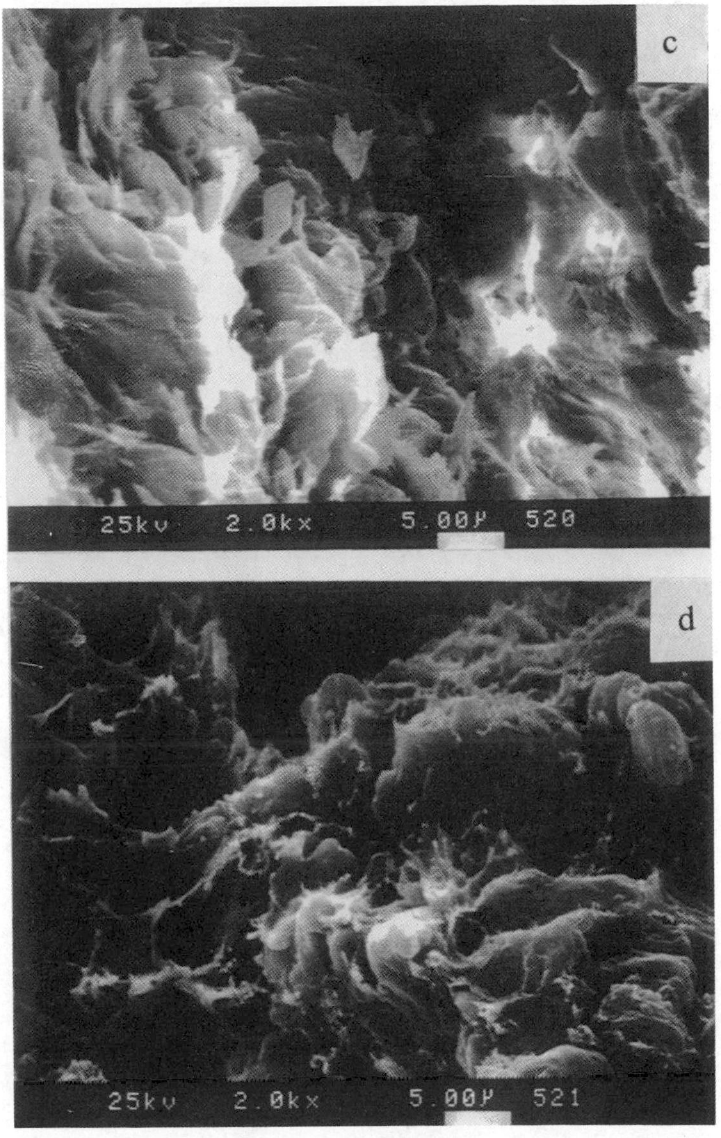

Figure 14. Continued.

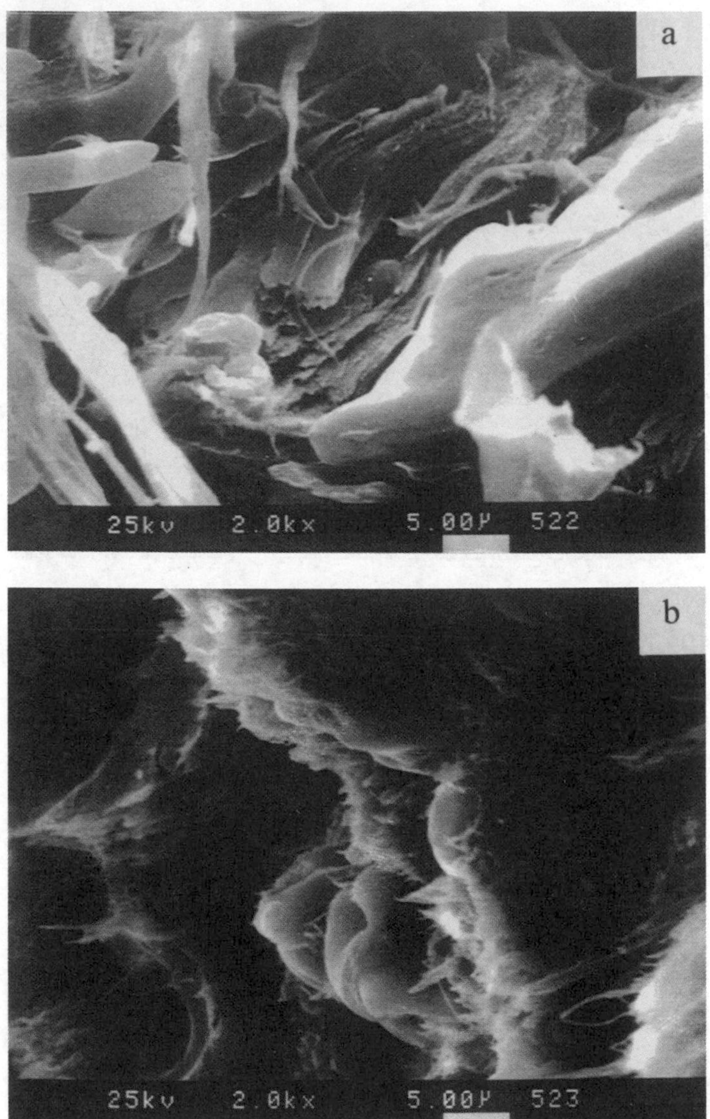

Figure 15. SEM photographs of uncured (a,b) and cured (c,d) 50/50 PAI-1/LCP MTB samples: (a,c) skin, (b,d) core.

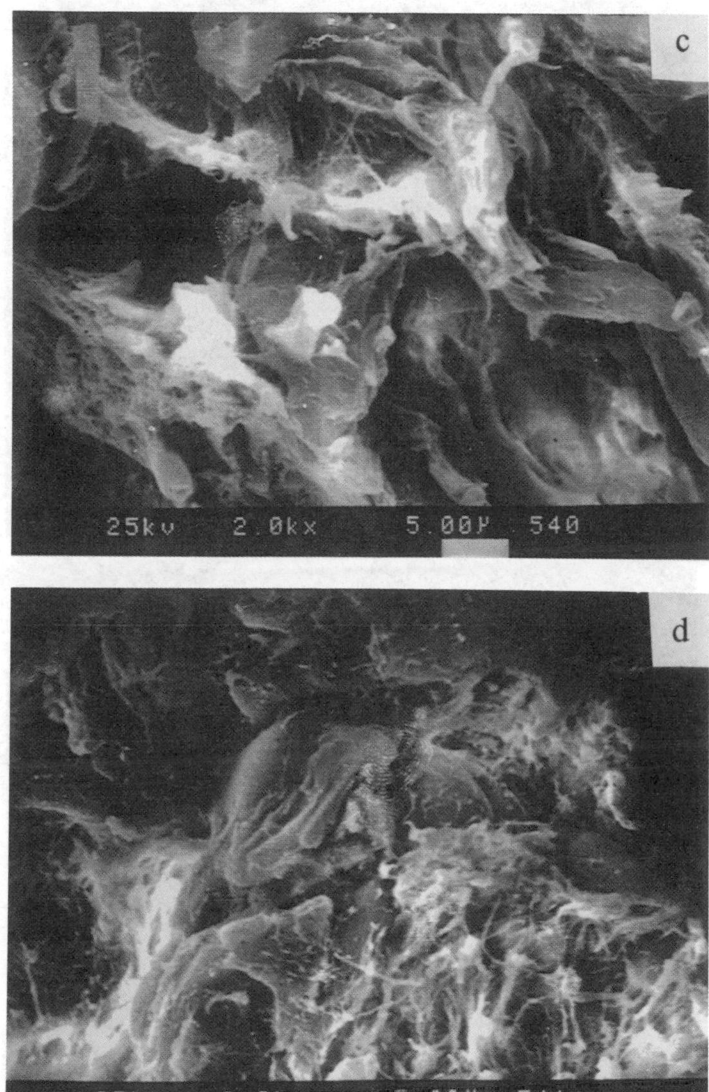

Figure 15. Continued.

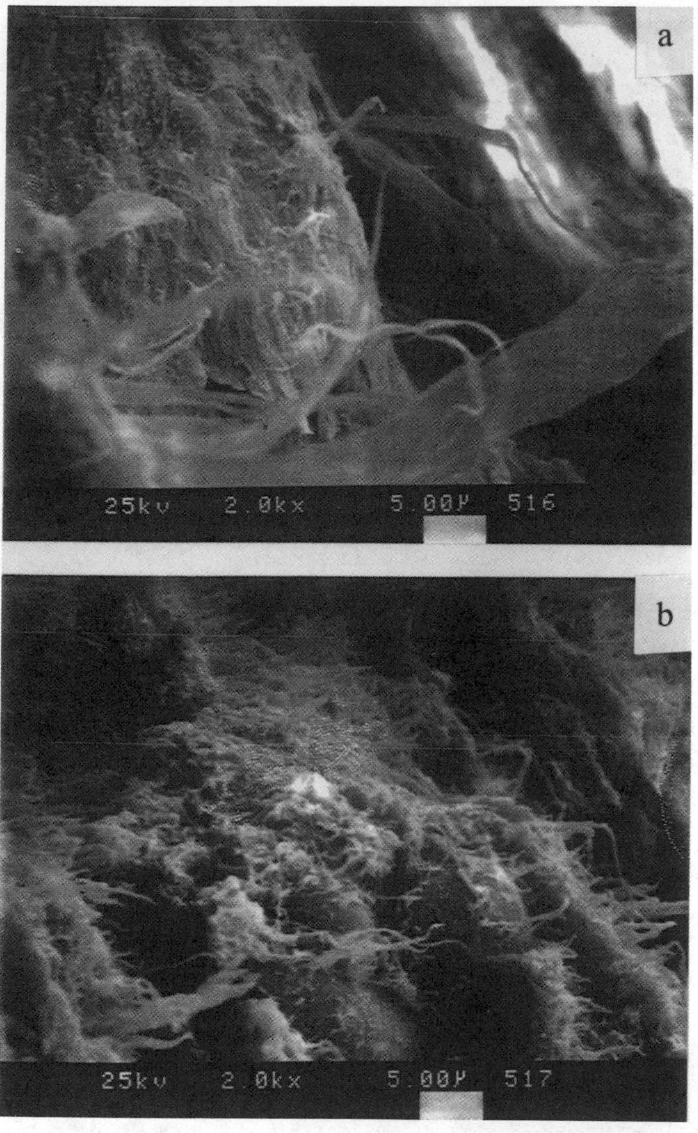

Figure 16. SEM photographs of uncured (a,b) and cured (c,d) 75/25 PAI-1/LCP MTB samples: (a,c) skin, (b,d) core.

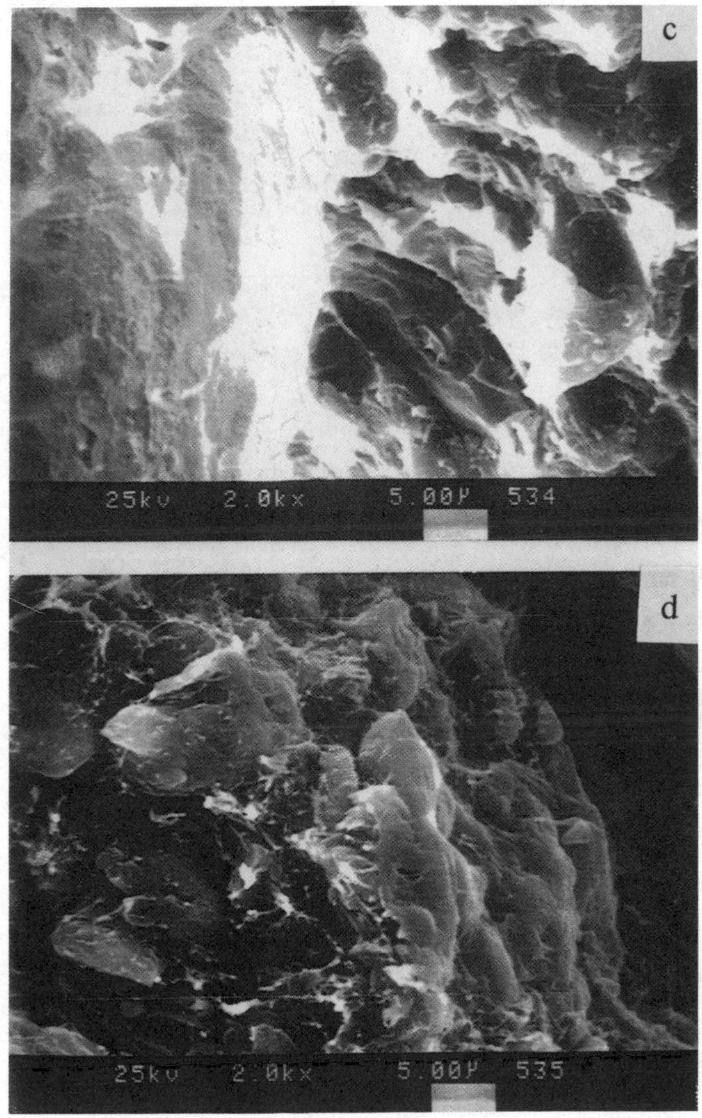

Figure 16. Continued.

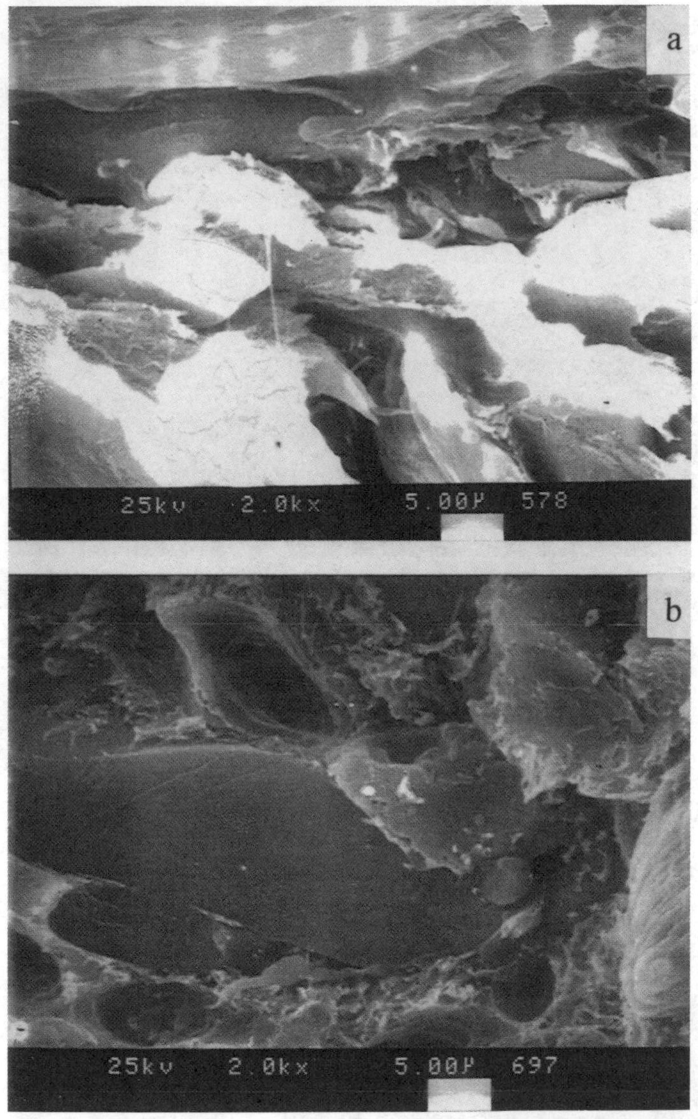

Figure 17. SEM photographs of uncured (a,b) and cured (c,d) 75/25 PAI-2/LCP MTB samples: (a,c) skin, (b,d) core.

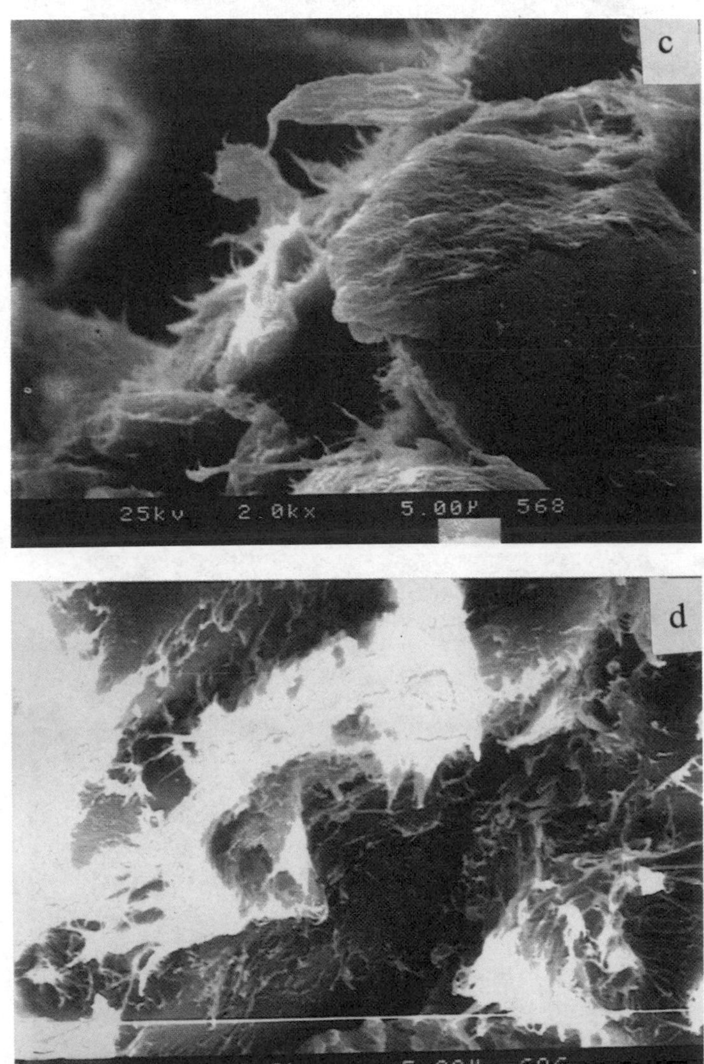

Figure 17. Continued.

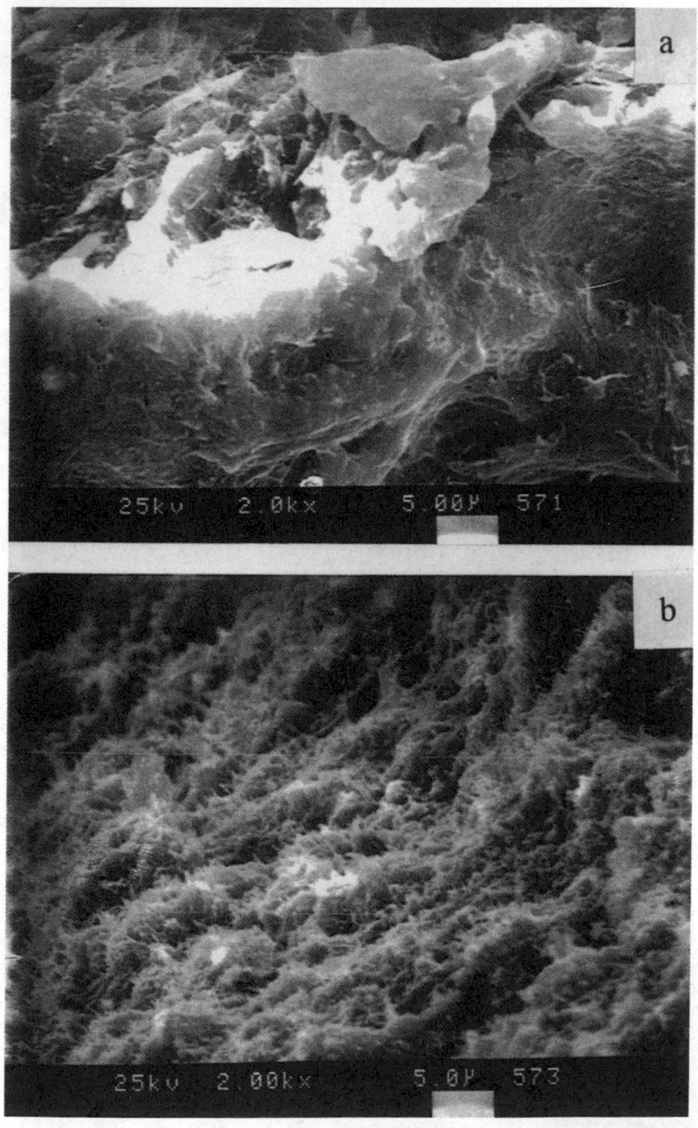

Figure 18. SEM photographs of uncured (a,b) and cured (c,d) 90/10 PAI-1/LCP MTB samples: (a,c) skin, (b,d) core.

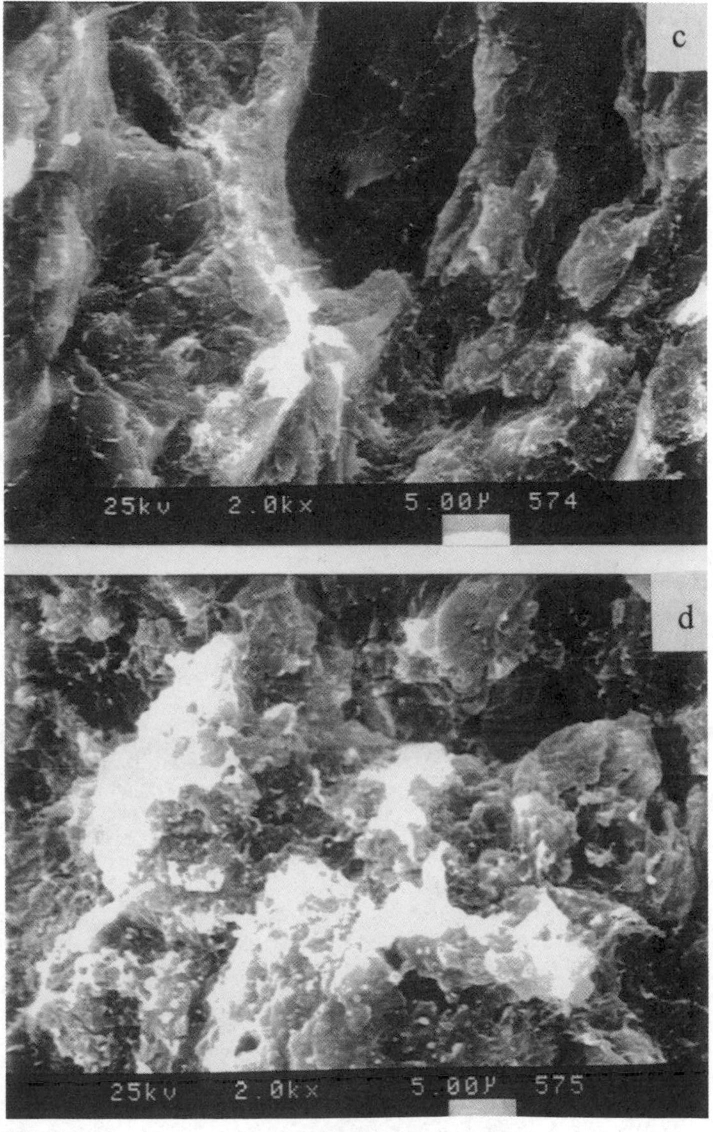

Figure 18. Continued.

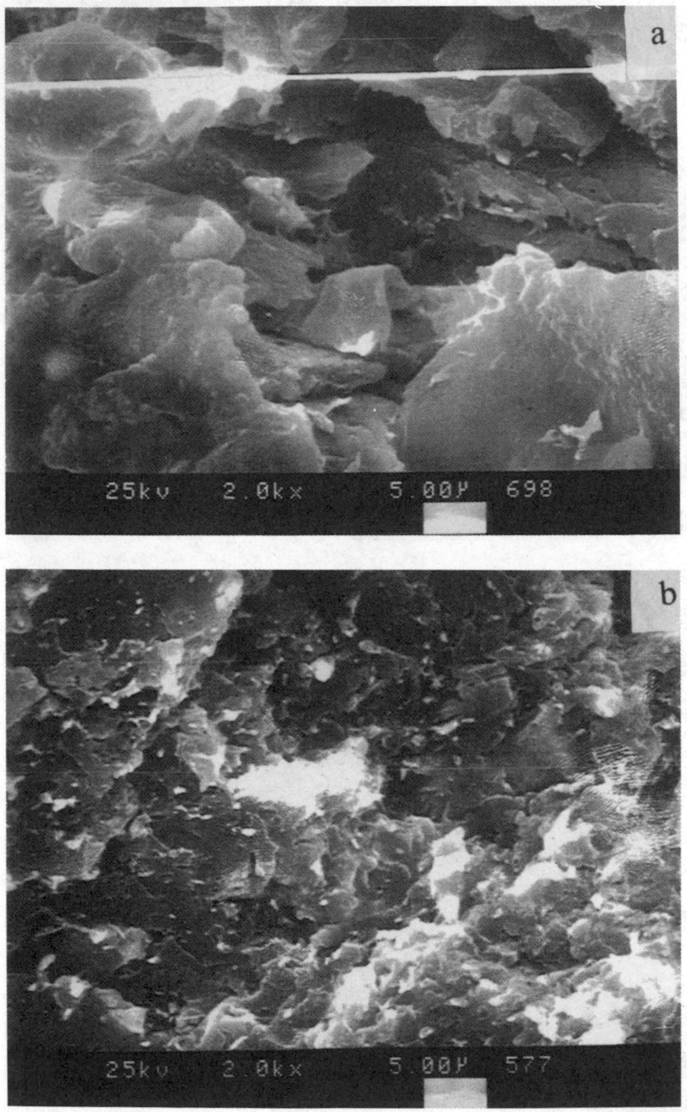

Figure 19. SEM photographs of uncured (a,b) and cured (c,d) 90/10 PAI-2/LCP MTB samples: (a,c) skin, (b,d) core.

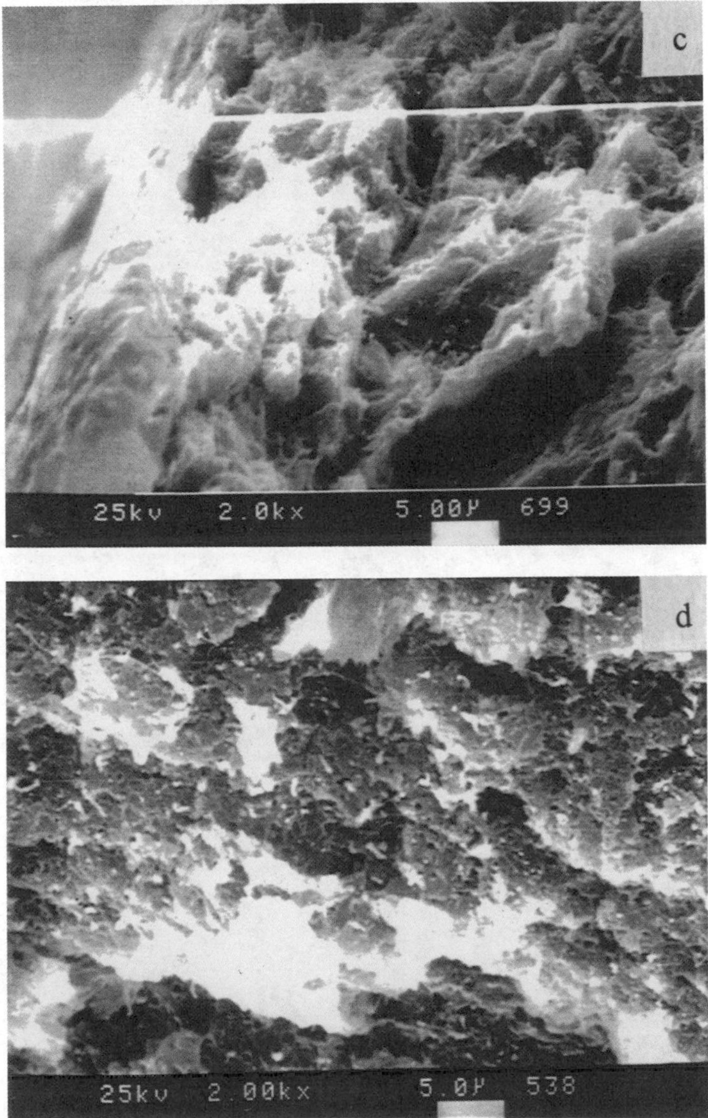

Figure 19. Continued.

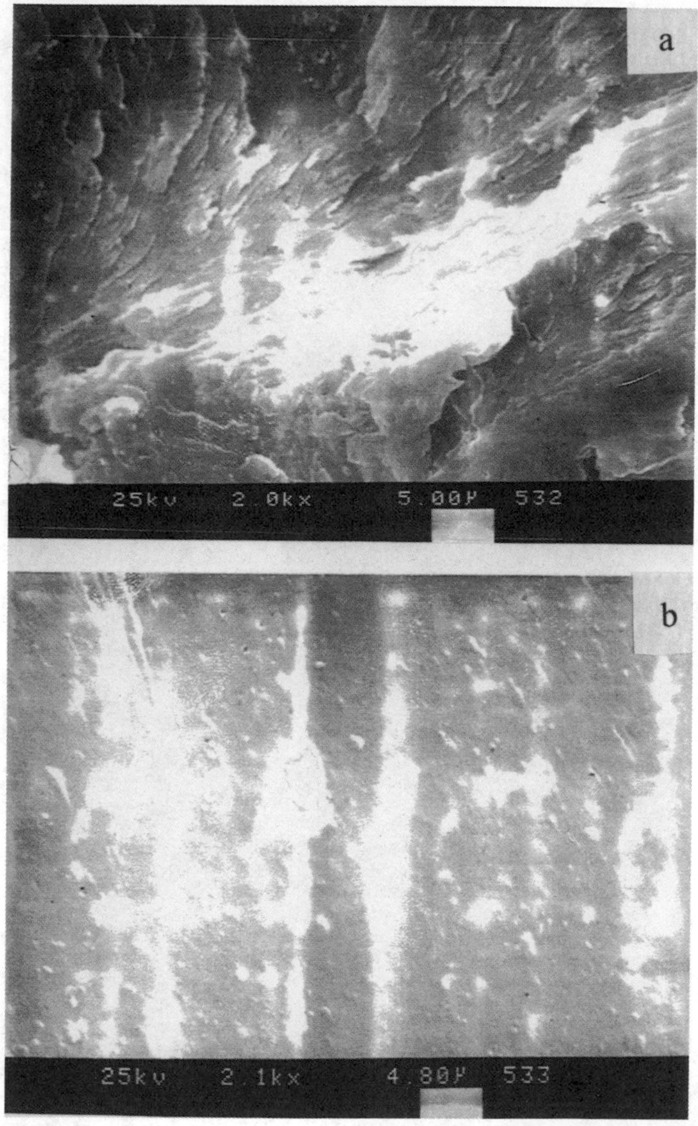

Figure 20. SEM photographs of uncured (a,b) and cured (c,d) PAI-2 MTB samples: (a,c) skin, (b,d) core.

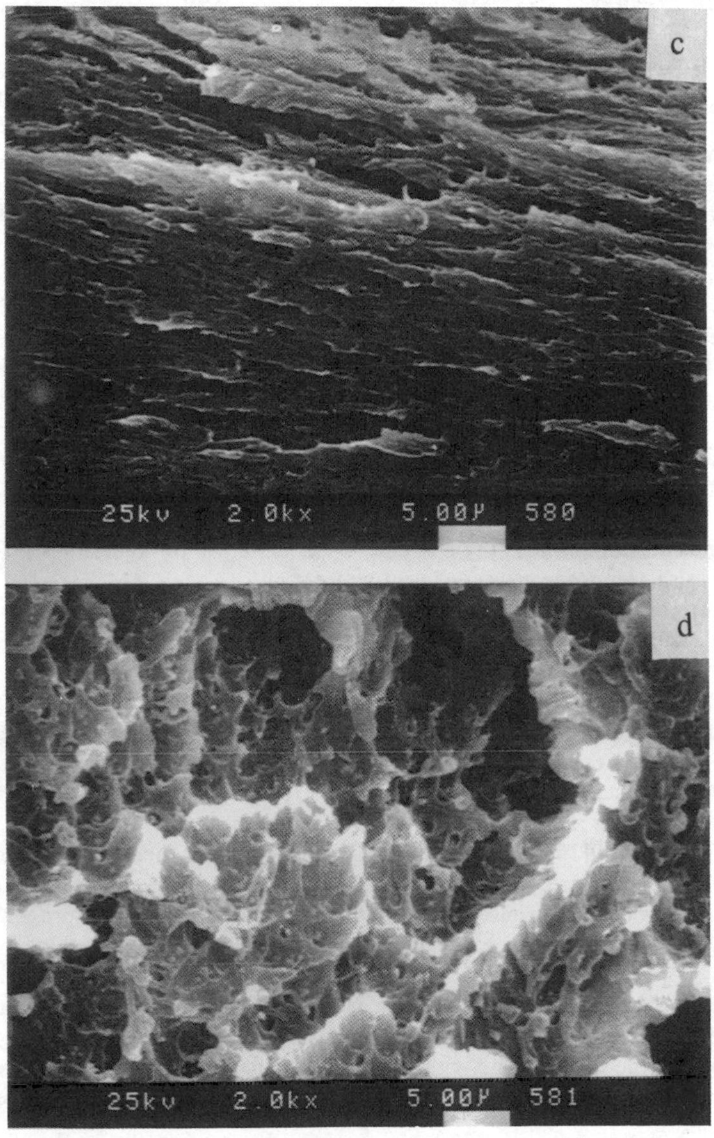

Figure 20. Continued.

contributors in the detriment of the physical properties of the blends with respect to the pure components.

Concluding Remarks

Thermal studies (DSC) indicate a very high moisture absorption rate for pure PAI-1. The amide linkage in PAI, as in case of polyamides, is known to be very moisture sensitive and undergoes hydrolysis and chain scission when in the melt state in the presence of moisture. PAI-2 does not indicate the same moisture absorption properties observed in PAI-1. Addition of only 10% LCP decreases the moisture absorption of the blend dramatically. The T_g of the pure components and the blends increases on heat treatment. The presence of LCP is seen to inhibit the imidization of PAI. The major factors affecting imidization are temperature, time and diffusion of the water of imidization through the molded part. The presence of LCP which has good barrier properties, inhibits the diffusion of the NMP solvent and water of imidization which in turn inhibits the reaction. The T_g transition for PAI in DSC scans is changed to a melting peak which is shifted to a higher temperature on heat treatment. This is due to the molecular aggregation within the amorphous regions after heat treatment observed by Jellinek et al (29).

The dynamic mechanical studies (DMTA) indicate that the blends are incompatible. The molecular stiffness and the T_g are also observed to increase on heat treatment, though not by much due to the inhibition of imidization. The loss tangent is also seen to decrease on heat treatment.

From the rheological studies, the pure PAI-2 showed shear thinning in the shear rate range studied. The studies also indicate that LCP acts as a lubricating agent for the blends by bringing down the viscosity of the blends dramatically. 90/10 PAI-2/LCP blends mixed using the twin screw extruder showed lower shear stress values than the corresponding static mixer mixed blends. The opposite is the case for the 75/25 PAI-2/LCP blends. There is a dramatic drop of one order of magnitude in the shear stress values between 90/10 and 75/25 PAI-2/LCP blends. A similar drop in shear stress values can be seen in case of pure PAI-1 and 90/10 PAI-1/LCP blends. In all blend compositions there is a drop in shear stress values with increase in temperature from 330° to 350 °C. There is no indication of any interaction between the phases. Taylor's equation applied to the present study indicates incompatibility since the blend viscosities are additive and no maxima exists.

The blend viscosities fall within the envelope of the Heitmiller prediction when applied to the present study. According to the theory this indicates complete segregation of the components with the lower viscosity component (LCP) forming the outer layer. Comparison with the Hashin model also leads us to the same conclusions.

Tensile modulus and elongation of the blends and pure PAI and LCP increases on heat treatment whereas the tensile strength decreases in case of the pure LCP. The latter case indicates that the flexibility and toughness of the LCP increases on heat treatment due to the loss of orientation whereas in the former case chain extension due to the loss of water molecules decreases the flexibility of the PAI chains. The properties of the pure components are better than that of the blends hence no reinforcement due to fiber formation is evident. Similar results were obtained by the Izod impact test. Here too, no reinforcement was evident. Morphological studies showed fiber formation in some cases. But these fibers were of very low aspect ratios and hence cannot provide reinforcement to the blend. Voids and the destruction of fiber structure on heat treatment can be observed. Agglomeration and segregation can be observed indicating poor mixing. This phenomenon combined with the lack of interaction between the phases, lack of reinforcing fibers and the inhibition of the imidization reaction of PAI by the LCP contributes to the poor mechanical properties of the blends.

Acknowledgement

Authors wish to thank the Amoco Performance Products for supplying PAI samples.

References

1. Bessonov, M.I., M.M. Koton, V.V. Kudryavtsev, and L.A. Laius, Polyimides: Thermally Stable Polymers, Consultants Bureau, NY, 1987.
2. Brydson, J.A., Plastics Materials, Butterworths, London, 1982.
3. Throne, J.L., in "Handbook of Plastic Materials and Technology", I.I. Rubin, ed., Wiley Interscience, New York, 1990, p.225.
4. Charbonneau, L., U.S. Patent 4, 024,108, 1977.
5. Chen, Y.T., U.S. Patent 4 224,214, 1980.
6. Brooks, G.T, and B.W. Cole, EP 97434, Patent to Standard Oil Co., 1983.
7 Brooks, G.T., EP 161053, Patent to Amoco Corporation, 1985.
8. Calundann, G.W., and M. Jaffe, Proc. R. A. Welch Conf. Chem. Res., Houston, 246, 1982.
9. Ide, Y., and Z. Ophir, Polym. Eng. Sci., 23, 261, 1983.
10. Isayev, A.I., and M. Modic, Polymer Composites, 8,158, 1987; U.S. Patent 4,728,698, 1988.
11. Blizard, K.G., and D.G. Baird, SPE ANTEC, 32, 311, 1986.
12. Huh, W., R. Weiss, and L. Nicolais., SPE ANTEC, 32, 306, 1986.
13. Isayev, A.I., and S. Swaminathan, Advanced Composites III, Expanding Technology, ASM, 259, 1989; U.S. Patent 4,835,047, 1989.
14. Kiss, G., Polym. Eng. Sci., 27, 410, 1987.
15. Brostow, W., Kunstoffe German Plastics, 78, 5, 15, 1988.
16. Isayev, A.I. and T. Limtasiri, in "The International Encyclopedia of Composites", S. M. Lee, ed., VCH Publishers New York, v.3, 1990, p55.
17. Dutta, D., H. Fruitwala, A. Kohli, and R.A. Weiss, Polym. Eng. Sci., 30, 1005 (1990).
18. Handlos, A.A., and D.G. Baird, J.Macromol. Sci., C35, 183, 1995.
19. Lai, X.Y., D.F. Zhou, and F.S. Lai, SPE ANTEC, 38, 382, 1992.
20. Lai, X.Y., D.F. Zhou, and F.S. Lai, SPE ANTEC , 39, 2676, 1993.
21 Taylor, G.I., Proc. Roy. Soc. London, A138, 431, 1932.

22. Taylor, G.I., Proc. Roy. Soc. London, A146, 501, 1934.
23. Chuang, H.K., and C.D. Han, J. American Chemical Society, 106, 171, 1984.
24. Han, C.D., Multiphase Flow in Polymer Processing Academic Press, NY, 1981.
25. Heitmiller, R.F., R.Z. Maar, and H.H. Zabusky, J. Appl. Polym. Sci., 8, 873, 1964.
26. Hashin, Z., Cited by H. Van Oene, Polymer Blends, D.R. Paul, and S. Newman eds., Vol. 1, Academic Press, NY., 1978, p.339.
27 Everage, Jr., A.E., Trans. Soc. Rheol., 19, 509, 1975.
28. Skross, R.E. and R.H. Walker, Plast. Eng. 35, January 1976.
29. Jellinek, H.H., R. Yokota, and Y. Itoh, Polym. J., 4, 601, 1973.
30. Adrova, N.A., L.A. Laius, A.P. Rudakov, and M.I. Bessonov, Polyimides: A New Class of Thermally Stable Polymers, Technomic Publishing, Stamford, 1970.
31. Joseph E., G.L. Wilkes, and D.G. Biard, ACS Polymer Preprints, 25, 2, 94, 1984.
32. Bigg, D.M., Polymer Composites, 6, 20, 1985.

Polymer-Dispersed Liquid Crystals

Chapter 11

Materials for Polymer-Stabilized Liquid Crystals

L.-C. Chien, M. N. Boyden, A. J. Walz, and C. M. Citano

Liquid Crystal Institute and NSF Advanced Liquid Crystalline Optical Materials Center, Kent State University, Kent, OH 44242

The synthesis, characterization and mesomorphic behavior of diacrylates based on polymerized liquid crystals are reported. Several types of polymer stabilized liquid crystal display devices were prepared from the dispersions of low concentration of diacrylates in liquid crystals and subsequently the prepolymers were polymerizaton by ultraviolet radiation toorm polymer networks. The morphology studies show that the orientation of polymer networks induced by the surface treatment of the substrate has led to the preferential liquid crystal alignment.

Liquid crystal display (LCD) technology is not a traditional field of chemical research, but it is chemically intensive and overlaps with physics, electronics and manufacture engineering. Advances in LCD technologies rely on new materials and materials engineering to produce light weight, low power consumption and high information content LCDs. The scope of advanced materials applied in LCDs are low-viscosity liquid crystals, color filters, polarizers, alignment layers, retardation films and various plastic substrates as well as dispersions of liquid crystals and polymers.

The introduction of dispersions of liquid crystals and polymers has grown into a broad class of materials important in electrically controlled, scattering-based light shutters and bistable reflective displays. In liquid crystal and polymer dispersions the weight of polymers used can be varied from 80% to as low as 0.5%, depending on the application and type of polymer used. The systems containing a polymer of 20% or higher has been extensively studied are referred to as polymer dispersed liquid crystals (PDLC). (*1*) Currently, the systems of most interest are those with polymer concentrations less than 10%.

The technique of incorporating a low concentration of polymers in cholesteric materials was brought to light due to the low solubility of a reactive diacrylate in cholesteric liquid crystals. The diacrylates can be either mesogenic or isotropic with a rigid core for the refractive indices matching. For low polymer concentrations, the mixtures of liquid crystals and the reactive monomers normally exhibit a liquid crystalline phases. Therefore, before photopolymerizations, the reactive monomers are aligned by the liquid crystals either via the surface treatment or an applied external field. After polymerization, the polymer network retain the alignment and can be used to control the orientation of the liquid crystal. Haze-free light shutters, operating at the scattering-mode, were discovered from the dispersions of polymer networks in cholesteric liquid crystals; there were no refractive indices matching problems

0097–6156/96/0632–0182$15.00/0

between the host LMWLCs and dispersed anisotropic networks. The dispersions of polymer networks in the liquid crystals present unique electro-optical properties which are well suited for applications such as overhead projection displays, head-up displays, large area flexible displays and hand-held video displays. Such technology is named as polymer stabilized cholesteric textures (PSCTs) with respect to the formulation of the materials. (*2-7*) A bistable reflective-mode display was developed using a polymer stabilized cholesteric liquid crystal display where the pitch of cholesteric liquid crystals was selected to reflect the visible light. (*8*)

Experimental

Materials. A series of diacrylates, **1** to **4**, were synthesized and characterized for polymer-stabilized liquid crystals. Reactive monomer (RM) **1**,4,4'-bis(acryloyloxy)-1,1'-biphenylene (BAB) was prepared from a biphenylene core and acryloyl chloride. The incorporation of hexamethylene spacer groups resulted in a low melting diacrylate monomer **2**, 4,4'-bis[6-(acryloyloxy)hexyloxy]-1,1'-biphenylene (BAB-6). RMs **3**, 4,4'-bis{4-[6-(acryloyloxy)-hexyloxy]benzoate}-1,1'-biphenylene (BABB-6) and **4**, 4,4'-bis{4-[6-(methacryloyloxy)-hexyloxy]benzoate}-1,1'-biphenylene (BMBB-6), were synthesized according to scheme 1. All the reagents were obtained from Aldrich and used as received. Liquid crystals, 4-pentyl-4'-cyanobiphenyl (K15) and 4-[2-methylbutyl-4'cyanobiphenyl (CB15), were purchased from E MercK). Benzoin methyl ether (BME), the photoinitiator was obtained from Polysciences Inc. The polyimide coated and anti-parallel rubbed cells, separated by 10 μm spacer, were purchased from Crystaloid Electronics, Hudson, Ohio. All the synthesized compounds were characterized by Varian XL 200 ^{1}H NMR, HPLC and Elemental analyses. Mesomorphic properties were studied using both the polarizing optical microscopy and the differential scanning calorimetry. The scanning electron microscope (SEM) was used to study the morphology of the polymer networks.

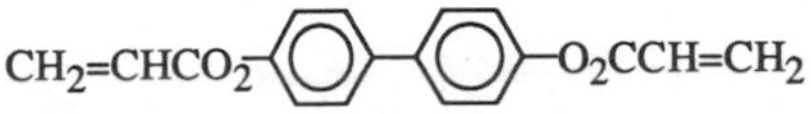

CH_2=CHCO_2–⟨⟩–⟨⟩–O_2CCH=CH_2

1

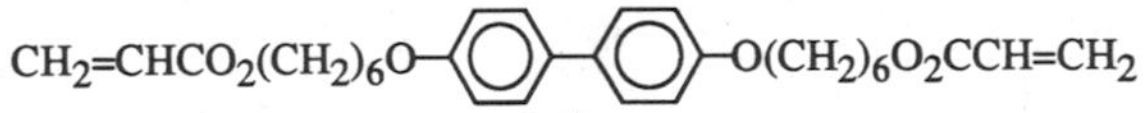

CH_2=CHCO_2(CH_2)$_6$O–⟨⟩–⟨⟩–O(CH_2)$_6$$O_2$CCH=$CH_2$

2

CH_2=CHCO_2(CH_2)$_6$O–⟨⟩–CO_2–⟨⟩–⟨⟩–O_2C–⟨⟩–O(CH_2)$_6$$O_2$CCH=$CH_2$

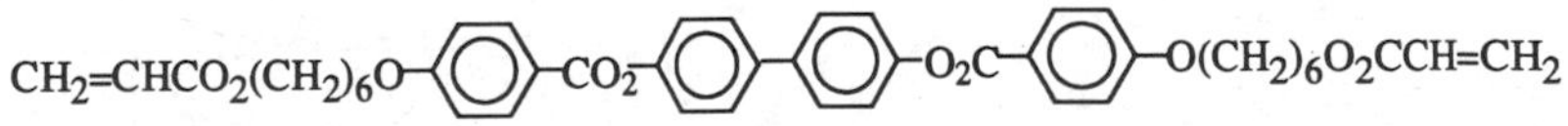

3

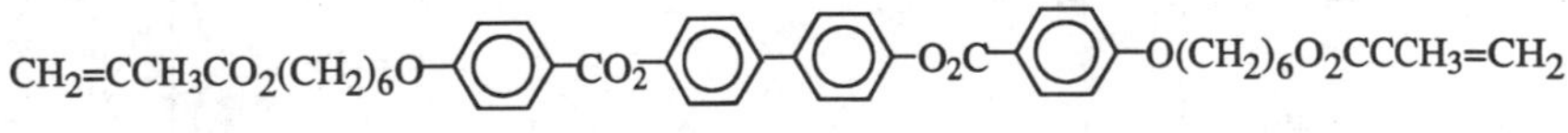

CH_2=CCH_3CO_2(CH_2)$_6$O–⟨⟩–CO_2–⟨⟩–⟨⟩–O_2C–⟨⟩–O(CH_2)$_6$$O_2CCCH_3$=$CH_2$

4

Synthesis of 4,4'-bis{6-[acryloyloxy)hexoxy]benzoate}-1,1'-biphenyl (BABB-6). (**3**). This reaction was carried out under dry conditions as shown in Scheme 1. The 4,4'-bis{6-[acryloyloxy)hexoxy]benzoate}-1,1'-biphenyl, **3**, was prepared from the reaction of 4-(acryloyloxyhexyloxy)benzoic acid chloride and 4,4'-biphenol. The crude was purified by flash chromatography using a mixture of methylene chloride/ethyl acetate (98:2) as the eluent. The ^{1}H NMR is shown in Figure 1a.

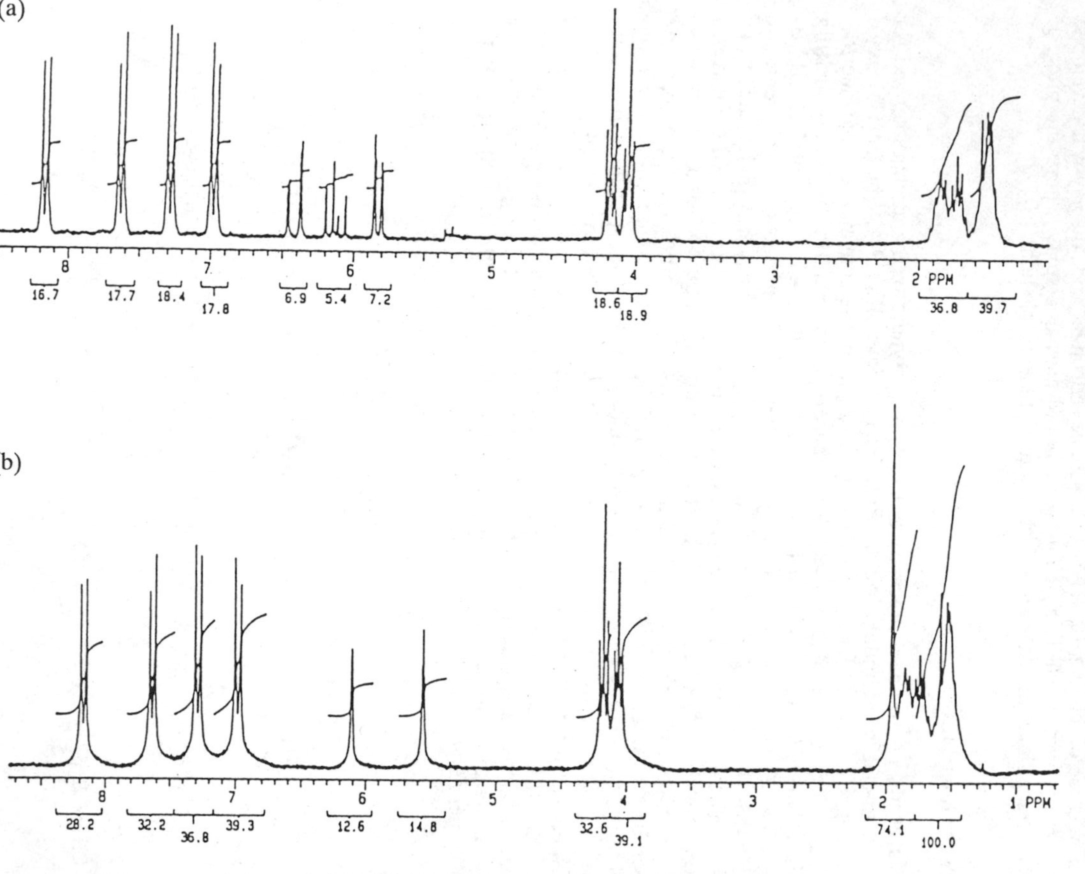

Figure 1. The ^1H NMR spectra of (a) RM **3** and (b) RM **4**.

^{1}H NMR ($CDCl_3$, δ ppm) 1.38-2.05 (m, 16H, alkyl spacer) 4.06 (t, 4H, J=6.35, Ph-O-C$\underline{H_2}$) 4.19 (t, 4H, J=6.56, =CH-COO-C$\underline{H_2}$) 5.82 (dd, 2H, J=1.71, 10.26, *cis*-$\underline{H}$CH=CH-COO-) 6.10 (dd, 2H, J=10.30, 17.21, =CH-COO-) 6.42 (dd, 2H, J=1.75, 17.26, *trans*-HC$\underline{H}$=CH-COO-) 6.98 (d, 4H, J=8.96, aromatic H ortho to Ph-O-) 7.28 (d, 4H, J=8.58, aromatic H ortho to COO-Ph-) 7.63 (d, 4H, J=8.67, 2,2,2'2'-aromatic H of Ph-Ph) 8.17 (d, 4H, J=8.95, aromatic H ortho to Ph-COO-) Elemental analysis: Cal. C: 71.92%; H: 6.31%; Exp. C: 70.79%; H: 6.11%.

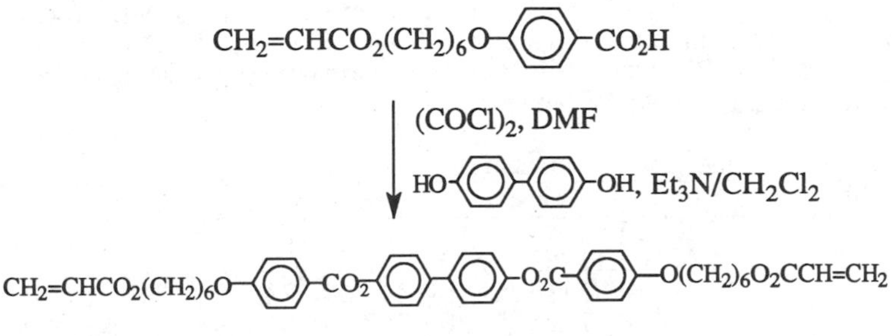

CH_2=$CHCO_2(CH_2)_6$O–⟨Ph⟩–CO_2H

↓ $(COCl)_2$, DMF

HO–⟨Ph⟩–⟨Ph⟩–OH, Et_3N/CH_2Cl_2

CH_2=$CHCO_2(CH_2)_6$O–⟨Ph⟩–CO_2–⟨Ph⟩–⟨Ph⟩–O_2C–⟨Ph⟩–$O(CH_2)_6O_2CCH$=CH_2

Scheme 1

Synthesis of 4,4'-bis{6-[methacryloyloxy)hexoxy]benzoate}-1,1'-biphenyl (BMAB-6). **(4)**. This reaction was carried out under dry conditions. The 4,4'-bis{6-[acryloyloxy)hexoxy]benzoate}-1,1'-biphenyl, **4**, was prepared from the reaction of 4-(methacryloyloxyhexyloxy)benzoic acid chloride and 4,4'-biphenol. The crude was purified by flash chromatography using a mixture of dichlomethane/ethyl acetate (98:2) as the eluent. The ^{1}H NMR of BABB-6 is shown in Figure 1b. ^{1}H NMR ($CDCl_3$, δ ppm) 1.38-1.90 (m, 16H, alkyl spacer) 1.95 (d, 3H, J=1.18, C$\underline{H_3}$-C=CH_2) 4.06 (t, 4H, J=6.35, Ph-O-C$\underline{H_2}$) 4.19 (t, 4H, J=6.56, =CH-COO-C$\underline{H_2}$) 5.56 (t, 2H, J=1.61,*E*-$\underline{H}$CH=CCH_3-COO-) 6.11 (t, 2H, J=1.79, Z-HC$\underline{H}$=CCH_3-COO-) 6.98 (d, 4H, J=8.96, aromatic H ortho to Ph-O-) 7.28 (d, 4H, J=8.58, aromatic H ortho to COO-Ph-) 7.63 (d, 4H, J=8.67, 2,2,2'2'-aromatic H of Ph-Ph) 8.17 (d, 4H, J=8.95, aromatic H ortho to Ph-COO-) Elemental analysis: Cal. C: 72.42%; H: 6.61%; Exp. C: 72.40%; H: 6.73%.

Characterizations. The textures of RMs were studied using a polarizing optical microscope fitted with a hot stage and controller. Also the thermal behavior was studied using a differential scanning calorimeter with the heating and cooling rate at 10°C/min.

Preparations of liquid crystal/reactive monomer dispersions. The dispersion samples were prepared by weighing the liquid crystal, monomer and photoinitiator in a vial, dissolving the mixture in dichloromethane, and evaporating the solvent in a vacuum oven. The display cells were vacuum filled. The formulation for sample number 1 contains a mixture of 97.45 wt% of K15, 2.5 wt% of BAB and 0.05 wt% of BME. The mixture number 2 is of 97.45 wt% of K15, 2.5 wt% of BAB-6 and 0.05 wt% of BME; sample number 3 is of 95.95 wt% of K15, 1.5 wt% of CB15, 2.5wt% of BABB-6 and 0.05wt% of BME.

Preparations of SEM samples. SEM samples were prepared by fracturing the display cells, and the cells were immersed in a mixture of hexane/dichloromethane solution. After being dried under reduced pressure, the cells were separated and the polymer network were sputtered with a thin layer of palladium. The SEM pictures of networks taken with the incident electron beam normal to the substrate.

Results and Discussion

Phase behavior of reactive monomers. The phase transition temperatures of reactive monomers, studied by both the polarizing optical microscope and DSC, are summarized in Table I. BAB melts from crystal to isotropic liquid at 150°C, whereas the BAB-6 melts at lower temperature due to the incorporation of flexible spacers in between the biphenylene and the diacrylate. By extending the length of rigid core, both the BABB-6 and BMBB-6 exhibit the smectic C phase besides the the higher ordered smectic phase. A typical Schliren texture of smectic phase was observed by the polarizing microscopy (see Figure 2a). Figure 2b shows the DSC heating and cooling curves of BABB-6 and BMBB-6. The thermal polymerizations occurred at the temperature below 180°C for both RMs.

Table I. The transition temperatures of reactive monomers **1** to **4**

RM	Transition Temperatures (°C) (Kcal/mol)						
1	K	150.0	I				
2	K	83.13 (63.72)	I				
3	K	102.5 (19.30)	S_X	114.0 (0.78)	S_C	121.1 (7.54)	I
4	K	82.7 (15.26)	S_X	107.3 (0.61)	S_C	111.3 (6.93)	I

Polymer stabilized liquid crystals. When a diacrylate monomer is polymerized in a liquid crystal solvent, the orientation and order of the resultant network depends on the orientation and order of the liquid crystal. The polymer networks formed are anisotropic, depending on the orientation of the liquid crystals during polymerization. After polymerizations, these polymer networks were found to affect the orientation of the liquid crystals. Typical polymerizations of reactive monomers in liquid crystals were examined by scanning electronic microscopy (SEM). Small amount of diacrylate monomers, 2.5% of BAB or BAB-6 by the weight of low molecular liquid crystal solvent K15, were homogenously dissolved in nematic state before polymerization. In Figure 3, the photopolymerization of BAB monomer took place in a polyimide-coated and homogeneously aligned nematic liquid crystal cell. The polymer network exhibits a globule-like morphology. In contrast, the network formed from BAB-6 has a dramatically different morphology (see Figure 4) of fibril-like network. After polymerization, the network is aligned along the rubbing direction. The flexible spacers provide the monomer molecules more conformational freedom to organize the molecules in the nematic solvent during the network formation.

Cholesterol liquid crystals in the planar texture possess the unique optical feature of separating incident white light into its left- and right-hand circular components by reflecting one component and transmitting the other. The wavelength of the reflected component is governed by the Bragg formula $\lambda = np$, where n is the average refractive index and p is the pitch length of the cholesteric helix. When the λ is in the visible light range, the material provides beautiful iridescent colors. The wavelength of the reflected light is easily controlled by adjusting the chemical composition which changes the chirality of the material and therefore the pitch. The reflected wavelength can also be made temperature independent over wide temperature ranges by using suitable chemical mixtures and polymer networks. In order to understand the formation of helicoidal polymer network, we prepared a cholesteric liquid crystal sample consisting of a diacrylate monomer BABB-6, a nematic liquid crystal mixture E48 and a chiral dopant CB15. The polymerization took place in the planar texture of such cholesteric mixture, which can be achieved by using a polyimide coated and anti-parallel rubbed cell. The SEM photo (see Figure 5) exhibits a long fibril-like network with a helicoidal superstructure, obtained with a mesogenic diacrylate BABB-6. Similar fiber-like morphology is obtained from reactive monomer **4**.

(a)

(b)

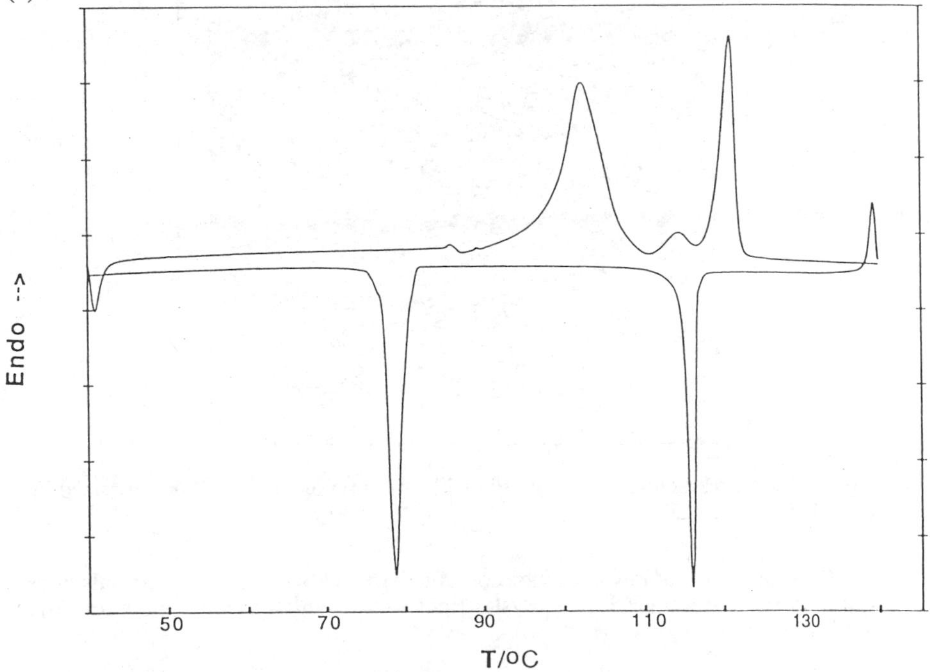

Endo -->

50 70 90 110 130

T/°C

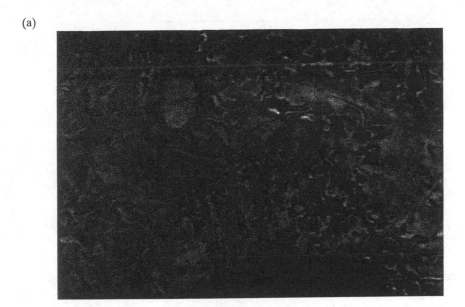

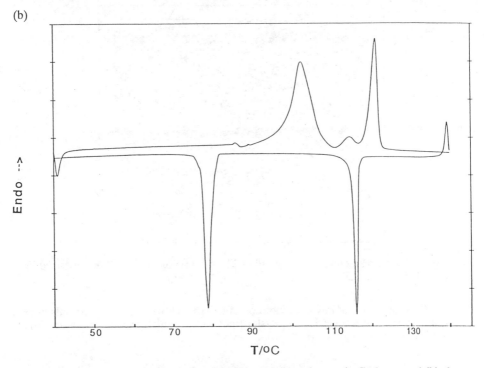

Figure 2. (a) The microphotograph of Schliren texture of smectic C phase; and (b) the DSC curves at the heating and cooling rate of 10°C/min of RM **3**.

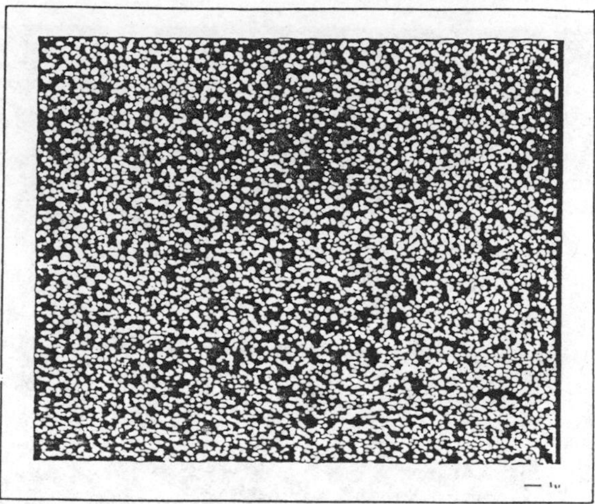

Figure 3. The SEM picture shows the bead-like morphology of RM **1** polymerized in a nematic liquid crystal (K15).

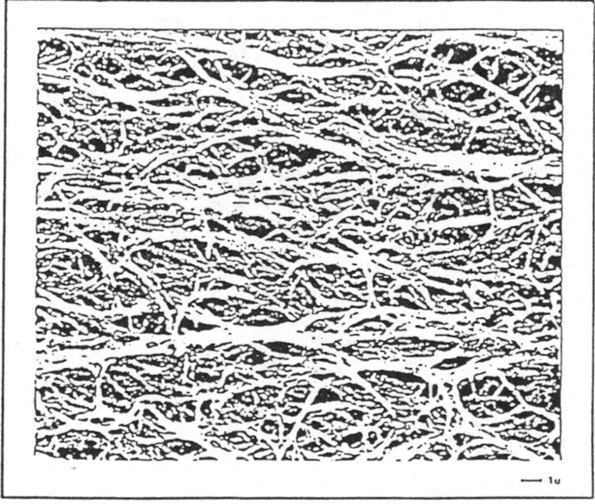

Figure 4. The SEM picture shows the fiber-like morphology of RM **2** polymerized in a nematic liquid crystal (K15).

The dispersions of low concentration of polymer network in liquid crystals have resulted in several unique liquid crystal light valves without using polarizers and providing the wide angle of view.

Conclusions

Several reactive monomers have been designed, synthesized and characterized for the polymer stabilized liquid crystals. The orientation of anisotropic polymer networks formed in different types of liquid crystals have been studied

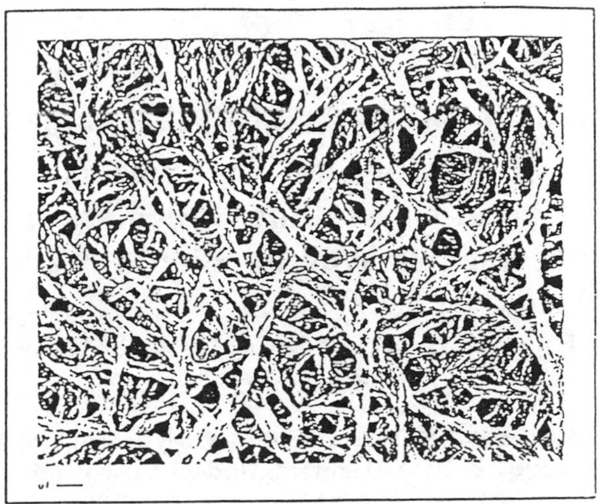

Figure 5. The SEM picture shows the helicoidal morphology of RM **3** polymerized in the planar texture of a cholesteric liquid crystal.

morphologically. We have explicated that the nature of liquid crystals controls the orientation of the polymer networks. The resultant polymer networks influence and thus, stabilize the liquid crystals.

Acknowledgment

This research is supported in part by ARPA/Low Power Displays N61331-94-K-0042 and the NSF ALCOM Center Grant DMR89-20147. The authors thank the ALCOM Characterization Center for obtaining SEM pictures and useful discussion from Dr Christine M. Hudson.

Literature Cited

1. Doane, J.W. in *Liquid Crystals, Applications and Uses*, Bahadur, B. Ed.; World Scientific Press, London, 1990, Vol. 1; pp 362.
2. Yang, D. K.; Chien, L. -C.; Doane, J. W.*Conference Record, International Display Reserach Conference* **1991,** pp. 49.
3. Doane, J. W.; Yang, D. K.; Chien, L. -C.*Conference Record, International Display Reserach Conference*, **1991**,pp. 175.
4. Yang, D. K.; Chien, L. -C.; Doane, J. W.*Appl. Phys. Lett*., **1992,** *60*, 3120.
5. Yang, D. K.; West, J. L.; Chien, L. -C.; Doane, J. W.*J. Appl. Phys*., **1994,** *76*, 1331.
6. Fung, Y. K.; Yang, D. K.; Sun, Y.; Chien, L. C.; Zumer, S.; Doane, J. W.*J. Appl. Phys*., to appear, **1995**.
7. Sun, Y, "Polymer Stabilized Cholesteric Textures--A reverse Mode Light Valve," Thesis, Kent State University, December, **1994**.
8. St.John, W. D.; Fritz, W.J.; Lu, Z. J.; Yang, D. K.; Doane, J. W.*Phys. Rev*., **1995,** *51*, 1191.

Chapter 12

Enhancement in Response Speed of Bistable Switching for Composites of Liquid-Crystalline Polymers and Low-Molecular-Weight Liquid Crystals

Tisato Kajiyama, Hirokazu Yamane, Hirotsugu Kikuchi, and Jenn Chiu Hwang[1]

Department of Chemical Science and Technology, Faculty of Engineering, Kyushu University, 6–10–1 Hakozaki, Higashi-ku, Fukuoka 812, Japan

The ternary composite was prepared from a miscible mixture of nematic liquid crystalline polymer (LCP) with a weak polar terminal unit in the mesogenic side group, nematic low molecular weight liquid crystal (LC) with a strong polar cyano end group (LC1) and LC with the similar chemical structure to the mesogenic side group of LCP (LC2). The reversible and bistable turbid-transparent states were successfully realized for the binary (LCP/LC1) and ternary (LCP/LC2/LC1) composites in an induced smectic phase upon the application of a.c. electric fields with low and high frequencies, respectively. The response speed for reversible light switching could be remarkably improved by an addition of LC2 to the binary composite at room temperature. The rise response time and decay one were a few seconds and several hundred ms, respectively. The molecular weight dependence of the electro-optical effect for the (LCP/LC) composite system was investigated by using LCPs with 12 or 40 of n. The magnitude of τ_D exhibited minimum for LCP with n=12, indicating that an appropriate length of polymeric main chain is required to realize an optimum condition for the balance between an electric field effect and an electric current effect

The preparation methods and the novel permselective characteristics of (polymer/LC) composite films have been extensively studied (*1-3*). Also, various types of (polymer/LC) composite systems have been reported as large area and flexible light-intensity controllable films (*4-11*). Since thermotropic liquid crystalline polymers (LCPs) with mesogenic side chain groups exhibit both inherent mesomorphic properties of LC and excellent mechanical characteristics of polymeric materials, LCPs

[1]Current address: Department of Chemical Engineering, Yuan-ze Institute of Technology, 135 Yuan-tung Road, Nei-li Chung-li 320, Taiwan

0097–6156/96/0632–0190$15.00/0

have attracted a major attention due to their promising applications as electro-optical devices (*12*). However, since LCPs in a mesophase state are more viscous than LCs, the response time of LCPs to an external stimulation such as an electric or a magnetic field is fairly longer than that of LCs. Therefore, (LCP/LC) mixtures in which LC takes a role of solvent or diluent to LCP has been studied in order to reduce the magnitude of viscosity of LCP, in other words, to reduce the magnitude of response time for the electro-optical devices of LCP (*13-16*).

It was reported that the binary mixtures of nematic LCs with both a strong polar terminal group (cyano or nitro) and a weak polar one formed an induced smectic phase (ISP) (*17-20*). Then, this concept was applied to the binary mixture composed of nematic LCP with a weak polar end group in the mesogenic side chain and nematic LC with a strong polar end group, resulting in the formation of an induced smectic phase (*21-24*). Then, the (LCP/LC) mixture in an induced smectic state is expected as a novel type of "light valve" which exhibits bistable and reversible light switching characteristics, that is, a memory effect.

In this paper, the phase transition behavior , aggregation states, and also, a bistable and reversible light switching behavior based on light scattering have been investigated for the ternary (LCP/LC2/LC1) composite system in an induced smectic phase.

Experimental

Materials. The chemical structures of LCPs and LCs are shown in Figure 1. PS3EM is polymethylsiloxane with a weak polar terminal group in the mesogenic side group. The molecular weight of PS3EM is $1.1\text{x}10^4$. 5OCB and HPPB were used as LCs, that is, LC1 and LC2, respectively. 5OCB has a strong polar cyano terminal group and HPPB does the similar chemical structure to the mesogenic side chain group of PS3EM. The mesomorphic phase of the (PS3EM/HPPB) composite is nematic. The mixture of PS3EM/HPPB/5OCB was dissolved in acetone and the ternary composite film was prepared by a solvent-casting method.

To investigate the role of LCP in the composite on the electro-optical properties, PS[4BC/DM]s with the degree of polymerization of 12 and 40 were used as LCP. In order to investigate the role of LCD in the composite on the electro-optical effect, low molecular weight smectic LC, S2 was employed as a sample of composite in which any LCP was not included. S2 is a binary smectic mixture of low molecular weight liquid crystals as shown in Figure 1. The ionic impurities (Pt catalyst) of about 1000 ppm was added to S2 to unify the condition of electric current effect with the (PS[4BC/DM](n=12, 40)/E7) composites which contain about 1000 ppm of Pt catalyst.

Characterization of the composite system. The phase transition behaviors and the aggregation states of the binary and ternary (LCP/LC(s)) composites were investigated on the basis of differential scanning calorimetry (DSC), polarizing optical microscopic (POM) observation and X-ray diffraction studies. The heating and cooling rates for DSC study and POM observation were 5 and 1 K/min, respectively. X-ray diffraction

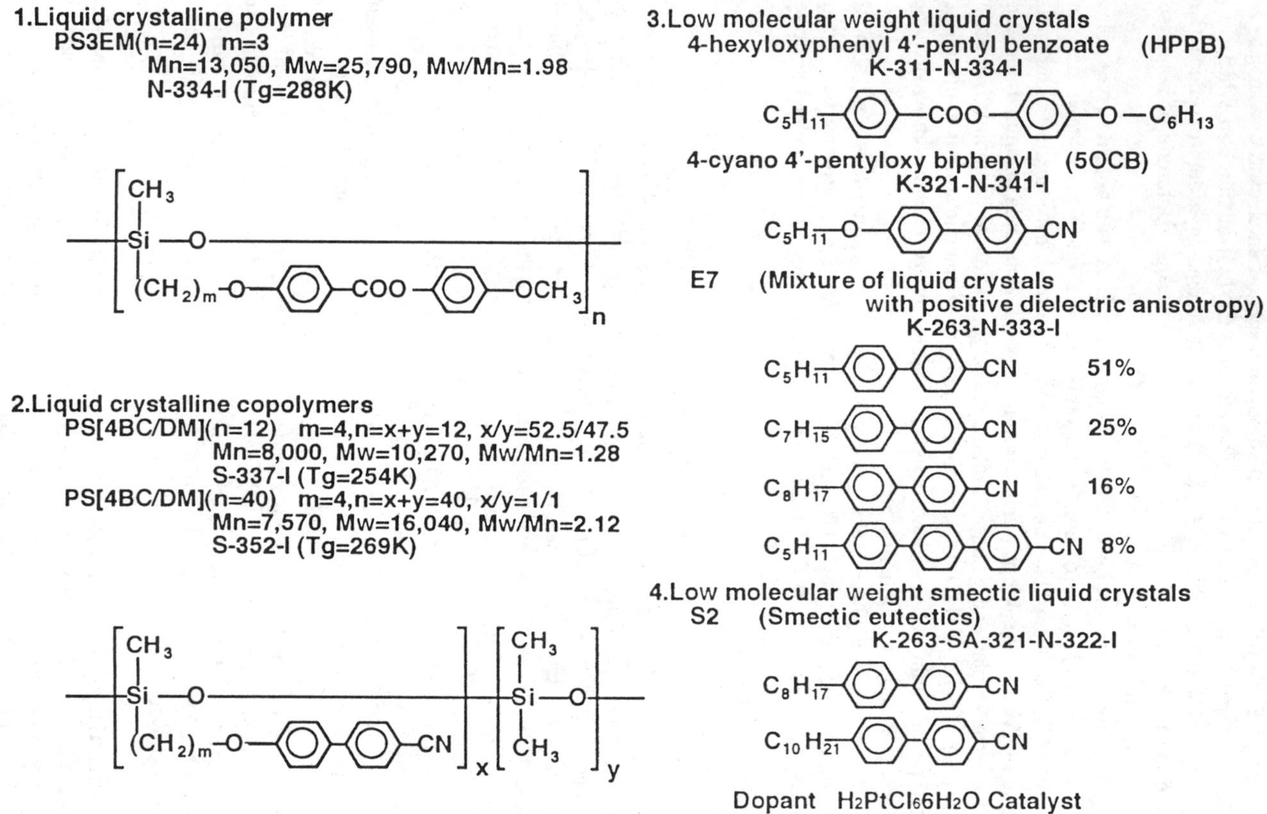

Figure 1. The chemical structures of liquid crystalline polymer and low molecular weight liquid crystals.

patterns were taken with a flat plate camera collimated with toroid mirror optics, using Ni-filtered CuK_{α} (λ=0.1542nm).

Measurement of electro-optical effect. In order to measure the electro-optical properties, the binary and ternary (LCP/LC(s)) composites were sandwiched between two transparent ITO-glass electrodes, which were separated by the PET spacer of 10 μm thick. The surface alignment treatment on the ITO-glass surface was not carried out, so that molecular orientation of LC in the composite was random. The electro-optical effects of the binary and ternary (LCP/LC) composites were studied under the application of an a.c. electric field. A generalized experimental scheme set up to investigate the electro-optical effects of the (LCP/L) composite systems is shown in Figure 2. Transmission intensity of He-Ne laser light through the cell was detected with a photodiode and was recorded with a digital storage oscilloscope. The distance between the cell and the photodiode was 305mm. The measuring temperature for electro-optical switching was controlled with accuracy of ±0.1K.

Results and Discussion

How to enhance the response speed. In the case that a sufficiently large electric filed, E above the threshold one, E_c is imposed to the (LCP/LC) composite system in a compatible smectic state, the rise time, τ_R is written by the following equation, because the composite system is homogeneous (one phase blend) (*23*).

$$\tau_R=\eta/\Delta\varepsilon(E^2-E_c^2)\approx\eta/\Delta\varepsilon E^2$$

where η and $\Delta\varepsilon$ are the viscosity and the dielectric anisotropy, respectively. Then, a decrease of viscosity in the (LCP/LC) composite system is apparently effective to increase the rise response speed for a bistable light switching. Since LCPs in a mesophase state is, in general, more viscous than LCs, the magnitude of η increases with an increase of the LCP fraction in the (LCP/LC) composite system. Therefore, this indicates that the magnitude of η of the (LCP/LC) composite system in a mesophase state must be reduced in order to realize the faster response speed for a bistable and reversible light switching.

Though the (PS3EM/5OCB:50/50 mol%) composite system in an induced smectic phase exhibited very stable memory effect at T_{NI} -5K ($T-T_{NI}$=-5K), the magnitude of τ_R was several hundred milliseconds-a few seconds, depending on the magnitude of an applied electric field. In order to realize the faster response time for the (LCP/LC) composite system, it is apparent that a decrease in the LCP fraction in the composite system is fairly effective, since η of LCP is generally higher than that of LC, as mentioned above. However, if the molar ratio of PS3EM/5OCB deviates much from 50/50 in an induced smectic phase, the stability of memory effect is lost. The response speed and the stability of memory have the opposite tendencies to the component fraction. Therefore, so long as in the binary composite system, it is difficult to realize simultaneously a fast response and a stable memory. Then, the low molecular weight liquid crystal, HPPB was added as the third component to the

(PS3EM/5OCB) composite system, in order to reduce the molar fraction of LCP, without any change in a smectic stability. Since HPPB has the corresponding similar chemical structure to that of the side chain mesogenic group of PS3EM, the (PS3EM/HPPB) mixture with a weak polar end in the mesogenic side group may become the counter part to 5OCB with a strong polar end to form an induced smectic phase with a fairly low magnitude of η. That is, the addition of HPPB can make the molar fraction of PS3EM effectively reduced in order to decrease the magnitude of η in the ternary composite system, maintaining similar induced smectic characteristics, such as the response speed, memory stability and so on to those of the (PS3EM/5OCB) composite system.

Phase transition behavior and aggregation states of the ternary (PS3EM/HPPB/5OCB) composite As reported previously (*23*), the (PS3EM/5OCB: 50/50 mol%) composite exhibited an excellent bistable and reversible light switching, though the magnitude of τ_R was too large at room temperature. From the view point of the sufficient memory stability, the molar percentage of the PS3EM/HPPB mixture and 5OCB was maintained to be 50/50. This means that the ratio of PS3EM/HPPB is 50-X/X, when the molar percentage of HPPB is X ($0 \leq X \leq 50$). Figure 3 shows the phase diagram for the (PS3EM/HPPB/5OCB:50-X/X/50 mol%) composite, which was obtained on the basis of DSC, POM and X-ray studies. The fan-shape texture being characteristic of a smectic phase was observed under POM for the ternary composite system. This indicates that a smectic phase was induced over a wide range of both mixing concentration and temperature as shown in Figure 3. Even in the extreme cases of the (PS3EM/HPPB/5OCB) composites of (50/0/50) and (0/50/50), an induced smectic phase was also observed. Then, this indicates that the mol% of LCP can be reduced by the addition of HPPB to the (PS3EM/5OCB) binary composite system, maintaining the similar bistable and reversible electro-optical switching characteristics in the (PS3EM/5OCB:50/50) binary composite. The narrow biphasic nematic-isotropic region of about 3K wide, in which the nematic and isotropic phase coexisted, was recognized by POM observation.

Electro-optical effects of the (PS3EM/HPPB/5OCB) composite in an induced smectic phase. The electro-optical effect based on light scattering was investigated under various conditions of an induced a.c. electric field at $T-T_{NI}$=-45K(in the range of room temperature). The transmittance of He-Ne laser through the composite cell without any optical polarizers was measured with a photodiode. Figure 4 shows the transmittances of PS3EM, HPPB, 5OCB and the (PS3EM/HPPB/ 5OCB:30/20/50 mol%) composite. In the cases of PS3EM, HPPB and 5OCB, no distinguishable transmittance changes were observed between a.c. electric field off- and on-states. As shown by the curve 4 in Figure 4 the (PS3EM/HPPB/5OCB:30/20/50 mol%) composite exhibited a bistable and reversible light switching driven by a.c. electric fields with low and high frequencies. Since the application of a low frequency electric field may induce an ionic current throughout the composite film, it is expected that an induced turbulent flow of LCP polymer main chains, being caused by an ionic current, collapses a fairy well-organized large smectic layer into many small smectic fragments

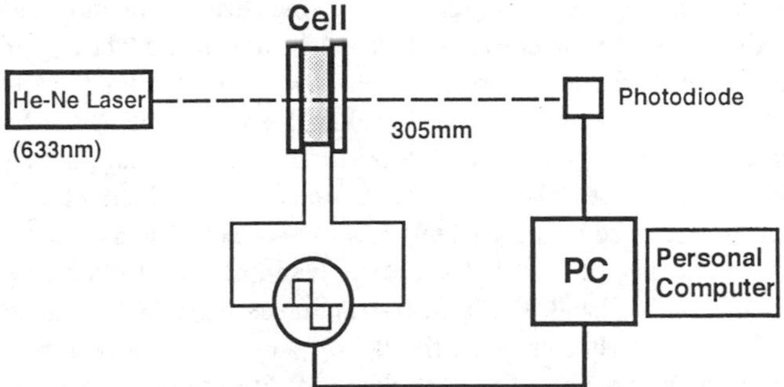

Figure 2. The generalized experimental scheme se up for the investigation of the electro-optical effect of a composite system.

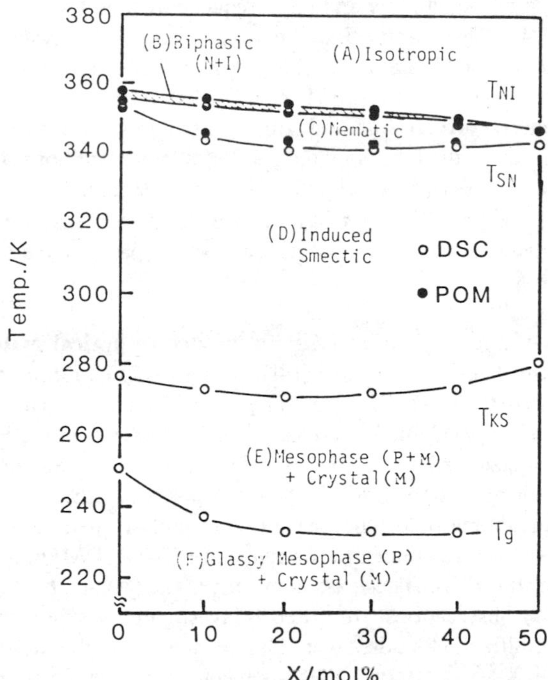

Figure 3. Phase diagram of the (PS3EM/HPPB/5OCB:50-X/X/50 mol%) ternary composite system.

(*15,16,21,22*). As reported previously, there are several possible supports to explain the origin of light scattering for the (LCP/LC) composite system upon the application of an a.c. electric field with low frequency ; ① in the case of low molecular weight LC molecules with a positive dielectric anisotropy in a smectic phase, light scattering could not be effectively induced with a low electric field, ② a light scattering state of the (LCP/LC) composite was more enhanced with an increase in the fraction or the dimension of ionic materials and also, ③ a purification of LCP and LC reduced a light scattering effect. Therefore, it is reasonable to consider that the random orientation of smectic directors, occurred by a turbulent flow of polymeric main chains induced by an ionic current, can be one of origins for optical heterogeneity to generate strong light scattering. On the other hand, the light transmittance increased up to 99% upon the application of a high frequency electric field. Any ionic current is not generally generated upon the application of an a.c. electric field with high frequency but a large scale of homeotropic alignment of smectic layers is easily formed due to a positive dielectric anisotropy of the smectic layer being composed of LC molecules and the side chain group of LCP. Also, turbid and transparent states of the (PS3EM/HPPB/5OCB) ternary composite system were remained unchanged even though a.c. electric fields with low and high frequencies were removed, respectively as shown in Figure 4. This means that the ternary composite system in an induced smectic phase has a bistable light switching characteristic, that is, a memory effect. The response speed for the (PS3EM/HPPB/5OCB) ternary composite system was remarkably enhanced several hundred times compared with that for the binary composite system under the corresponding experimental conditions, that is, the same reduced temperature ($T\text{-}T_{NI}$=-45K) and the same magnitude of an electric field ($16.7MV_{p\text{-}p}\ m^{-1}$). Especially, the magnitude of decay response time, τ_D ranges in several hundred ms even though the composite system has a stable memory effect at room temperature.

Effect of Degree of polymerization LCP on the electro-optical switching. Figure 5 shows the phase diagram of the (PS[4BC/DM](n=12)/E7) composite which was obtained on the basis of DSC, POM and X-ray studies. The glass transition temperature, Tg of PS[4BC/DM] decreased with an increase in the E7 fraction, may be due to the plasticizing effect of E7. Since a single endothermic peak attributed to the mesophase-isotropic transition, T_{MI} was observed, the composite forms a homogeneously mixed mesomorphic phase (molecularly dispersed state). Therefore, E7 is miscible over a whole concentration range of PS[4BC/DM] in both isotropic (A) and mesomorphic states (C). These assignments for the (A) and (C) regions were also confirmed by POM observations. In order to investigate the aggregation state of the composite, X-ray diffraction studies were also carried out in the mesophase (C). The sharp small angle X-ray scattering corresponding to a smectic layer structure was observed over a weight fraction range of E7 below 60 %. Also in this region a fan-shape texture being characteristic of smectic state was observed with POM. Therefore, the smectic phase certainly exists in a mesophase region with a E7 fraction at least below 60 %. In the region above 70 wt% of E7, sharp small angle X-ray scattering disappeared and a diffuse scattering was observed. It was also confirmed by

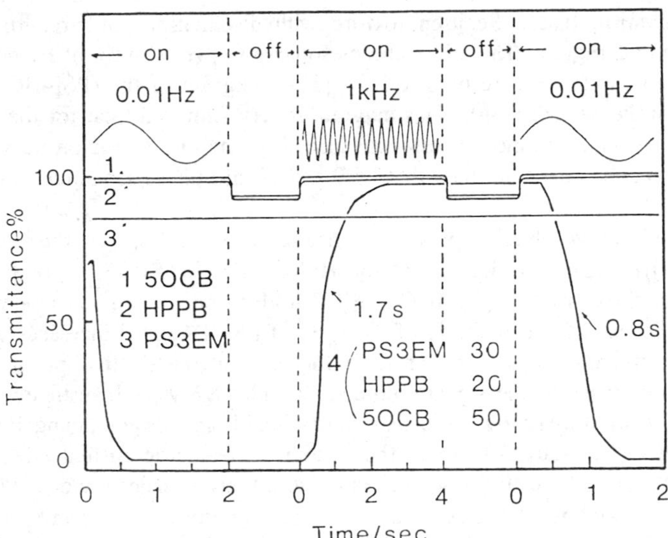

Figure 4. Electro-optical effect of PS3EM (nematic), HPPB (nematic), 5OCB (nematic) and the PS3EM/HPPB/5OCB composite (30/20/50 mol%) in an induced smectic state upon a.c. electric fields with low (0.01Hz) and high (1kHz) frequencies under the conditions of an applied electric field of 16.7MV$_{p-p}$ m^{-1}, at T-T$_{NI}$=-45K(nearly room temperature) and the cell thickness of 16 μm.

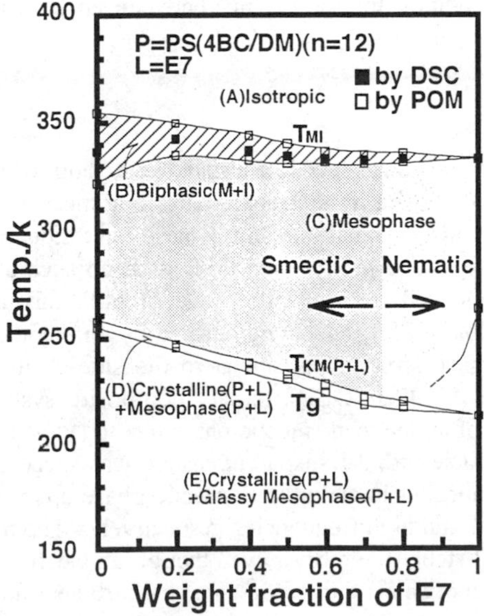

Figure 5. Phase diagram of the (PS[4BC/DM](n=12)/E7) composite.

POM observation that a Schliren texture with disclinations of strength of ± 1/2 appeared in that region. Therefore, the mesophase above 70 wt% of E7 corresponds to the nematic state. Figure 6 shows the phase diagram of the (PS[4BC/DM]/ E7) composite in the case of n=40. This diagram is very similar to that for the composite with n=12. A miscible smectic state was also confirmed in the region below 70 wt% of E7. It has been also confirmed that LCP and LC are phase-separated in the case of n=120.

Figure 7 shows electro-optical response times, τ_R and τ_D, vs. the magnitude of applied electric field. τ_R and τ_D of the (PS[4BC/DM](n=12)/E7) composite were shorter than those of the (PS[4BC/DM](n=40)/E7) composite. These results are reasonable because the magnitude of viscosity of PS[4BC/DM] increases with n. In order to discuss the role of LCP in the composite on the electro-optical effect, the switching properties of the smectic eutectic sample, S2 were investigated. S2 is a binary smectic mixture of low molecular weight liquid crystals containing 1000 ppm of ionic impurity (Pt catalyst) to unify the condition of electric current effect with the (PS[4BC/DM](n=12, 40)/E7) composites because there remained about 1000 ppm of Pt compound which was used as a catalyst. The magnitudes of τ_R and τ_D of S2 were smaller and larger, respectively, in comparison with the response times of (PS[4BC/DM](n=12, 40)/E7) composites. The result of the decrease in τ_R is also due to the small viscosity of S2. The electro-optical decay, transparent to turbid, process is attained through a hydrodynamic turbulent flow induced by ionic current upon application of an electric field with low frequency. These results on τ_D for (PS[4BC/DM](n=12, 40)/E7) composite and S2 indicate that the LCP plays a role of an agitator to enhance the turbulent flow. The magnitude of τ_D exhibited minimum for LCP with n=12. There might be an appropriate length of polymeric main chain to realize an optimum condition for the balance between an electric field effect and an electric current effect.

Conclusions

In order to enhance the response speed of bistable switching for (LCP/LC) composite system, the ternary composite was prepared from a miscible mixture of nematic PS3EM with a weak polar terminal unit in the mesogenic side group, nematic 5OCB with a strong polar cyano end group and HPPB with no such a group. An induced smectic phase was observed over a wide range of both mixing concentration and temperature. The mol% of PS3EM can be reduced by mixing the third component of HPPB, of which chemical structure is similar to the side chain mesogenic group of PS3EM. The (PS3EM/HPPB/5OCB) ternary composite system can maintain the similar characteristics of an induced smectic phase to the (PS3EM/5OCB) binary one. The reversible and bistable turbid-transparent change was successfully realized for the binary and ternary composites in an induced smectic phase upon the application of a.c. electric fields with low and high frequencies, respectively. The transparent and turbid states were maintained stably (memory), even though an electric field was turned off. The random orientation of smectic directors due to a turbulent flow of polymeric main chains being induced by an ionic current can be one of origins for optical heterogeneity

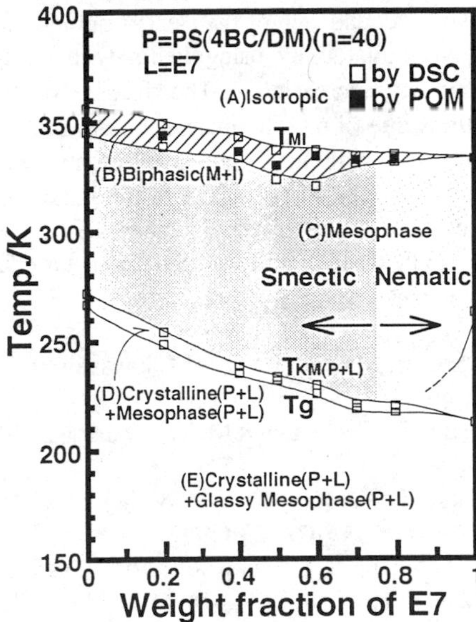

Figure 6. Phase diagram of the (PS[4BC/DM](n=40)/E7) composite.

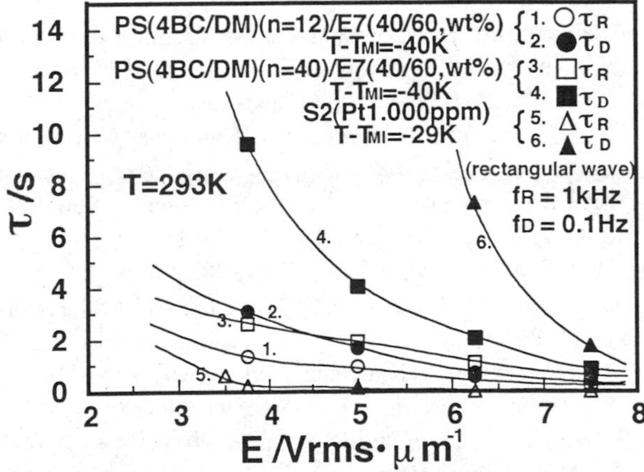

Figure 7. Electro-optical response times vs. the magnitude of applied electric field.

to generate strong light scattering. It is reasonably concluded that a bistable and reversible light switching of the (PS3EM/HPPB/5OCB) ternary composite system in an induced smectic phase can be realized by the balance between an electric current effect based on the electrohydrodynamic motion of the LCP main chain and an electric field effect based on the dielectric anisotropy of the smectic layer being composed of LC molecules and the side chain part of LCP. The response speed for bistable and reversible light switching could be remarkably improved by an introduction of HPPB. The magnitudes of τ_R and τ_D are a few seconds and several hundred ms, respectively.

The molecular weight dependence of the electro-optical effect for the (LCP/LC) composite system was investigated by using LCPs with 0, 12 or 40 of n. The magnitude of τ_R increased with increasing n. The viscosity of the composite system is strongly dependent on the value of n. The increase in τ_R might be due to an increase in the magnitude of viscosity with increasing n. The magnitude of τ_D, however, exhibited minimum for LCP with n=12. This result indicates that an appropriate length of polymeric main chain is required to realize an optimum condition for the balance between an electric field effect and an electric current effect.

References

1) T. Kajiyama, Y. Nagata, E. Maemura and M. Takayanagi, *Chem.Lett.*, **1979**, 679 (1979).
2) T. Kajiyama, Y. Nagata, S. Washizu and M. Takayanagi, *J. Membrane Sci.*, **11**, 39(1982).
3) H. Kikuchi, A. Kumano, T. Kajiyama, M. Takayanagi and S. Shinkai, *J. Chem. Soc. Jpn, Chem. Ind. Chem.*, **1987**, 423 (1987).
4) H. G. Craighead, J. Cheng and S. Hwackwood, *Appl. Rhys. Lett.*, **40**, 22 (1982).
5) J. L. Fergason, *SID Int. Symp. Dig. Tech.*, **16**, 68 (1985).
6) H. Kikuchi, A. Miyamoto, A. Takahara, T. Furukawa, T. Kajiyama, *Prepr. 2nd SPSJ Int. Polym. Conf.* 33 (1986).
7) P. S. Drazic, *J. Appl. Phys.*, **60**, 2142 (1986).
8) J. W. Doane, A. Golemme, J. L. West, J. B. Whiteneat, B-G. Wu, *Mol. Cryst. Liquid Cryst.*, **165**, 511 (1988).
9) T. Kajiyama, A. Miyamoto, H. Kikuchi and Y. Morimura, *Chem. Lett.*, **1989**, 813.
10) T. Kajiyama, H. Kikuchi, A. Miyamoto and Y. Morimura, "Frontiers of Macromolecular Science", IUPAC 505 (1989).
11) A. Miyamoto, H.Kikuchi, Y. Morimura and T. Kajiyama, *New Polym. Mater.*, **2**, 1 (1990).
12) C.B. McARDLE Ed., Side Chain Liquid Crystal Polymers, New York (1989).
13) A. I. Hopwood, H. J. Coles, *Polymer.*, **26**, 1312 (1985).
14) M. S. Sefton, H. J. Coles, *Polymer.*, **26**, 1319 (1985).
15) T. Kajiyama, H. Kikuchi, A. Miyamoto, S. Moritomi and J. C. Hwang, *Chem. Lett.*, **1989**, 817 (1989).
16) H. Kikuchi, J. C. Hwang and T. Kajiyama, *Polym. Adv. Tech.*, **1**, 297 (1991).
17) A. C. Griffin, J. F. Johnson, *J. Am. Chem. Soc.*, **99**, 4859 (1977).
18) B. Engelen, G. Heppke, R. Hopf and F. Schnider, *Ann. Phys.*, **3**, 403 (1978).
19) M. Domon, J. Billard, J. Phys., Paris , **40**, C3-413 (1979).
20) F. Schneider, N. K. Sharna, *Z. Naturforsch.*, **36**, 62(1981).
21) T. Kajiyama, H.Kikuchi, A. Miyamoto, S. Moritomi and J. C. Hwang, *Mat. Res. Soc. Sym. Proc.*, **171**, 305 (1990).
22) T. Kajiyama, H. Kikuchi, J. C. Hwang, A. Miyamoto, S. Moritomi and Y. Morimura, *Prog. in Pacific Polym. Sci.*, **1**, 343 (1991).
23) J. C. Hwang, H. Kikuchi and T. Kajiyama, *Polymer*, **33**, 1821 (1992).

Chapter 13

Thermal-Induced Phase Separation in a Mixture of Functional Poly(methyl methacrylate) and Low-Molar-Mass Liquid Crystals

Thein Kyu, I. Ilies, C. Shen, and Z. L. Zhou

Institute of Polymer Engineering, University of Akron, Akron, OH 44325–0301

Miscibility phase diagram and phase separation dynamics of a mixture of a flexible polymer and a monomeric liquid crystal, hereafter called a polymer dispersed liquid crystal (PDLC), have been investigated. A theoretical calculation was carried out for predicting an equilibrium phase diagram of a PDLC system by taking into consideration Flory-Huggins (FH) free energy of mixing of isotropic liquid phases in combination with Maier-Saupe (MS) free energy of nematic ordering of the liquid crystal. The combined FH-MS theory predicts a *"tea-pot"* phase diagram in which liquid-liquid phase separation is overlapped with an isotropic-nematic coexistence region. A temperature-composition phase diagram of a mixture of hydroxyl functionalized polymethyl methacrylate (PMMA-OH) and eutectic nematic liquid crystals (E7) was established by light scattering and differential scanning calorimetry. This phase diagram can be characterized as an upper critical solution temperature exhibiting liquid-liquid phase separation between the polymer and the isotropic phase of the liquid crystal. The nematic-isotropic transition occurs at high liquid crystal compositions. The calculated phase diagram was found to conform well with the cloud point phase diagram. The dynamics of phase separation was investigated by means of light scattering and optical microscopy. The time-evolution of structure factor was analyzed in terms of a power law.

Inhomogeneous composite films comprising of a polymer binder and low molar mass liquid crystals, hereafter termed polymer dispersed liquid crystals (PDLC), are of interest because of their potential for applications in optical devices such as optical switches, flat panel display screens, and privacy windows (1-4). The PDLC devices are operated on the principle of electrical switching from an opaque to a transparent state (4). For a successful application of PDLC devices, the optimization of various parameters including maximum opacity in the off-state of the electric field, high transparency in the on-state, fast electrical switching and response times, is of paramount importance. The electro-optical properties depend on various parameters such as size, shape, uniformity of liquid crystal dispersions in the polymer matrix. In general, polymer dispersed liquid crystals are produced via phase separation of initially homogeneous binary mixtures of polymer and liquid crystals either by thermal quenching or chemical reactions such as thermal or photo-polymerization (4-7). The

0097–6156/96/0632–0201$15.00/0

understanding of mechanisms of phase separation and pattern formation driven by thermal or chemical processes is essential for the elucidation of the relationship between the structure and electro-optical properties of PDLC.

Our effort in this area has been directed to the elucidation of phase equilibria and dynamics of phase decomposition of various PDLC systems (5-7). Recently, we have derived a theoretical scheme by combining the Flory-Huggins free energy (FH) for describing isotropic mixing and the Maier-Saupe free energy (MS) for the nematic-isotropic phase transition of a nematic liquid crystal (8). We found that this combined FH/MS theory is capable of predicting an equilibrium phase diagram of PDLC systems and tested favorably well with the cloud point phase diagrams of polymethyl methacrylate/E7 (eutectic liquid crystals) and polybutyl methacrylate/E7 (8,9).

In this paper, phase behavior of a polymer dispersed liquid crystal consisting of functional polymethyl methacrylate having hydroxy groups (PMMA-OH) and eutectic main chain liquid crystals (E7) have been investigated without the involvement of cross-linking. The emphasis is placed on mutual interference between an upper critical solution temperature (UCST) associated with liquid-liquid phase separation and the nematic-isotropic transition of the liquid crystals. The experimental phase diagram has been tested with the theoretical prediction. Pattern formation and kinetics of phase decomposition driven by thermal quenching into a liquid-liquid region and a nematic-liquid region have been investigated by time resolved small angle light scattering and optical microscopy. The growth behavior is analyzed in the framework of a power law.

Theoretical Model

A theoretical phase diagram for mixtures of polymer and liquid crystals has been calculated by combining Flory-Huggins free energy for isotropic mixing (10) and Maier-Saupe free energy for nematic ordering (11,12) as follows;

$$g = \phi \ln \phi + \frac{1-\phi}{r} \ln(1-\phi) + \chi\phi(1-\phi) - \phi\Sigma(s) - \frac{1}{2}\nu\phi^2 s^2 \quad (1)$$

where $\Sigma(s)$ represents entropy of mixing of liquid crystals which is defined as

$$\Sigma(s) = -\int f(\theta)\ln[4\pi f(\theta)]d\Omega \quad (2)$$

where ϕ represent the volume fraction of component 1 (liquid crystal) and $d\Omega = \sin\theta d\theta d\psi$. r is the ratio of the numbers of lattice sites or segments occupied by a single polymer chain with respect to that by one liquid crystal molecule. χ represents the intermolecular interaction, often known as the Flory-Huggins interaction parameter, which is generally assumed to be inversely proportional to absolute temperature (T) (10,13), viz.,

$$\chi = a + b/T \quad (3)$$

where a and b are constants. The last two terms in equation 1 represent the entropic and enthalpic contributions from the nematic ordering. Here, ν is the Maier-Saupe quadrupole interaction parameter which has the inverse temperature dependence (8) as described by the following equation:

$$\nu = 4.541\frac{T_{NI}}{T} \quad (4)$$

where T_{NI} is the nematic-isotropic (NI) transition temperature of liquid crystals and s represents an orientational order parameter defined as:

$$s = \frac{1}{2}(3 < \cos^2\theta > -1) \tag{5}$$

The directional cosine can be determined by averaging the director orientations in terms of an orientation distribution function, f(θ), as follow (14):

$$\left\langle \cos^2\theta \right\rangle = \int \cos^2\theta \, f(\theta) \, d\cos\theta \tag{6}$$

while

$$f(\theta) = \frac{1}{4\pi Z} \exp\left[-\frac{u(\theta)}{kT}\right] \tag{7}$$

in which the partition function, Z, is given by

$$Z = \int \exp\left(-\frac{u(\theta)}{kT}\right) d\cos\theta \tag{8}$$

where the psuedo-potential of director orientations, u(θ), may be expressed in terms of a second order Legendre polynomial by the following equation (12):

$$\frac{u(\theta)}{kT} = -\frac{1}{2} m(3\cos^2\theta - 1) \tag{9}$$

where m is a mean field parameter characterizing strength of a potential field. It is essential to determine the orientational order parameter for various blend compositions by means of the free energy minimization approach, and subsequently the free energy of the nematic ordering may be evaluated.

At equilibrium, the chemical potentials of the individual component are, by definition, equivalent in every phase, i.e., the polymer rich-phase (α) and liquid crystal rich-phase (β) have the same chemical potential (10,12):

$$\mu_1^\alpha = \mu_1^\beta \tag{10}$$

$$\mu_2^\alpha = \mu_2^\beta \tag{11}$$

where the chemical potentials of the LC rich- and poor-phases may be expressed as

$$\mu_1 = \ln\phi + \left(1 - \frac{1}{r}\right)(1-\phi) + \chi(1-\phi)^2 - \ln Z + \frac{1}{2}\nu\phi^2 s^2 \tag{12}$$

$$\mu_2 = \ln(1-\phi) + (1-r)\phi + \chi r\phi^2 + \frac{1}{2} r\nu\phi^2 s^2 \tag{13}$$

By solving equations 10 and 11 simultaneously, equilibrium temperature - composition phase diagrams for various PDLC systems may be constructed (8).

Experimental Section

Functional polymethyl methacrylate (PMMA-OH) with hydroxyl groups were obtained from Rohm and Haas Co. Unfortunately, the detailed chemical structure is not disclosed to us. This PMMA-OH is in a liquid form as typical for a low molecular weight polymer. The dispersing liquid crystal is a mixture of nematic liquid crystals, commercially known as E7 (purchased from EM Industries). E7 consists of various cyanobiphenyl (CB), oxycyanobiphenyl (OCB) and cyanoterphenyl (CT) derivatives in the following proportion: 5CB (51%), 7CB (21%), 8OCB (16%) and 5CT (12%) by weight. Although the composition of E7 is seemingly complex, E7 exhibits an eutectic behavior, i.e., it displays a single nematic-isotropic transition at 60 °C and a single crystal-nematic transition at -30 °C. Thus the PMMA-OH/E7 mixture may be regarded as a pseudo binary system. The preference of choosing E7 is due to the broad nematic range (-30 to 60 °C) as opposed to a few degrees for a single component liquid crystal (cf. 5 °C for 5CB).

The nematic-isotropic transitions of neat E7 as well as PMMA-OH/E7 mixtures were investigated by means of differential scanning calorimetry (Model 9900, Du Pont) with a heating module (Model 910). Indium standard was used for temperature calibration. The DSC experiments were conducted under flowing dry nitrogen. The heating rate was 10 °C/min unless indicated otherwise.

The cloud point phase diagram was established by using a light scattering technique described elsewhere (15). The light scattering apparatus consists of a He-Ne laser light source (2 mW, LSR2R Aerotech), a sample hot stage controlled by a programmable temperature controller (Omega CN2012) interfaced with a personal computer (IBM PS2/30). The detection system was slightly modified by replacing the two dimensional (2-D) Vidicon camera with a Reticon 1-D silicon diode array detector for dynamical studies and with a conventional photo-diode for the cloud point measurements.

Various proportions of PMMA-OH and E7 were mixed by stirring rigorously in a flask without using any solvent. Trapped bubbles were removed by vacuuming the mixtures. These mixtures were spread on glass slides and held at 120 °C for 5 min and then covered with a microscope cover glass. The thickness of the sample was 20 μm. A cloud point temperature versus composition phase diagram was established by means of a light scattering method reported elsewhere. The angle of observation can be varied, but it was set at 20° in the present study. In the homogeneous state, the observed scattered intensity was low, it increased substantially when the system phase separated into the inhomogeneous two-phase state during cooling. The temperature at which the scattered intensity increased abruptly was designated as a cloud point. The heating and cooling rate was 0.5 °C/min unless indicated otherwise. In time resolved light scattering experiments, a couple of sample hot stages were employed; one was controlled at an experimental temperature and the other was used for preheating at a predetermined temperature typically in the single phase region (cf. 80 °C). Several temperature (T) quenches were performed from a single phase temperature (80 °C) to a two-phase region (62, 57, and 53 °C) by rapidly transferring the blend samples. Structure evolution during various T quenches was investigated by time resolved optical microscopy.

Results and Discussion

Figure 1 depicts the experimental cloud points and the nematic-isotropic transition temperatures in comparison with the theoretical phase diagram of the PMMA-OH/E7 PDLC system. The phase diagram calculation was carried out based on the combined Flory-Huggins (FH) and Maier-Saupe (MS) free energies using r = 2.25 and a = -4.0. The b value was estimated from the critical temperature using $\chi = a + \{(\chi_c - a)T_c\}/T$.

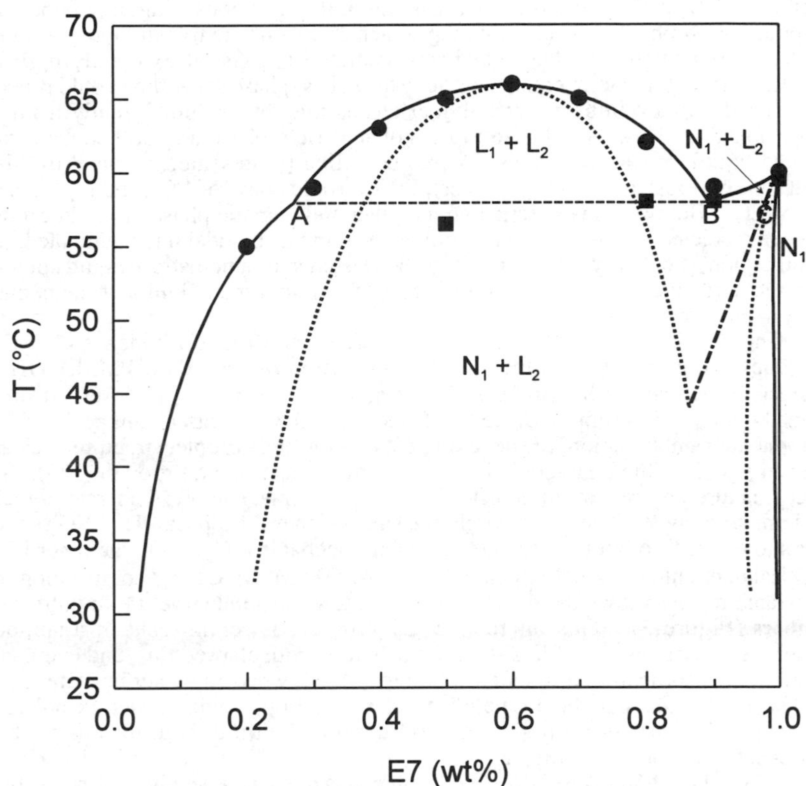

Figure 1. The predicted "tea-pot" phase diagram in comparison with the experimental cloud points (●) by LS and nematic-isotropic transition temperatures (■) by DSC. Line ABC represents the peritectic line. The solid and dotted lines represent the binodal and spinodal, respectively.

The fit between the experiment (filled circles) and the theory (a solid line representing a binodal curve) was found remarkably good. The combined FH/MS theory predicts several distinct regions consisting of isotropic, liquid-liquid, liquid-nematic, and pure nematic phases. Such a complex phase diagram of a PDLC system may be termed a "tea-pot" phase diagram. The liquid-liquid region is essentially an upper critical solution temperature (UCST) type having a convex maximum approximately at the 60% E7 and a critical temperature of 66 °C. At elevated temperatures above the liquid-liquid coexistence line, the system is isotropic and homogeneous. At very high liquid crystal concentrations (> 90%), there is a narrow immiscibility gap consisting of polymer-rich and nematic-rich phases associated with the nematic-isotropic transition of the liquid crystals. A pure nematic phase has been predicted to exist at extremely high LC concentrations. At low temperatures or deep quenches, phase separation could proceed to this nematic phase limit where polymer chains may be excluded totally from the nematic domains. However, the isotropic (polymer-rich) phase may still contain some dissolved liquid crystal molecules. A peritectic line (represented by the ABC line) signifies the co-existence of L_2 (polymer), L_1 (isotropic phase of LC), and N_1 (nematic phase of LC). Below this peritectic line, polymer and nematic phase of liquid crystals have been predicted to co-exist. In addition to the regular spinodal line within the liquid + liquid region, the theory further predicts the existence of a nematic + liquid spinodal line in the narrow coexistence region of the isotropic polymer fluid and the nematic liquid crystal phase.

The predicted liquid + liquid region may be confirmed by means of optical microscopy. Figure 2 shows the structure evolution of the 60/40 PMMA-OH/E7 mixture (off-critical) following a quench from 80 °C to 62 °C as observed under polarized optical microscopy. Since the T quench at the off-critical composition falls within the metastable region, the development of a multiple droplet structure is clearly discerned which is the characteristic of a typical nucleation and growth (NG). The droplet structure grows in size (diameter) with elapsed time, namely, the larger domains (one dominant length scale) grow while the smaller ones disappear. This NG process appears to occur through the Oswald's ripening mechanism (16). On the other hand, the critical quenching (from 80 to 62 °C) at the 40/60 PMMA-OH/E7 composition into the unstable region shows the development of interconnected structure initially for a few hours (Figure 3). This interconnected structure is reminiscent of a spinodal structure. The length scale of this structure increases with elapsed time and eventually the pattern transforms into droplet morphology probably driven by surface tension. It should be emphasized that there is no identifiable anisotropic region under the polarized microscopic investigation indicating a true liquid-liquid equilibrium without involvement of the nematic ordering.

When the 40/60 PMMA-OH/E7 mixture was quenched deeply into the unstable region within the spinodal boundary and below the peritectic line (e.g. 53 °C), a very tiny modulated structure develops and grows with time (Figure 4). Unlike the droplet structure of the T quenches to 62 °C, this percolated structure is birefringent, suggesting the occurrence of nematic ordering. After a few hours, the bicontinuous structure transforms to a droplet morphology. Within these LC-rich domains, some Schlieren textures develop, and subsequently annihilation of the LC defects takes place. This temperature gap can be identified as the nematic-liquid coexistence region which is consistent with the above theoretical prediction.

At the T quench of the same composition to 57 °C which is very close to the border of the peritectic line, the initial structure as viewed under unpolarized configuration resembles the spinodal structure displaying some interconnected structure having some birefringent entities (Figure 5). The average initial domain (grain) size is intermediate between those of the above two quenches (62 and 53 °C). The domain size increases with elapsed time. At later times, the percolated structure transforms to droplet morphology suggesting a transition from the percolation to a cluster regime. It was noticed that some of the droplets are seemingly larger than the film thickness which could be a consequence of a change of growth dimension from three to two. Such a

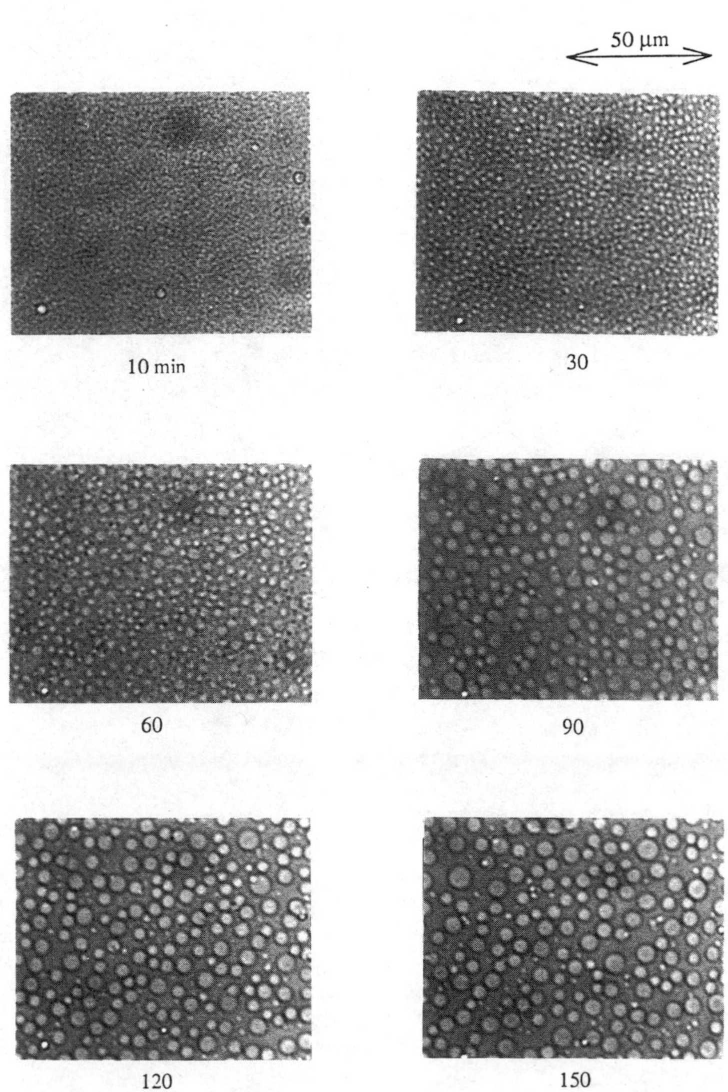

Figure 2. Time evolution of NG structure following a T quench from 80 °C to 62 °C at the off-critical 60/40 PMMA-OH/E7 mixture.

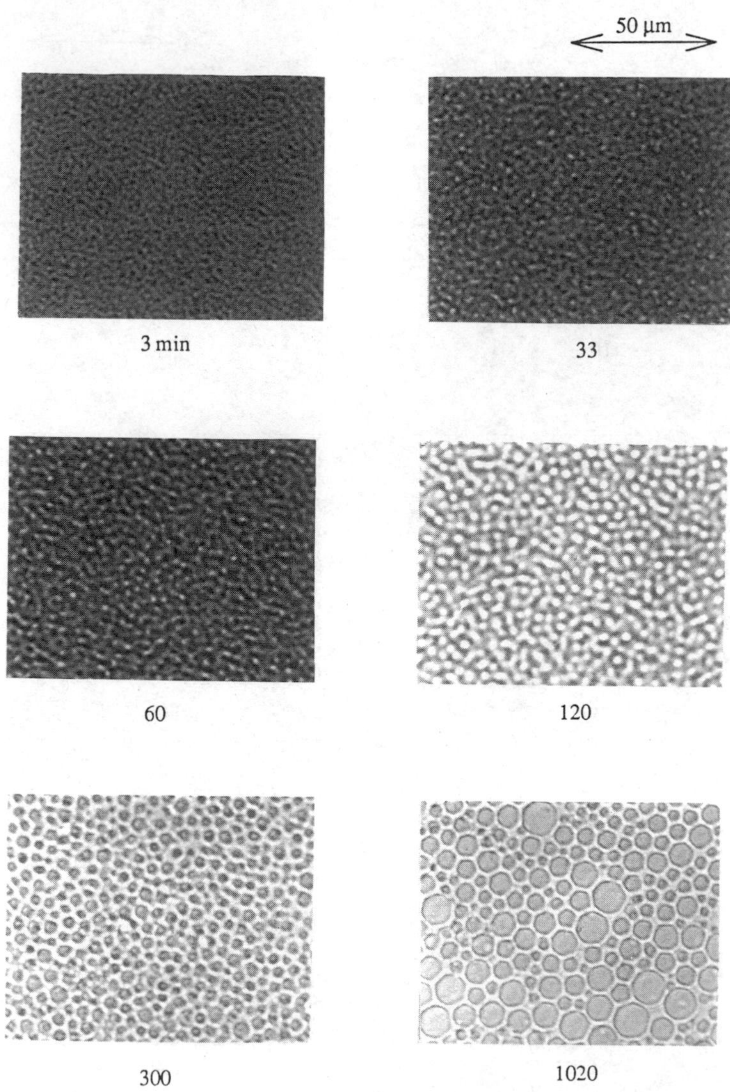

Figure 3. Time evolution of SD structure following a T quench from 80 °C to 62 °C at the critical 40/60 PMMA-OH/E7 mixture.

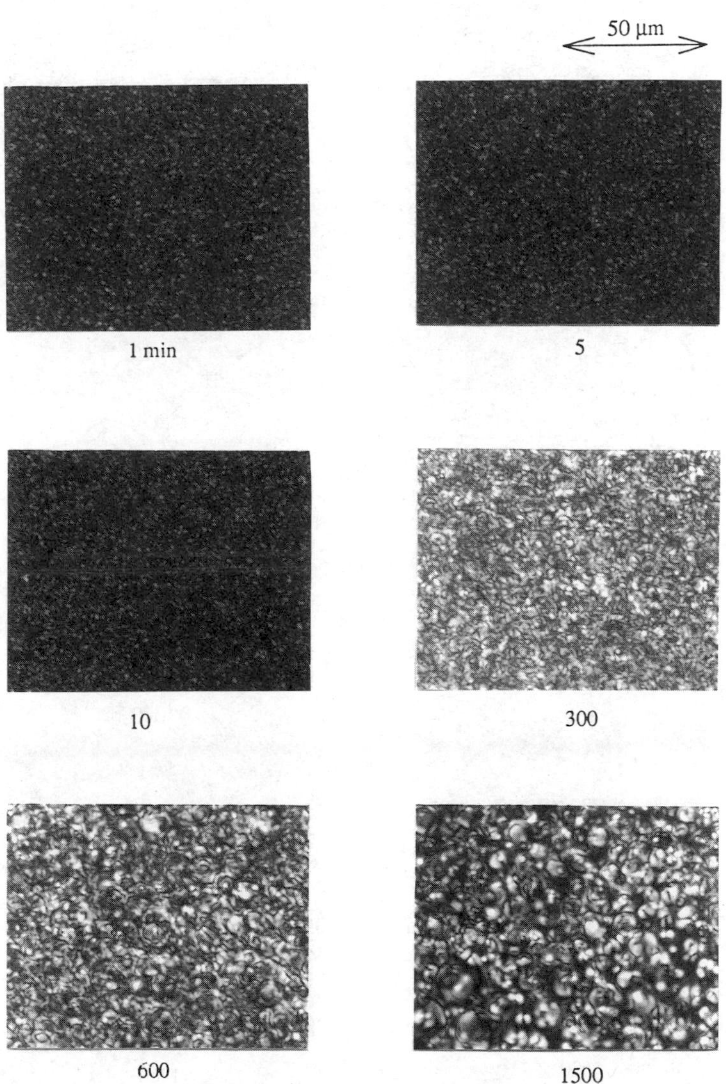

Figure 4. Time evolution of SD structure following a T quench from 80 °C to 53 °C at the critical 40/60 PMMA-OH/E7 mixture.

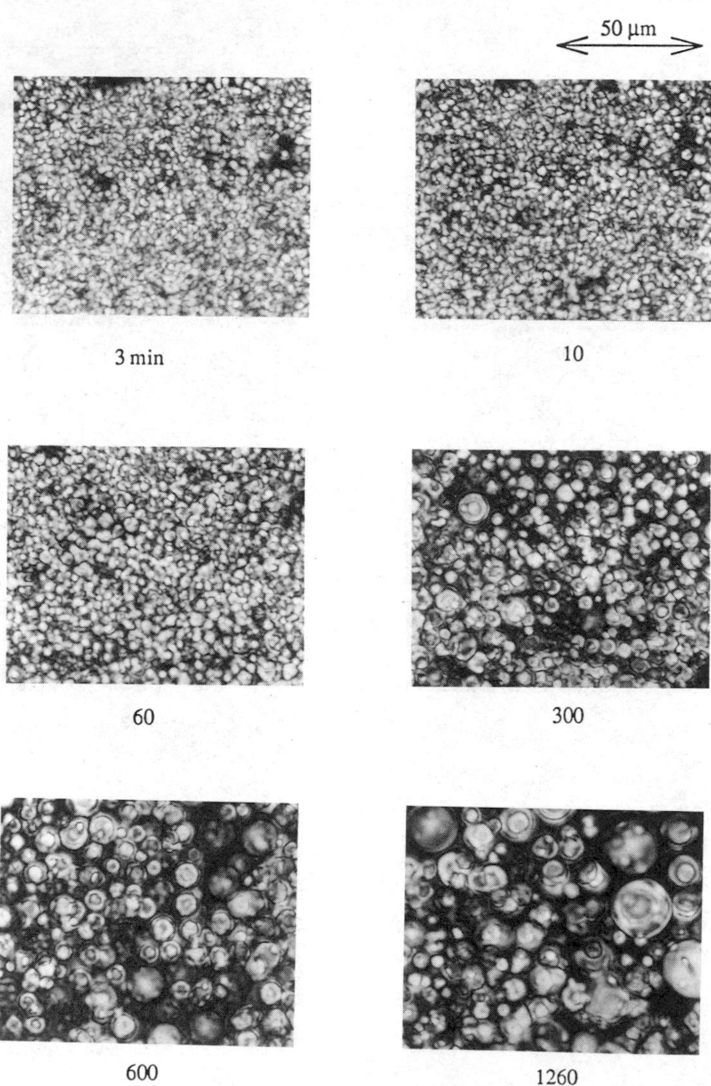

Figure 5. Time evolution of SD structure following a T quench from 80 °C to 57 °C at the critical 40/60 PMMA-OH/E7 mixture.

cross-over phenomenon has been readily observed in conventional polymer blends when a cover glass was in use (17). The dimensional change from there to two can be observed in the off-critical quench to 60 °C with or without the cover glass, thus this phenomenon is not an artifact. The cross-over of percolated SD structure to the cluster regime was also discerned in the off-critical but deep quench to 53 °C at the 60/40 PMMA-OH/E7 composition. Concurrently, Schlieren texture develops within the preformed domains. These observations confirmed the existence of the nematic-liquid region below the peritectic line at both compositions.

We further examined dynamics of spinodal decomposition (SD) in various binary blends extensively by time resolved light scattering. The early stage of SD is observable if the phase separation process is very slow. Most studies showed a typical growth behavior of late stages of SD with an growth exponent of 1/3. This value often crossed over to 1 at very late stages depending on the quench depths and compositions such as the critical or the off-critical composition. At this point, the dynamics of SD in liquid-liquid phase separation of binary polymer blends, polymer-oligomer, prepolymer mixtures is fairly understood and quite predictable. Therefore, we focused on the phase separation dynamics of PMMA-OH/E7 mixtures in other regions: (i) liquid-liquid phase separation in the metastable region and (ii) spinodal decomposition in the nematic-liquid regime. A temperature quench was performed from the isotropic phase (80 °C) to a two-phase liquid-liquid region (62 °C) at the off-critical composition of 40 % E7 composition. This region corresponds to the metastable region where multiple droplets are formed. The size (length scale) and the distribution of the droplets appear uniform seemingly suggesting a homogeneous nucleation and growth. According to Cumming et al. (18), such uniform size and distribution of the droplets could give rise to multiple scattering maxima. In general, the homogeneous NG is a rare occurrence, and a majority of cases result in heterogeneous nucleation. Hence, the scattering maximum can be washed out due to the heterogeneity of the system. As can be seen in Figure 6, a scattering maximum first develops at a large wavenumber, q, defined as $q=(4\pi/\lambda)\sin(\theta/2)$, where λ and θ are the wavelength of incident light and the scattering angle measured in the medium. The peak moves to a lower scattering angle with elapsed time while the intensity increases, indicative of domain growth. A cross-over of scattering curves can be noticed at larger wavenumbers which implies that larger concentration fluctuations (or domains) grow at the expense of smaller fluctuations. Such cross-over behavior of the scattering curves has been regarded as a signature of SD (19) which is contradictory to the present case where the cross-over of scattering profiles indeed could occur in the metastable region.

The growth of phase separated domains may be analyzed by examining the shift of wavenumber maximum (q_m) and the corresponding peak intensity (I_m) as a function of elapsed time. A power law scheme has been customarily employed for characterizing the phase growth as follow:

$$q_m(t) = \xi(t)^{-1} \sim t^{-\alpha} \tag{14}$$

and

$$I_m(t) \sim <\eta(t)^2> \xi(t)^3 \sim t^{\beta} \tag{15}$$

where $<\eta(t)^2>$ and $\xi(t)$ are the mean square fluctuations of dielectric constants and the length scale of the domains, respectively. The growth exponents, α and β, have been predicted theoretically to be 1/3 and 1 for binary systems and also borne out experimentally in a number of polymer blends (16). The classical 1/3 law has been predicted by the condensation-evorporation model of Lifshift and Slyozov (20) and the cluster dynamics by Binder and Stauffer (21). As shown in Figure 7, the plot of q_m

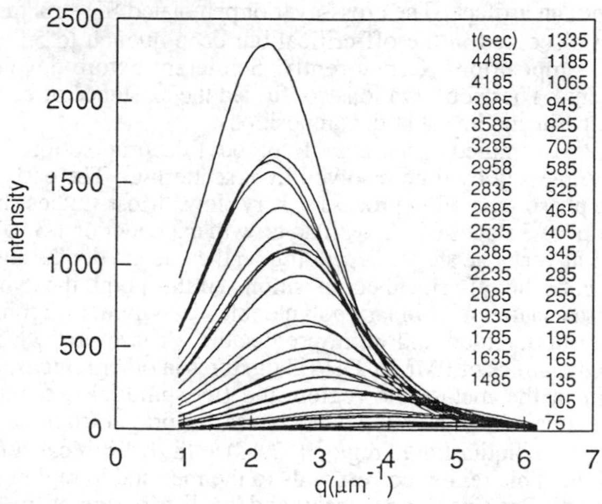

Figure 6. Time evolution of sacttering profiles of the 60/40 PMMA-OH/E7 composition following a T quench from 80 °C to 62°C.

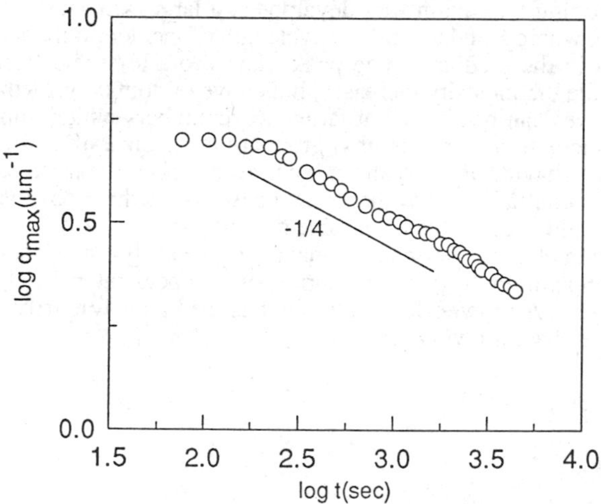

Figure 7. Time dependence of q_{max} in the nucleation and growth regime following a T quench from 80 °C to 62 °C at the 60/40 PMMA-OH/E7 composition.

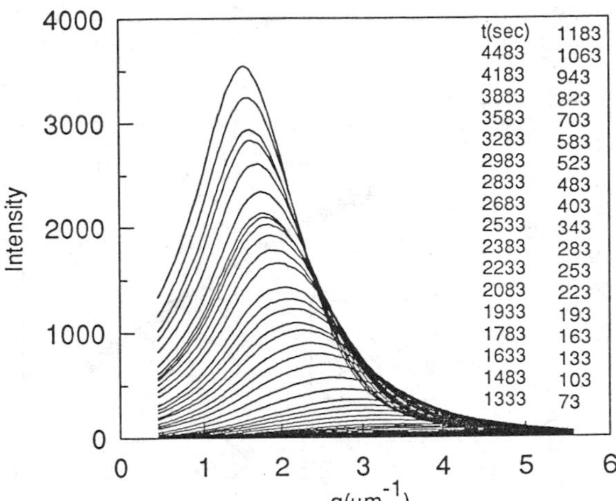

Figure 8. Time evolution of scattering profiles of 40/60 PMMA-OH/E7 following a T quench from 80 °C to 57 °C.

versus t in the double logarithmic form gives a very small slope in some initial period then the growth exponent crosses-over to 1/4 at later times. This exponent is smaller than the predicted 1/3 law and also smaller than the value of 1/2 obtained by Cumming et al. (18) for the NG in the blends of polyisoprene and polyethylene-propylene with low but narrowly distributed molecular weights. This T jump falls within the metastable limit of the liquid + liquid coexistence region, therefore nematic ordering is not involved.

Next, several temperature quenches were undertaken at the 40/60 PMMA-OH/E7 mixture from a single phase temperature (80 °C) to a two-phase unstable region (57 and 53 °C) where interconnected morphology was discerned. Figure 8 shows time evolution of scattering profiles of the 40/60 PMMA-OH/E7 mixture, following the T quenches to 57 °C as obtained by small angle light scattering. The scattering maximum first occurs at a large scattering angle and shifts to low angles with elapsed time. A cross-over of the scattering curves is also discerned suggesting that the smaller fluctuations decay while the larger fluctuations grow in time.

Figure 9 shows the variation of q_m versus t in the log-log plots for two T quenches at the 40/60 PMMA-OH/E7 composition into the nematic + liquid region (53 and 57 °C). In the deep T quench to 53 °C, the initial change of q_m is subtle, but the slope becomes -1/3 at longer times. The 57 °C quench shows the same -1/3 slope over the time scale investigated, thus the exponent is consistent with that of the classical liquid-liquid phase separation in binary polymer blends. It appears that the α=1/3 law is valid even when the liquid crystal in the PDLC is in the nematic phase. It should be pointed out that the dynamics of annihilation of nematic disclinations in liquid crystalline polymers has shown to obey the same 1/3 law (22) while the coarsening dynamics of low molar mass nematic liquid crystals (23) has revealed the exponent value of 1. Hence, the observed growth law of 1/3 in the nematic-liquid region appears reasonable.

Conclusions

We have demonstrated that the combination of Flory-Huggins theory and Maier Saupe theory predicts a complex "tea pot" phase diagram comprising of liquid-liquid

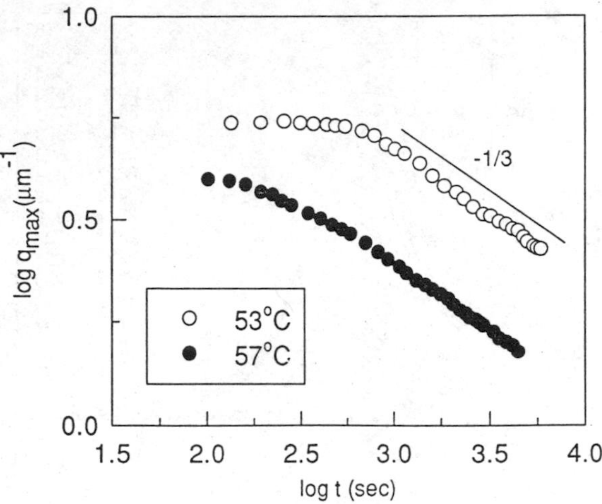

Figure 9. Time dependence of q_{max} in the spinodal regime following two different quenches from 80 °C to indicated temperatures at the 40/60 PMMA-OH/E7 composition.

coexistence region and a narrow nematic-liquid coexistence region. Below the peritectic line, nematic ordering takes place within the preformed LC-rich domains. The combined FH/MS theory was found to accord well with the cloud point phase diagram of the PMMA-OH/E7 mixture. The dynamics of liquid-liquid phase separation in the metastable region shows the droplet formation with seemingly uniform size and distribution, suggestive of homogeneous nucleation and growth. The growth exponent of 1/4 was estimated which is smaller than the predicted value for the NG. At deep quenches into unstable region, the growth exponent of 1/3 was observed, suggesting that the nematic ordering has little or no influence on the growth dynamics. This observation is not surprising in view of the fact that the dynamics of liquid-liquid phase separation and the annihilation of the disclinations shows the exponents of 1/3 or 1 in liquid crystalline polymer systems.

Acknowledgment: Support of this work by the National Science Foundation - Science and Technology Center for "Advanced Liquid Crystal Optical Materials: ALCOM" # DMR89-20147 is gratefully acknowledged.

Literature Cited

1. Doane, J.W.;. Vaz, N.A.P; Wu, B.G; Zumer, S. *Appl. Phys. Lett.*, **1986,** 48, 269.
2. Kajiyama, T.; Miyamoto, A.; Kikuchi, H.; Morimura, Y. *Chem. Lett.*, **1989,** 813, 1989.
3. Bahadur, B. Ed. "*Liquid Crystals: Their Applications and Uses,*" **1992,** World Sci. Publ., Teaneck, New Jersey, .
4. West, J. in "*Polymer Liquid Crystals,*" Weiss, R.A.; Ober, C. Eds., ACS Symp. Ser. # 435, **1990,** Ch. 32, p. 475
5. Kim, J.Y.; Cho, C.H.; Palffy-Muhoray, P.; Mustafa, M.; Kyu, T.; *Phys. Rev. Lett.*, **1993,** 71, 2232.
6. Kyu, T.; Ilies, I.; Mustafa, M.; *J. de Physique IV*, **1993,** 3, 37.
7. Kim, W.K.; Kyu, T. *Mol. Cryst. Liq. Cryst.*, **1994,** 250, 131.
8. Shen, C.; Kyu, T. *J. Chem. Phys.*, **1995,** 102, 556.

9. Shen, C., *Ph. D. Dissertation*, University of Akron, **1995.**
10. Flory, P.J. "*Principles of Polymer Chemistry,*" Cornell Univ. Press, Ithaca, N.Y., **1953.**
11. Brochard, F.; Jouffroy, j.; Levison, P. *J. de Physque*, **1984,** 45, 1125.
12. de Gennes, P. "*The Physics of Liquid Crystals*," **1974**, Oxford Sci. Pub., Oxford.
13. Olabisi, O.; Robeson, L.M.; Shaw, M.T. "*Polymer-Polymer Miscibility,*" *1979*, Academic Press, New York, N.Y.
14. Maier, W.; Saupe, A. *Z. Naturforsh.*, **1959,** 14a, 882; ibid., *Z. Naturforsh.*, **1960,** 15a, 287.
15. Kyu, T.; Saldanha, J.M.; *J. Polym. Sci. Polym. Lett. Ed.,* **1988,** 26, 33.
16. Gunton, J.D.; San Miguel, M.; Sahni, P.S. in "*Phase Transitions and Critical Phenomena,*" Domb. C.; Lebowitz, J.L., Eds., Academic Press, New York, N.Y., **1983,** ch. 8, p. 267.
17. Tanaka., H. *Europhys. Lett.*, **1993,** 24, 665
18. Cumming, A.; Wiltzius, P.; Bates, F.S. *Phys. Rev. Lett.*, **1990**, 65, 863.
19. Xie, Y.; Gallagher, P.D.; Gupta, A.; Morales, G.; Ludwig, K.; Bensil, R.; Konak, C. *Bull. Am. Phys. Soc.*; Annual Meeting, **1994,** 39, F31 10, 344.
20. Lifshiftz, I.M.; Slyozov, V.V. *J. Phys. Chem.,* **1961**, 19, 35.
21. Binder, K; Stauffer, D. *Phys. Rev. Lett.*, **1973,** 33, 1006.
22. Shiwaku, T.; Nakai, A.; Hasegawa, H.; Hashimoto, T. *Polym. Commun.*, **1987,** 28, 174.
23. Yurke, B.; Pargellis, A.N.; Turok, N. *Mol. Cryst. Liq. Cryst.*, **1992**, 222, 195.

Chapter 14

Polymer-Dispersed Liquid Crystals: Compositional Dependence of Structure and Morphology

Andrew J. Lovinger, Karl R. Amundson, and D. D. Davis

AT&T Bell Laboratories, 600 Mountain Avenue, Murray Hill, NJ 07974

We have investigated the structure and morphology of polymer-dispersed liquid crystals (PDLCs). These materials are commonly prepared by UV-induced crosslinking of a compatible mixture of a prepolymer with a liquid-crystal (LC) eutectic and are of high interest for flat-panel display applications. We have found that the morphology of these materials varies greatly with UV-irradiation temperature and composition. Within a range of ca. 20-70 wt.% LC, lower irradiation temperatures and higher LC contents favor a two-phase dispersion of bipolar droplets within the polymeric matrix. At low LC contents and high irradiation temperatures a new, space-filling spherulitic morphology is seen. These spherulites have a tangential orientation of the nematic LC molecules and are characterized by a highly unusual radial proliferation of surface inversion walls. We have found these defects to be initiated consistently at s=+1/2 disclinations and to be terminated at s=–1/2. The spherulites as well as the disclinations survive heating above the nematic-isotropic transition with little change.

Polymer-dispersed liquid crystals (PDLCs) are μm-sized dispersions of nematic liquid-crystalline droplets within a polymeric matrix (1-3). They are finding intense interest for applications such as light switches in flat-panel displays and windows (3, 4). They are usually prepared by UV-induced phase separation and cross-linking of a prepolymer containing a compatible blend of liquid crystals (LCs), although solution-casting from a common solvent and cooling from the melt below the upper-critical-

0097–6156/96/0632–0216$15.00/0

solution temperature are also occasionally used (5). Contrary to the usual twisted nematic liquid-crystal displays, PDLCs operate on a scattering principle and therefore do not require polarizers (which absorb over 50% of all transmitted light). The LC molecules are selected so that their ordinary refractive index matches that of the host polymer. Since they have a positive dielectric anisotropy, they are aligned with an applied electric field, causing the initially scattering, randomly oriented bipolar droplets to give rise to a transparent film.

Detailed calorimetric studies of the phase separation and curing processes in UV-irradiated PDLC materials have been given by Smith (6, 7). The morphology of PDLCs has been observed by scanning electron microscopy of blends of nematics in aqueous emulsions of polymers (8-10). Very recently, we have published a detailed morphological investigation of UV-cured PDLCs (11). Here, we summarize our recent findings on the compositional dependence of phase separation and morphology, and extend them to the description of the temperature dependence of the new morphologies that we reported earlier (11).

Experimental Section

The polymeric matrices were prepared by photopolymerization of commercial mixtures containing primarily trimethylolpropane diallyl ether, trimethylolpropane tristhiol and a diisocyanate ester (6). They were obtained from Norland Products, Inc., New Brunswick, NJ, under the designation NOA65. Laboratory-made mixtures of different acrylates were also studied. The liquid crystals were in most cases commercial eutectic mixtures of cyanobiphenyls and -triphenyls having a nematic-isotropic transition at 59-60°C and obtained from Merck Industrial Chemicals under the designation E7. Other, halogenated LCs were also occasionally used. The materials (which had been stored in the dark and handled under diminished light) were blended and sandwiched between thin glass cover slips to yield films of thickness ca. 12-18 µm. Photopolymerization was performed with a 100-W Hg lamp at selected temperatures and with doses of 4.5 J/cm^2 or greater. Specimens were examined in the polarizing optical microscope during heating and cooling using a Mettler microscope oven. In some cases, the interior of the films was visualized by scanning electron microscopy of properly prepared surfaces.

Results and Discussion

Typical PDLC morphologies of ≥50% LC blends of LC:polymer (E7:NOA65) cured at ambient temperature consist of a bimodal distribution of droplets having sizes of 4-8 μm and 0.2-1 μm, respectively (see Figure 1a). This figure is a scanning electron micrograph from a blend containing 65% LC, which is close to the maximum that can be incorporated at room temperature within this widely-used prepolymer. The smaller droplets are seen to be concentrated in the interstices between the larger ones and are probably generated at the later stages of curing when the matrix has become quite viscous and highly cross-linked. This morphology, which incorporates large amounts of polymer matrix between droplets, obviously affects the transparency of the thin films, resulting in the well-known off-axis haze, which is associated with the refractive index mismatch between polymer and LC. One way to minimize this is through mixtures that can accommodate larger amounts of LC. Such materials may contain, for example, over 80% of a halogenated multiphenyl liquid-crystal mixture in a blend of alkyl acrylates (such as TL205:PN393 from Merck). We have found (12) that their morphology is strikingly different (see Figure 1b): the polymer wall thickness is much smaller and the droplets are now arranged in a polygonal "foam texture". As a result, scattering occurs overwhelmingly *among droplets* rather than between LC and matrix, leading to favorable reduction in turbidity (13).

Returning now to the E7:NOA65 blends, we examine the compositional dependence of their morphologies in the optical microscope. The droplet morphology is by no means exclusive or even prevalent, but depends on both composition and UV irradiation temperature (11). For example, Figure 2 shows that the two-phase PDLC morphology in 50:50 blends is replaced by a spherulitic one at irradiation temperatures between 40 and 50°C. The spherulites are space-filling and are seen to have profuse radial striations. Overall, they exhibit a striking similarity to the typical spherulites of crystalline polymers. In addition to this morphological similarity, we also found that they have the same molecular orientation (i.e. tangential), as evidenced by their *negative* birefringence upon insertion of a first-order red plate at 45° to the crossed polars (see figure 3). This is somewhat surprising because the elastic constant for splay distortions in this liquid crystal is smaller than that for bend (K_3/K_1=1.54), which would then favor a radial director orientation. However, these constants are provided from the manufacturer for the LC component alone, and it is not known how the presence of large amounts of flexible polymer molecules might modulate the elastic distortions experienced by the LC rods.

Polymeric spherulites have been known (14, 15) to grow with a radially directed and non-crystallographically branched lamellar orientation. The molecules are generally perpendicular to the lamellar

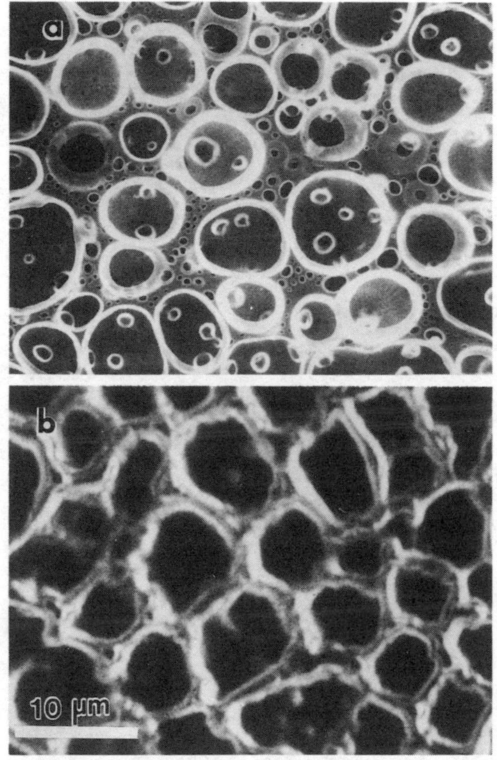

Figure 1. Scanning electron micrographs (secondary electrons) showing the typical morphologies of PDLCs after extraction of the liquid crystal components. (a) 65:35 blend of LC:polymer, (b) 80:20 blend of (a different) LC:polymer. See text for further details.

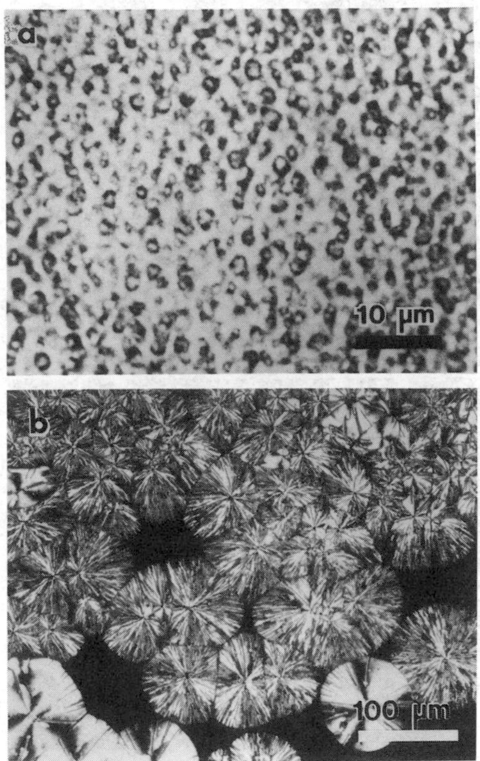

Figure 2. Typical morphologies of 50:50 blends of LC:polymer recorded in the polarizing optical microscope. Both were UV-irradiated with 4.5 J/cm^2 at (a) 40°C and (b) 50°C.

crystallites, which in turn are separated from each other by amorphous regions (see Figure 4a). In the novel LC spherulites described here, the rod-like molecules are again tangentially disposed, but there is no analogue of radially continuous lamellae. Instead, the molecules are expected to be aligned to each other within a network of cross-linked amorphous polymeric chains (see Figure 4b). The centers of the polymeric vs. nematic spherulites are also different. In crystalline polymers, nucleation occurs most commonly heterogeneously (e.g. on impurities) or quasi-homogeneously (see Figure 5a). In our LC case, the central region must involve a singularity: either an s=+1 disclination line anchored at the two interior glass surfaces or an escaped point singularity (see Figure 5b).

The above general behavior of 50:50 LC:polymer mixtures can now be extended to other compositions and temperatures. We found that higher concentrations of the nematic component favor the phase-separated PDLC, e.g. for 65% LC samples the spherulitic morphology is not seen at any irradiation temperature. On the other hand, lower LC contents favor formation of spherulites. For example, a 25% LC mixture no longer yields PDLC droplets above room temperature: only by irradiation at low temperatures (e.g., 10°C) is such a morphology obtained, but it is not stable upon return to ambient, leading instead to a continuous nematic phase (11). We can summarize our findings with the aid of a "morphological map" as in Figure 6, which shows the regions of PDLC versus spherulitic phases. Beyond ca. 72%, the liquid crystal component is no longer miscible with the initial photopolymerizable mixture. Since electrical and optical properties improve with LC content (13), it is desirable to be able to incorporate more than the ca. 72% that is possible with this mixture. It is for this reason that matrix polymers and liquid crystals of the general types referred to in Figure 1b are becoming increasingly more useful, since they allow LC contents as high as ca. 84% and also lead to the advantageous scattering mechanism mentioned above (12, 13).

Having discussed the general morphological map of these PDLC materials we now concentrate on the new spherulitic morphologies, specifically their internal microstructural features. Of these, the most obvious are the radial striations that are already discernible in Figure 2b. At high magnification between crossed polarizers, these striations generally exhibit closed dark loops; the material inside the loops has opposite birefringence to that outside. We therefore identify these features as surface inversion walls (16, 17): These are initiated and terminated by s=+1/2 or −1/2 disclination lines that are attached to the glass surfaces and are generally normal to them (see Figure 7). A related possibility consists of half-integral disclination lines running along the surface of one of the walls, but these have been associated

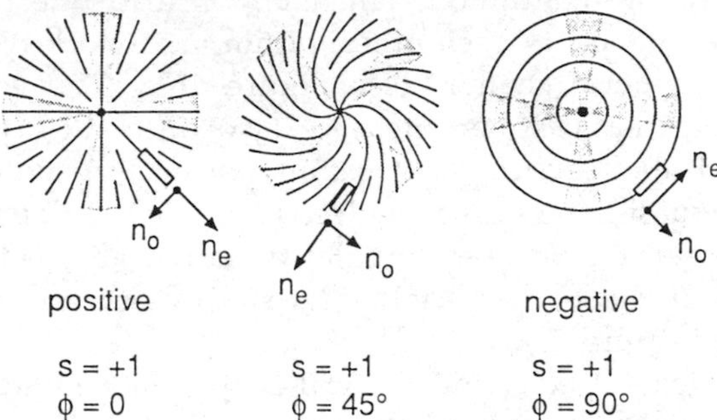

Figure 3. Schematic representations of the molecular orientations of liquid-crystal molecules in various types of spherulitic morphologies originating at an s=+1 disclination. The ordinary and extraordinary refractive indices are designated n_o and n_e, respectively. (Reproduced from Ref. 11; Copyright American Chemical Society).

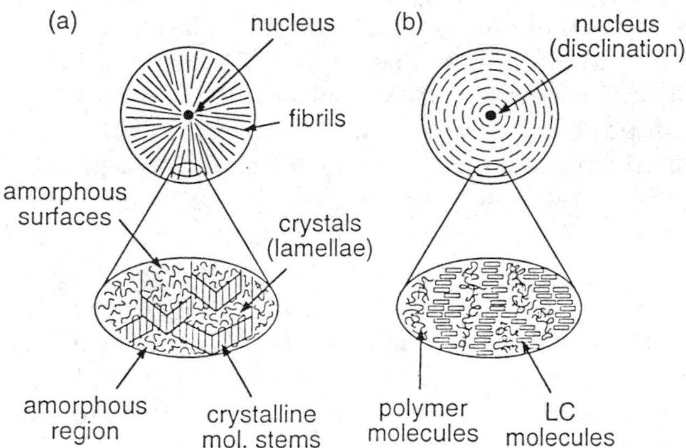

Figure 4. Comparison of (a) lamellar growth in spherulites of semi-crystalline polymers with (b) possible growth of liquid crystals in spherulites containing partly cross-linked amorphous polymeric molecules. (Reproduced from Ref. 11; Copyright American Chemical Society).

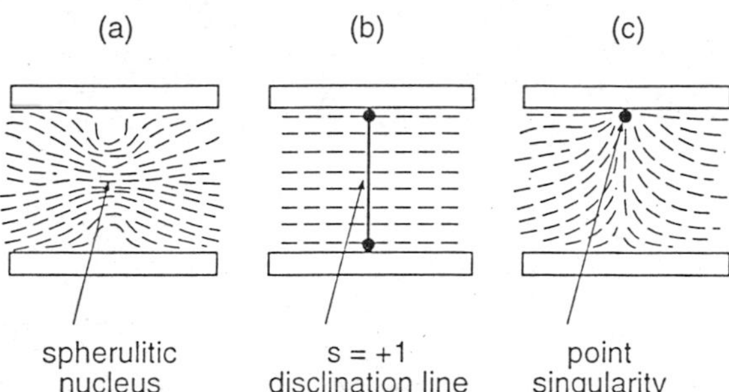

Figure 5. Comparison of spherulitic centers in (a) semi-crystalline polymers versus liquid crystals growing in a partly cross-linked amorphous polymeric matrix (b and c) for samples confined to thin films between two flat surfaces. (Reproduced from Ref. 11; Copyright American Chemical Society).

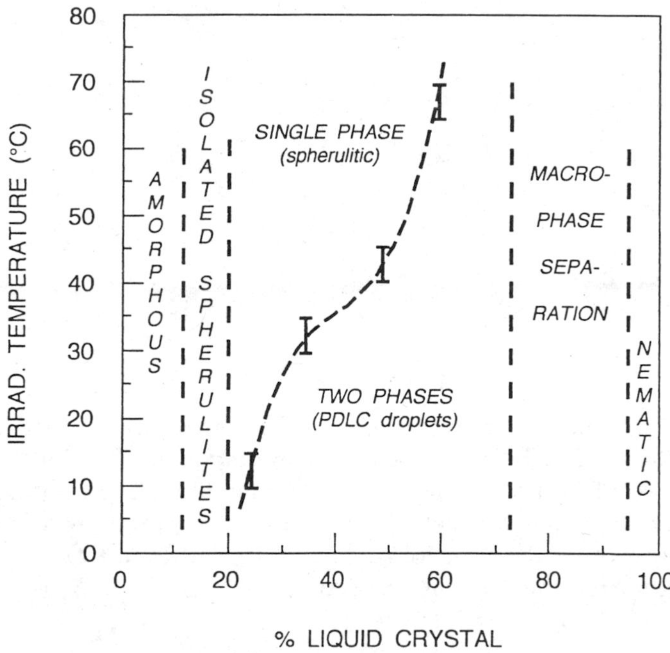

Figure 6. Approximate morphological map of the E7:NOA65 LC:polymer blend system as a function of composition and UV irradiation temperature. (Reproduced from Ref. 11; Copyright American Chemical Society).

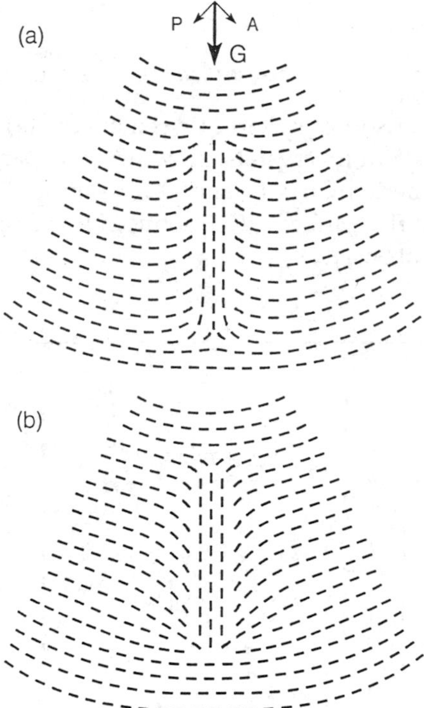

Figure 7. Schematic representations of inversion wall defects in LC spherulites and of their possible radial orientation modes. Initiation at (a) an s=+1/2 and (b) an s=–1/2 disclination line. The polarizer and analyzer directions are denoted by P and A, respectively, and the spherulitic growth direction by G. (Reproduced from Ref. 11; Copyright American Chemical Society).

with free surfaces (18) and exhibit difuseness at their s=–1/2 termination points that is not seen in our samples.

A number of different manifestations of these surface inversion walls is seen in the polarizing micrographs of Figure 8. Figure 8a shows the detailed morphology of these inversion wall defects. This is one case where successive proliferation of such defects occurs during growth, leading to crankshaft or zigzag appearance of extinction lines in the radial direction. Figure 8b demonstrates initiation of such inversion walls along a line that traverses obliquely through one spherulite. This shows that factors extrinsic to the growing LC/polymer mixtures can cause these defects to be introduced. Such extrinsic factors are probably scratches in the inner surfaces of the coverglass or other discontinuities. In Figure 8c we observe the detailed microstructure of inversion walls as they reach the interfaces between growing spherulites. The wall defects are seen to be reoriented along the interspherulitic boundaries instead of being terminated at the intersections. Remarkably, the orientation of the LC molecules remains the same within these walls (i.e., parallel to the original growth direction), as we ascertained through introduction of a first-order red plate into the optical path of the microscope, between sample and analyzer. Preservation of the overall nematic director within these regions is dictated by the packing requirements for the rod-like molecules. Finally, Figure 8d shows the typical morphology of LC/polymer spherulites growing in "confined" regions, i.e. those remaining after the surrounding material has already grown spherulitically. The inversion wall defects are now seen to form irregularly curved, coarse channels that imply growth from regions enriched in amorphous polymer. This is entirely analogous to the phenomena explained many years ago by Keith and Padden (14) in their classic studies of spherulites of crystalline materials (polymeric, organic, and inorganic) growing from "impure" melts.

A remarkable feature of these inversion wall defects involves their uniformity in generation and termination. As seen schematically in Figure 7, these defects could be initiated at s=+1/2 disclination lines and terminated at s=–1/2 ones (Fig. 7a), or vice versa (Fig. 7b). Through rotation of spherulites about the axis of a polarizing microscope and examination of the concurrent optical changes in their radially oriented inversion walls, we found a uniform behavior: Always the inner non-birefringent boundary (i.e., the one closest to the polarizer or analyzer direction) narrows as the defect approaches that polarizer or analyzer while the outer boundary becomes wider. This can be seen even in the stationary case of Figure 8b by comparing the widths of the dark brushes defining each radial defect in relation to their positions vis-à-vis the spherulitic Maltese cross. By examination of the schematics in Figure 7,

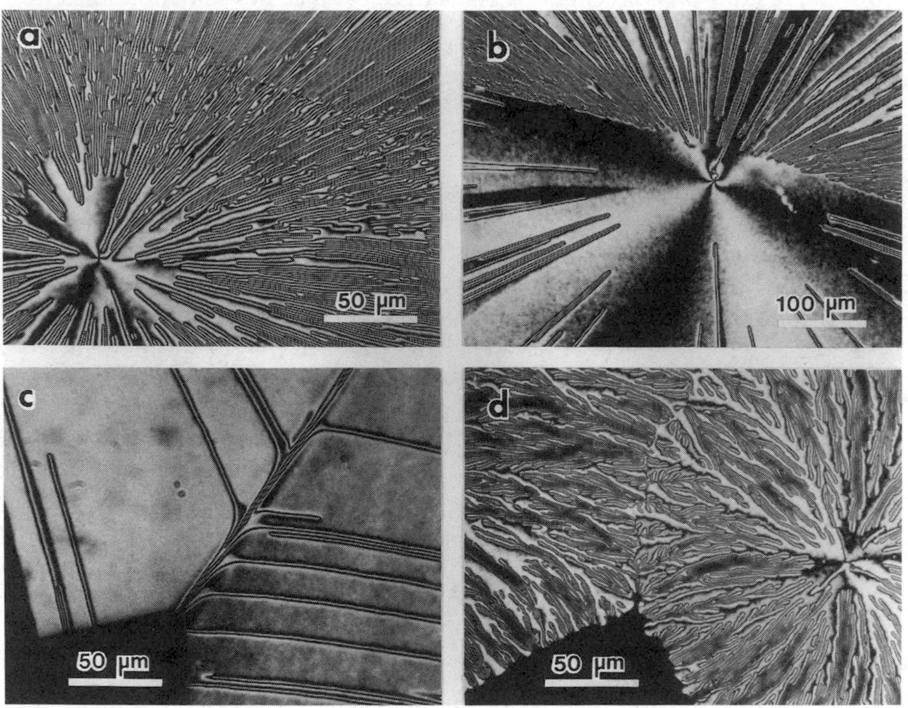

Figure 8. Appearance between crossed polars of various morphological features associated with the inversion wall defects in LC spherulites. (a) Successive radial initiation, (b) initiation by extrinsic factors, (c) rejection and reorientation of inversion walls in interspherulitic boundaries, and (d) spherulitic growth in confined regions. The material is a 25:75 E7:NOA65 LC:polymer blend UV-irradiated with 4.5 J/cm^2. (Reproduced from Ref. 11; Copyright American Chemical Society).

this narrowing and widening behavior is consistent with case (a), i.e. initiation at s=+1/2 disclination lines. Reasons for this propensity may have to do with the orientation of LC molecules with respect to polymeric cross-links or other singularities. The nematic rods are not expected to exhibit preferential wetting or other favorable interactions with such inhomogeneities (Figure 9b), and will thus adopt a radial orientation toward them (Figure 9a) leading to s=+1/2 disclinations.

The isotropization ("melting") of these nematic spherulites during heating is also remarkable. The general behavior is seen in Figure 10. The spherulites (and their inversion wall defects) survive essentially unchanged until ca. 61.5°C (Figure 10a). Above that temperature, isotropic droplets appear throughout the samples. In some cases (11) these are concentrated near the spherulitic boundaries; in some others isotropization is initiated at the spherulitic peripheries (see spherulites marked "1" and "2" in Fig. 10b and c) and grows inward. Upon further increase in temperature by only a few tenths of one °C the non-birefringent droplets coalesce and the spherulites "melt" completely at 62°C (Figure 10c). However, even after further heating to 65°C (Figure 10d), subsequent cooling below the isotropic-to-nematic transition temperature leads to re-formation of the very same spherulites. This includes reappearance of the original inversion walls with minimal or no disruptions. In Figure 10d, these disruptions are seen to be associated with the locations of the original isotropic droplets. All of these results indicate that a memory of the spherulitic morphology and substructure survives even above the isotropization temperature. Possibly, this is a result of an imprinting of the LC molecular arrangements onto the partly cross-linked polymeric matrix, as well as of the confinement of the nematic rods by the polymer in such a way that regeneration of their original orientations is highly favored.

Conclusions

Polymer-dispersed liquid crystals are seen to exhibit exceptionally complex morphologies ranging from LC droplets phase-separated within the polymeric matrix to nematic spherulites incorporating the polymer. The spherulites have been examined in detail and are seen to be characterized by radially oriented defects, which we attribute to inversion walls; these are initiated uniformly through s=+1/2 disclination lines. The molecular orientation in these LC spherulites appears to be tangential, similar to the case of semicrystalline polymers. The isotropization behavior of these spherulites is also complex, beginning with nucleation of non-birefringent regions at the interiors (and occasionally the peripheries) and followed by reappearance of the main morphological features upon subsequent re-cooling.

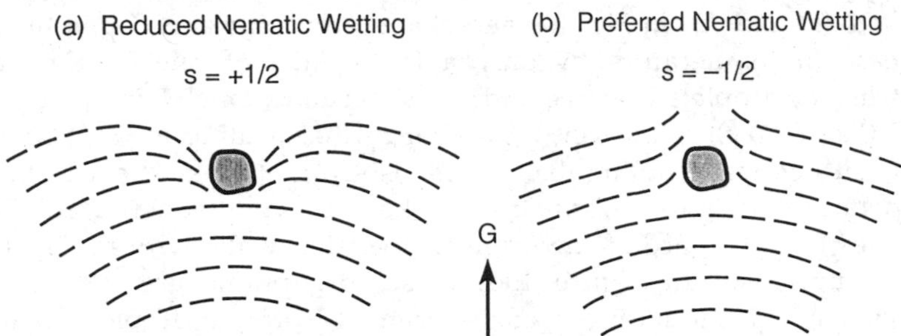

Figure 9. Hypothetical modes of initiation of (a) s=+1/2 and (b) s=−1/2 disclinations during radial growth in LC spherulites depending upon mutual orientations of the LC molecules on regions of polymeric cross-links or other inhomogeneities. (Reproduced from Ref. 11; Copyright American Chemical Society).

Figure 10. Polarizing micrographs depicting the isotropization behavior during heating of LC spherulites in a 25:75 E7:NOA65 LC:polymer blend UV-irradiated with 4.5 J/cm^2 at 25°C. (a) Heated to 61.5°C, (b) 61.6°C, and (c) 61.8°C. (d) Cooled to ambient after complete isotropization at 62°C and further heating to 65°C.

Literature Cited

1. Fergason, J. L. *SID Tech. Dig.* **1986**, *16*, 68.
2. Drzaic, P. S. *J. Appl. Phys.* **1986**, *60*, 2142.
3. Doane, J. W.; Vaz, N. A.; Wu, B.-G.; Zumer, S. *Appl. Phys. Lett.* **1986**, *48*, 269.
4. Doane, J. W. *MRS Bull.* **1991** (1), 22.
5. West, J. L. *Mol. Cryst. Liq. Cryst.* **1988**, *157*, 427.
6. Smith, G. W. *Mol. Cryst. Liq. Cryst.* **1991**, *196*, 89.
7. Smith, G. W. *Mol. Cryst. Liq. Cryst.* **1993**, *70*, 198.
8. Drzaic, P. S. *Liq. Cryst.* **1988**, *3*, 1543.
9. Drzaic, P. S.; Muller, A. *Liq. Cryst.* **1989**, *5*, 1467.
10. Havens, J. R.; Leong, D. B.; Reimer, K. B. *Mol. Cryst. Liq. Cryst.* **1990**, *178*, 89.
11. Lovinger, A. J.; Amundson, K. R.; Davis, D. D. *Chem. Mater.* **1994**, *6*, 1726.
12. Amundson, K. R.; Lovinger, A. J.; Davis, D. D. (unpublished results).
13. Drzaic, P. S.; Gonzales, A. M. *Appl. Phys. Lett.* **1993**, *62*, 1332.
14. Keith, H. D.; Padden, F. J., Jr. *J. Appl. Phys.* **1963**, *34*, 2409; **1965**, *35*, 1270, 1286.
15. Wunderlich, B. *Macromolecular Physics*; Academic, New York **1973**; vol. 1.
16. Nehring, J.; Saupe, A. *J. Chem. Soc., Faraday Trans.* **1972**, *68*, 1.
17. Kléman, M.; Williams, C. *Philos. Mag.* **1973**, *28*, 725.
18. Mazelet, G.; Kléman, M. *Polymer* **1986**, *27*, 714.

Structure of the Mesophase

Chapter 15

The Difference Between Liquid Crystals and Conformationally Disordered Crystals

Bernhard Wunderlich and Wei Chen

Department of Chemistry, University of Tennessee, Knoxville, TN 37996–1600 and Chemical and Analytical Sciences Division, Oak Ridge National Laboratory, Oak Ridge, TN 37831–6197

A series of criteria and signature properties that distinguish liquid crystals from condis crystals has been developed and will be discussed. Thermotropic liquid crystals achieve their special properties and structure *via* rod-, disk-, or board-like rigid mesogens. Amphiphilic liquid crystals, in contrast, result from the formation of interfaces where the junctions between incompatible parts within the molecule collect. In the liquid crystalline state, the liquid-like motion and disorder of translational (positional) and conformational type is largely maintained, while some of the orientational motion and disorder is decreased. Liquid crystals must be distinguished from conformationally disordered crystals (condis crystals). Condis crystals are more closely related to the crystalline state with translational and orientational long-range order, but with conformational disorder. In macromolecules which alternate in chemical nature along the chain one can find a full spectrum of substances ranging from thermotropic and amphiphilic liquid crystals to condis crystals and true crystals.

Liquid crystals (LC) were discovered about 100 years ago (*1*). Their structure is intermediate in order and mobility between crystal and melt, *i.e.* an LC is a mesophase (Gr. μέσος = middle). *Macromolecules* were recognized as a special class of molecules only in 1920 (*2*). They consist of more than 1000 atoms, in contrast to small molecules with fewer atoms [definition given by Staudinger (*3*)]. The macromolecules of concern in this paper are flexible and linear, to be distinguished from rigid macromolecules, such as diamond, poly-p-phenylene, metals, and salts. The identifiers of classical *thermotropic, liquid-crystal-forming molecules* are their mesogens, represented by rod- or disc-like segments (*4,5*). More recently board-like, sanidic liquid crystals were described (Gk. σανίς = board, plank) (*6*). Most thermotropic LCs have one or more flexible appendages on the edge of the mesogen. On ordering of the mesogens, a certain amount of phase-separation of the two, usually incompatible constituents occurs. The resulting phases are of nanometer size and may contribute to the driving force of LC formation (*nanophase separation*).

0097–6156/96/0632–0232$15.00/0

In *amphiphilic liquid crystalline phases* (*4*), incompatible molecular parts of flexible molecules cause an LC anisotropy without the presence of a distinct mesogen. The alignment of the junctions between the incompatible parts at interfaces produces the LC structure.

*Macromolecular LC*s were first observed in form of, lyotropic systems (*7*) and discussed in terms of segmental rigidity (*8*). Lyotropic liquid crystalline phases derive their anisotropy from interaction with a solvent. They are not further considered in this paper. When thermotropic, polymeric LCs that have an LC-to-isotropic phase transition at the isotropization temperature, T_i, were first mentioned in the scientific literature (*9*), it became of interest to find the similarities and differences to small-molecule LCs.

The structure-determining constituent of the molecules that form *thermotropic liquid crystals* can be a mesogen, as in small molecules. These mesogens of macromolecules must be separated by flexible links to maintain a liquid-like mobility. There are two such molecular structures. One has the mesogens placed at intervals in the otherwise flexible main chain (*10*), the other has the mesogens placed as side-groups that are attached to the macromolecule with flexible links (*11*). The side-chain LCs are closely related to the small molecules, just as normal macromolecules with sufficiently long side-chains resemble in many of their properties the analogous small molecules (*12,13*). The major difference to small-molecule LCs is, in both cases, the higher viscosity resulting from the macromolecular nature. The main-chain LCs have an additional difference relative to small molecules with similar mesogens. The flexible spacers enter, through changes in their conformation, more actively into the alignment of the mesogens. As the sizes of the different segments of the macromolecule increase, nanophase separation of the two incompatible constituents can contribute to the stability of the LC. This nanophase separation is enhanced by conformational ordering in the flexible spacer, needed to achieve an advantageous packing of the more rigid segments. This type of nanophase separation may in some cases become the main driving force in LC formation. The morphology of *macromolecular amphiphilics* varies with composition, molar mass, and conformation (*14,15*). Because of the large molecular size, macromolecular amphiphilics are usually multiphase structures with dimensions in the micrometer scale (*microphase separation*).

Crystals with conformational disorder (condis crystals) were proposed in 1985 to belong to a separate mesophase, characterized otherwise by orientational and positional order of the molecules as a whole (*16*). Their properties were reviewed on many examples from the literature in 1988 (*17*). The need to include conformational disorder into the causes for the formation of mesophases was already pointed out by Smith in 1975 (*18*). Furthermore, it was suggested by Leadbetter that the term liquid crystal should be reserved for phases lacking positional long-range order in at least one dimension (*19*). This distinguishes condis crystals from liquid crystals and places condis crystals as a separate mesophase between crystals and liquid crystals. Conformational contributions to the order in main-chain LCs may, in some cases, be sufficient to change their behavior from liquid-like to more plastic-like, as expected of condis crystals. The disorder in condis crystals occurs on a segmental scale, contrasting orientational disorder on a molecular scale in crystals. Orientational disorder (and mobility) on a molecular scale of close-to-spherical molecules is found in *plastic crystals* (*5*). The identification of LCs with conformational order in the flexible spacer and their distinction from condis crystals is the main topic of this paper.

Descriptions of the Melting of Crystals

A detailed study of *crystals of macromolecules* (*20,21*) and their melting under equilibrium conditions revealed that the entropy of fusion, ΔS_f, is often about 7–12 J/(K mol) per mobile unit or "bead" (*22*). This entropy is linked mainly to the conformational disorder (ΔS_{conf}) and mobility that is introduced on fusion. Sufficiently below the melting temperature, disorder and thermal motion in crystals is exclusively vibrational. While vibrations are *small-amplitude motions* that occur about equilibrium positions, conformational, orientational, and translational motions are of *large amplitude*. These types of large-amplitude motion can be assessed by their contributions to heat capacity (*23*), entropy (*22*), and identified by relaxation times of the nuclear magnetization (*24*).

Orientational and positional entropies of fusion ΔS_{orient} and ΔS_{trans} are of importance to describe the *fusion of small molecules*. They can be derived from the many data on fusion of the appropriate rigid, small molecules of nonspherical and spherical shapes [nonspherical molecules: Walden's rule (1908), $\Delta S_f = \Delta S_{orient} + \Delta S_{trans}$ = 20–60 J/(K mol); and spherical molecules: Richards' rule (1897), $\Delta S_f = \Delta S_{trans}$ = 7–14 J/(K mol)].. The contributions of ΔS_{orient} and ΔS_{trans} to the melting of macromolecules is negligible since these contributions to ΔS_f are counted per molecule and not per repeating unit (*16,22*). For a general molecule one can thus write:

$$\Delta S_f = \Delta S_{orient} + \Delta S_{trans} + n\Delta S_{conf} \tag{1}$$

where n is the number of bonds capable of producing rotational isomerism. Equation (1) can be used to estimate the entropy of fusion from the structure of the molecule, using the empirical rules for the three contributions (within the given, broad variations).

Liquid Crystal Formation

To change a melt to an *LC*, it is necessary to orient the mesogens. In case the molecule contains, as is usually the case, one or more flexible extensions on the mesogen, there must also be a certain amount of nanophase separation to bring the mesogens into preferential proximity. The large-amplitude motions remain otherwise little affected, accounting for the liquid-like nature. This is particularly important for the translational motion. As soon as the translational motion stops and positional, long-range order is achieved for the molecule, the phase should be called a true crystal (*19*). Thermodynamically the order in a liquid crystal is characterized by a small ΔS [often less than 2 J/(K mol) (*16–18*)] and an almost unchanged heat capacity (but note that many LCs show some degree of continuous ordering on cooling between the transitions and have, thus, an increased apparent heat capacity relative to the liquid).

Figure 1 displays the correlation between the mesophases and the classical phases: glass, melt, and crystal. The first-order transition LC-to-crystal is marked on the right side of the figure along with all other observed transitions. In case the LC can be quenched to low temperature avoiding further ordering, it undergoes a glass transition, as indicated on the left of Fig. 1. The decrease in heat capacity on going to the glass, a solid that shows only vibrational motion, can be used to estimate the loss in mobility on vitrification. It is about 11 J/(K mol) for small mobile units (beads) and double to triple this amount for larger beads, as found in the mesogens (*23*).

Orientational order in an LC is given by the parameter O = 1.5(<cos^2 θ> - ⅓) (*25*), where θ represents the deviation of the molecular axis from the director. The director, in turn, is the symmetry axis of the distribution function of the molecular axes. For nematic LCs (LCs with order in one direction only) O is often about 0.5. The order parameter may increase to above 0.8 for the various smectic LCs (*26*) (LCs ordered in two dimensions, in a lamellar arrangement).

The entropy of isotropization is a much more stringent measure of order. It is often only 10% or less than can be expected for full orientational order of the mesogen. Entropies of isotropization, ΔS_i, were collected earlier (*16*). They showed a major difference in ΔS_i for the nematic phases of small molecules and macromolecules with mesogens in the side-chain on the one hand, and macromolecules with mesogens in the main-chain on the other [ΔS_i = 2.1±1.5, 3.5±2.5, and 15.6±7.3 J/(K mol), respectively, for 11, 5, and 64 examples]. Smectic crystals may have somewhat larger ΔS_i, but it is still lower than expected for orientational disordering of the mesogen alone. For LCs of small molecule there seems to be no conformational order in the flexible appendages.

Amphiphilic LCs can be found, for example, among block copolymers, soaps, and lipids. The driving force towards establishment of an amphiphilic LC is governed by the incompatibility of the different parts of the molecule and the need to establish an interface of minimal interfacial free energy. The entropy of mixing is always favorable to the mixed state and must be overcome by the energetics. Figure 2 shows the typical liquid-crystalline-like lamellar structure of a diblock copolymer. The example is poly(styrene-*block*-butadiene) of overall molar mass of 88,000 Da with an approximate 50 to 50 ratio of the two molecular parts (*27*). The sizes of the incompatible segments within the molecule determine the phase structure (spherical or rod-like inclusions, lamellae, or more complicated structures with bicontinuous surfaces). In the example of Fig. 2, both components of the block copolymer are large and amorphous, so that no ordering parameter can be defined at some distance from the interface. If in such an amphiphilic LC the copolymer segments are marginally compatible, an isotropization transition would become possible above a lower or below an upper critical solution temperature (LCST or UCST, respectively).

A decrease in size of the two phases leads in the limit to molecules like soaps or lipids and the difference between amphiphilics and thermotropics may completely disappear. In neat (water free) soaps, for example, the smectic phase consists of layers of partially conformationally-ordered, nonpolar hydrocarbon segments, separated by layers of the polar molecular ends (*28*). The small diameter of the polar groups, being forced into close proximity by strong interactions, make the nonpolar groups take on a mobile, parallel, smectic structure. No large, rigid mesogen is needed for the alignment of the molecules. The order within the nonpolar segments is purely conformational. In the presence of a solvent these molecules arrange as micelles.

Figure 3 shows a phase diagram of a series of thallium soaps with longer alkyl chains (*29*). The neat phases shown are all liquid crystals with fan-type texture under polarized light. Their entropies of isotropization, ΔS_i, vary from 7 to 2.4 J/(K mol), typical for low molar mass liquid crystals. The mesophases stable at lower temperature show much larger entropies of transition, as is indicated in the figure. These phases were suggested to be condis crystals (*17*). The first member of the series, CH_3COOTl, displays a one-step melting with a ΔS_m of 43.5 J/(K mol). Only if the alkyl chain is

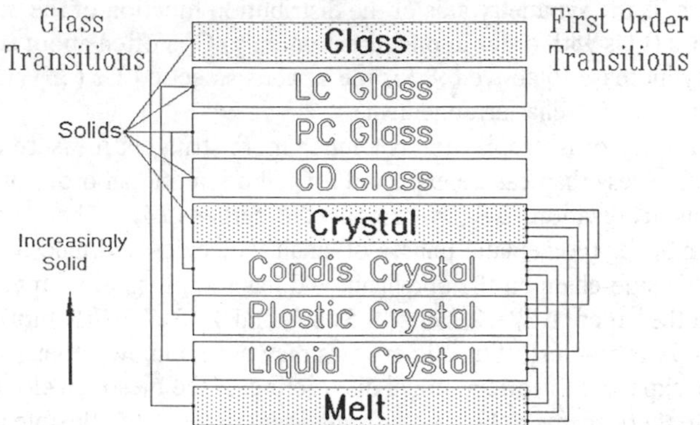

Figure 1: Schematic of phases, mesophases and transitions.

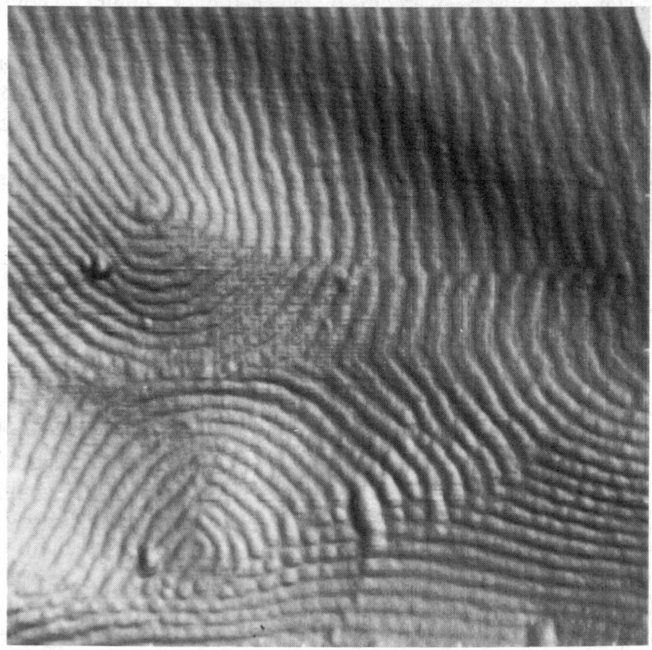

Figure 2: Atomic force microscopy image of a diblock copolymer of poly(styrene-*block*-butadiene). The length of the figure is 2.1 μm.

sufficiently long to produce a stable nanophase, is the liquid crystalline texture possible (at n > 4).

Formation of Condis Crystals

To go from the melt to a *condis crystal*, positional and orientational order must be established. In addition, partial conformational order may also be introduced. For molecules of low molar mass this leads to a much larger loss of entropy than the formation of an LC. Figure 4 illustrates a thermal analysis trace for OOBPD (*30*). The transition entropies are summarized in Fig. 5. The combined transitions from the melt to the first condis phase reach the Walden's rule entropy for the crystallization of the mesogen. Thus the K-phases are, as often in liquid-crystal-forming molecules with two flexible appendages, condis phases. With solid state ^{13}C NMR it could be shown that all nematic and smectic phases reorient their mesogen quickly on changes in the magnetic field direction and show liquid-like spectra (*i.e.* cross polarized, CP, spectra can be obtained without spinning about the magic angle, MAS). The condis phases, in contrast, behave like solids [no reorientation of the mesogen and sharp spectra can only be obtained with CP MAS) (*26*)]. The K_1-phase has only little conformational order, but even in K_3, the conformational order is incomplete. The remaining disorder is located in ten specific bonds (five in each end of the molecule):

$$CH_3-\mathbf{r}-CH_2-\mathbf{r}-CH_2-\mathbf{t}-CH_2-\mathbf{r}-CH_2-\mathbf{t}-CH_2-\mathbf{r}-CH_2-\mathbf{r}-CH_2-\mathbf{t}-O-\mathbf{r}-\text{phenylene}\cdots$$

$-\mathbf{r}-$ stands for a bond that randomly changes between *gauche* and *trans* conformations in the K_3 phase at sufficiently high temperature, $-\mathbf{t}-$ indicates a bond that is largely fixed in the *trans* conformation. The first bond, CH_3- is known to be mobile down to very low temperatures. The large-amplitude motion of the five $-\mathbf{r}-$ bonds freezes over a glass transition temperature range of about 100 K (*30*), as indicated in Fig. 6, and summarized numerically in Fig. 5, above. This example of the low-molar-mass OOBPD can serve as a model for main-chain LCs. The interpretation the OOBPD properties is easier since all first-order phase transitions occur close to equilibrium and no partial transitions that lead to fractional crystallinity are observed.

Typical condis crystals of flexible, linear macromolecules are hexagonal polyethylene (stable at elevated pressure), *trans* 1,4-polybutadiene (crystal phase II), polytetrafluoroethylene (polymorph I), and the high-temperature crystal forms of poly(diethylsiloxane) (*17*). Their structures are usually hexagonal, with the symmetry of the motif (asymmetric unit of the crystal) increased by the conformational mobility. Typical entropies of isotropization of the condis crystals are 1/3-2/3 of the total entropy of fusion of the fully ordered crystal, *i.e.* much larger than the ΔS_i found in liquid crystals. In all these polymers with rather simple repeating units, all mobile bonds seem to contribute equally to the disorder. The macromolecular condis crystals are usually of partial crystallinity, similar to the standard semicrystalline polymers. Their larger internal mobility permits, however, to anneal the crystals from the initial folded chain macroconformation to an extended chain macroconformation (*31*) with a mechanism of sliding diffusion (*32,33*). Linear polyethylene could, in addition, be crystallized to 99% perfection at pressures 400 MPa and display then equilibrium thermal properties even after the transformation from the condis phase to the orthorhombic phase on cooling. At

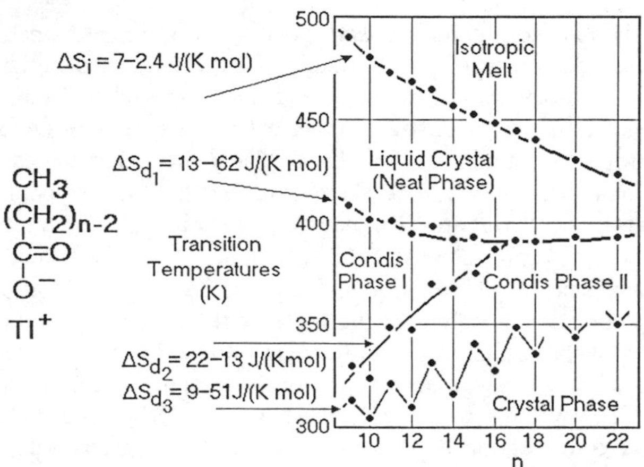

Figure 3: Phase relationships for a thallium soap.

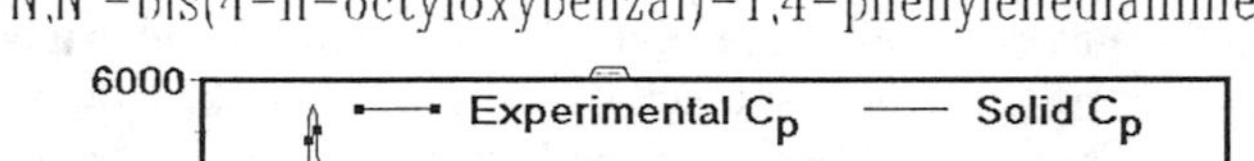

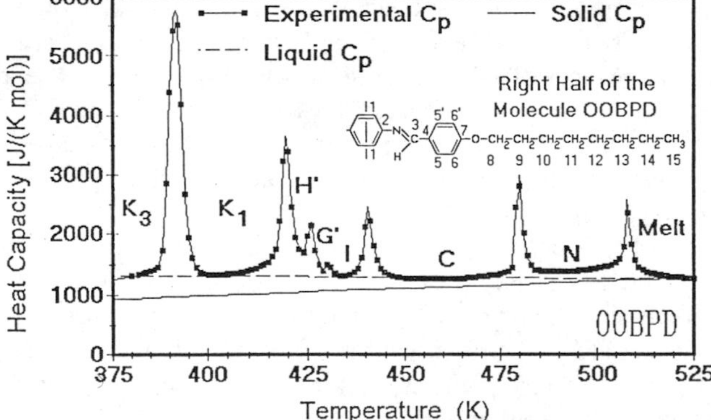

Figure 4: Differential scanning calorimetry of OOBPD. The heat capacity of the liquid is extrapolated from the melt, the heat capacity of the solid is evaluated at low temperatures as vibrational C_p and extrapolated to higher temperature.

N,N'-bis(4-n-octyloxybenzal)-1,4-phenylenediamine

$CH_3-CH_2-CH_2-CH_2-CH_2-CH_2-CH_2-CH_2-O-C_6H_4-CH{=}N-C_6H_4-N{=}CH-C_6H_4-O-CH_2-CH_2-CH_2-CH_2-CH_2-CH_2-CH_2-CH_3$

Measurements:

Transition	T (K)	ΔS [J/(K mol)]
Melt → N	504.8	8.5
N → C	476.6	11.0
C → I	436.8	8.8
I → G'	427.2	1.3
G' → H'	421.8	6.0
H' → K1	415.5	23.0
subtotal:		58.6
K1 → K3	387.0	47.4
grand total:		106.0

Analysis:

1 mesogen = 59 J/(K mol)
16 small beads = 152 J/(K mol)
total expected = 211 J/(K mol)
total measured = 106 J/(K mol)

outside of trans. = 75 J/(K mol)
remains at 250 K = 31 J/(K mol)

from NMR 10 mobile beads
or 3.1 J/(K mol) below T_g

[glassy polyethylene per CH_2
at 0 K = 3.0 J/(K mol)]

Figure 5: Summary of the transition parameters of OOBPD.

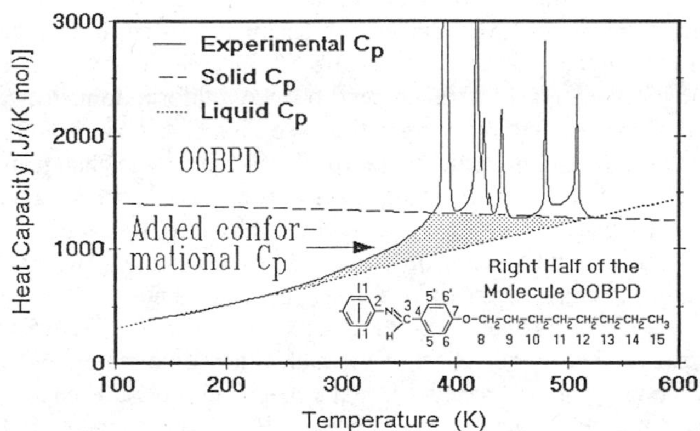

Figure 6: Thermal analysis of OOBPD, compare to Fig. 4. The shaded area is due largely to additional changes in the conformational motion (outside the transition regions).

atmospheric pressure the melting temperature of equilibrium orthorhombic polyethylene is 414.6 K, and the entropy of fusion 9.91 J/(K mol)] (*34*). Figure 7 shows a typical fracture surface of extended chain crystals of polyethylene with close-to-equilibrium properties.

Differences Between the Liquid and Condis Crystals

The differences between standard thermotropic LCs and macromolecular condis crystals are summarized in Fig. 8. The first three and the last two points make it easy to experimentally identify low molecular mass LCs. For macromolecules, however, the viscosity may be sufficiently large to lose the obvious liquid character; the birefringence does not always show the well-known LC texture (*35*); the small ΔS_i of LCs may be confused with partial crystallinity of the condis crystals; and in polymers, some larger main-chain rigid groups are not always easily identifiable as mesogens. This leaves points four and eight for differentiation between the two mesophases. Points five and six are more difficult to establish, and solid state NMR and detailed X-ray structure-determinations may be necessary for full characterization. Furthermore, borderline structures may be possible between thermotropic LCs, amphiphilic LCs, and condis crystals. A few examples and the resolution of their structures are discussed next, to illustrate the resolution of some of these problems.

The difference between liquid crystals and condis crystals is documented best for the low- molar-mass model-compound OOBPD (see Figs. 4-6). The instrumentation for the analysis included differential scanning calorimetry (heat capacity, transition parameter, and entropy measurement and calculation), optical microscopy (birefringence and morphology determination), thermomechanical analysis (evaluation of expansivity and volume change on transition), X-ray diffraction (observation of crystal structure changes), and solid state NMR (proof of chemical structure and estimation of intramolecular mobility) (*26,30,36*). The change of the smectic phase I to H' shows the most dramatic morphological rearrangement to large, uniform domains, accompanied, however, by the smallest entropy of transition (see Figs. 4 and 5) (*30*). The change to the condis state K_1 brings little change in morphology, but a large change in volume and enthalpy and the appearance of long-range X-ray order. The H'-to-K_1 transition is best described as a stopping of the remaining mesogen mobility (translational and residual orientational) coupled with a better ordering and lower volume. The static ^{13}C NMR spectra of all smectic phases showed axially-symmetric chemical-shift-anisotropy about the molecular director that is oriented parallel to the magnetic field. Reorientation of the magnetic field finds fast reorientation of the molecules (time scale minutes or faster). As soon as the condis phase is reached, the static spectrum is broadened and only magic-angle-spinning can produce sharp resonance lines. Points 3, 5, 6, and 8 in Fig. 8 permit the assignments given.

The superficial similarity between the hexagonal condis crystals of the flexible macromolecules, listed above, and liquid crystals leads frequently to wrong assignments (*37*). All of the listed flexible polymers are classified as condis crystals based on points 3, 4, 6, 7, and 8 of Fig. 8 (*16,17*). It is of particular importance to note that a hexagonal, columnar structure may exist not only for LCs, but also for condis crystals. The X-ray structure of a mesophases is often insufficient to identify the type of mesophase. Most of these condis crystals were earlier called smectic liquid crystals on the basis of X-ray

Figure 7: Fracture surfaces of extended chain crystals. The molecular chain direction is along the striations. Scale marker = 1 μm.

Liquid Crystals	Condis Crystals
1. "liquid"	1. "solid"
2. birefringence	2. birefringence
3. small ΔS_i, often 2-5 J/(K mol)	3. large ΔS_i, n x [7-12 J/(K mol)]
4. 100% crystalline for small and large molecules	4. limited crystallinity for macromolecules
5. no positional order, some orientational order	5. positional order, orientational order
6. full conformational disorder (as in melt) for small molecules and side-chain LC macromolecules	6. partial or full conformational disorder, largely parallel chains
7. mesogen shape: rod, disc, or board	7. mesogen may not contribute to ΔS_i, or may be absent
8. little or no super-cooling on phase change	8. "normal" super-cooling on phase change

Figure 8: Differences between liquid crystals and condis crystals.

data alone (*17*). In the direction of higher order, the condis states that many LCs develop at low temperatures by achieving positional order for the mesogen (such as K_1 and K_3 of OOBPD, see Figs. 4-6) were earlier called crystals. This is perhaps a natural situation when a new intermediate phase is introduced between two well-established phases. The need for the class of condis crystals is particularly obvious when attempting to describe orientationally disordered and mobile crystal phases (*19*). For flexible molecules, the orientational disorder is usually achieved through a segmental twist (*38*), and the phase is best called a condis crystal as long as the director of the chains has long-range positional order. This makes the smectic E phases of molecules which can change between different rotational isomers condis crystals. The classification of the smectic E phase of rigid molecules as an orientationally disordered crystal has already been suggested earlier (*19*). A rigid molecule that can have orientational disorder and mobility about less than all three axes of rotation should be classified as a subgroup of the plastic crystals (*5,18*). A typical example of such plastic crystal with rotation about only one axis of rotation is one of the mesophases of C_{70} (*39*). If one accepts smectic E as a separate type of phase, it is located between plastic crystals and condis crystals, overlapping with both. In the cases of orientational disorder without large-amplitude motion, the corresponding glasses would describe the phases.

More difficulties arise when groups that may be mesogenic are included in the flexible chain, permitting a possible nanophase separation and partial axial ordering of the mesogens in the more rigid phase. In these cases structures arise that are at the borders between thermotropic LCs, amphiphilic LCs, and condis crystals. By checking several or all of the appropriate characteristics of Fig. 8, a conclusive description may be possible, but borderline cases may also exist, that may even change from one class to another with temperature. Of particular value are often the easily determined criteria four and eight. Figure 9 illustrates the increasing enthalpy of isotropization of the high-temperature mesophases of three homologous series of liquid-crystal-forming molecules (*40*). Quite clearly, the major driving force for mesophase formation of the chains with longer CH_2-sequences comes from the conformational ordering in the flexible spacers and not the mesogen. An identification of these mesophases as LCs needs more information on the order and mobility in the two nanophase areas.

More detailed studies of the mobility and order were carried out for the MBPEs with the mesogen 1-(4-hydroxyphenyl)-2-(2-methyl-4-hydroxyphenyl)ethene, using thermal analysis (41) and ^{13}C NMR (42). The LC phase in MBPE-9 (nine CH_2-groups in the flexible spacer) is monotropic, *i.e.* it appears only on cooling, since it requires less supercooling to grow than the stable condis phase (point eight of Fig. 8). At lower temperatures, an additional transition yields a stable condis crystal, as in OOBPD. The condis crystal is easily recognized by its partial crystallinity (point four of Fig. 8). The formation of the LC is, thus, based largely on the conformational order in the nanophase-separated flexible spacers. This nanophase separation is either driven by the thermotropic conformational ordering of the CH_2-sequences, or it may already exist to some degree in the isotropic phase (amphiphilic nanophase separation, still in need of verification). Liquid crystalline behavior requires rather facile sliding diffusion of the CH_2-sequences to satisfy points one and five of Fig.8. The mesogens of MBPEs with odd numbers of CH_2-groups show little orientational ordering (very small entropy of transition, Fig. 9) and their remaining mobility in the mesophase is documented by NMR (*42*). The main effect of the mesogen seems to be the hindering of the flexible chain on crystallization, but, almost all orientation in the LC phase is based on the entropy decrease of the flexible spacer. As soon as the mobility within the flexible spacer is sufficiently low, a positionally ordered condis crystal with further phase separation arises (partial condis crystallinity, approaching microphase dimensions).

The thermal properties of the stable condis phase of MBPE-9 are shown in Fig. 10 (*41*). Using solid state NMR, it was possible to find that in the condis phase of MBPE-9 four flexible bonds are fixed in the *trans*-conformation (the O-C- bonds and the C-C-bonds that are two bonds removed from O-C) and all others keep their liquid-like disorder and mobility (*42*). Figure 10 shows besides the partial crystallinity a fraction of the mobile phase as being rigid amorphous. The considerable disorder in the condis state freezes on cooling over a broad temperature range as in the OOBPD example of Fig. 6.

An example of a molecule of similar amphiphilic structure as MBPE that yields only condis crystals is poly(oxy-1,4-phenylene-oxy-xylylene-oxy-1-4-phenylene-oxy-undecamethylene). It has two polymorphs with different numbers of flexible bonds fixed in the *trans*-conformation (2/3 and 1/3) (*43*). The condis phases were in this case identified by the large ΔS_i, partial crystallinity, absence of a true mesogen, and supercooling before crystallization (Points 3, 4, 7, 8 in Fig. 9).

A number of macromolecular mesophases have been described that are kept in a *metastable state*, either by being quenched into the glassy state before more complete crystallization is possible, or by being strained by crystals in a drawn fiber. A typical example of the first type is found in quenched polypropylene. The CD glass (see Fig. 1) can be identified by its X-ray diffraction, as shown in the bottom curve of Fig. 12 (*16*). Again, this structure was initially called a smectic LC (*44*). At about 380 K, somewhat above the CD glass transition, the stable monoclinic phase forms with an exotherm of about 600 J/mol (ΔH_{fusion} of the monoclinic phase = 8.7 kJ/mol at 460.7 K) (*45,46*). Figure 11 indicates the gradual development of the monoclinic crystals out of the condis phase. The main conformational disorder in the CD phase is in form of helix reversals (*47*), similar to the condis phase of polytetrafluoroethylene.

An X-ray diffraction pattern of drawn fibers of poly(ethylene terephthalate) (PET) is given in Fig. 12 in the upper left pattern (*48*). The known triclinic crystal structure of PET cannot account for this pattern, even after correcting for all possible crystal defects. Only by postulating a mesophase, as shown in the lower right pattern of Fig. 12, can one quantitatively reproduce the diffraction pattern. The 36% of mesophase in the PET fiber determine the mechanical properties through its degree of orientation (*49*). Because of the small size of the terephthaloyl and the oxyethyleneoxy groups, it is most likely that the mesophase is an oriented condis phase, rather than a liquid crystal. The sizes of the three phases range from nanophases to microphases. A similar mesophase has also been suggested for gel-spun polyethylene, where a mobile preferentially trans component has been found in addition to the much less mobile crystal (*50*).

Conclusions

Research of the last ten years has strengthened the case for the existence of condis crystals. The condis crystals have long-range orientational and positional order of the chain director, although chain diffusion in the axial direction (sliding diffusion) is more

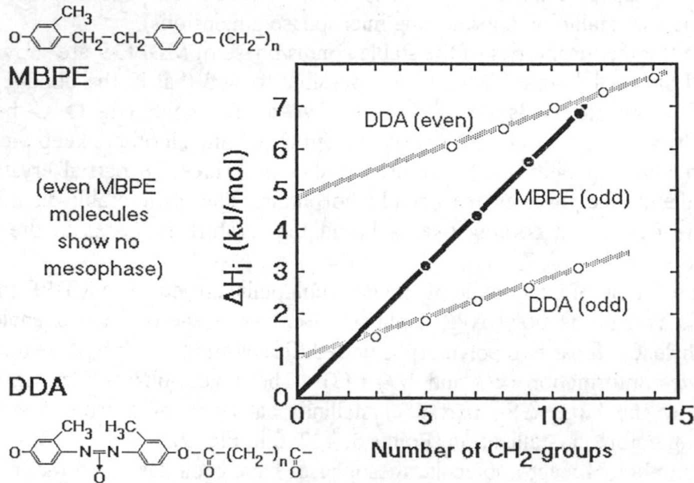

Figure 9: Enthalpies of isotropization for three sets of main-chain macromolecular liquid crystals. The ordinate at zero number of CH_2-groups indicates the mesogen contribution.

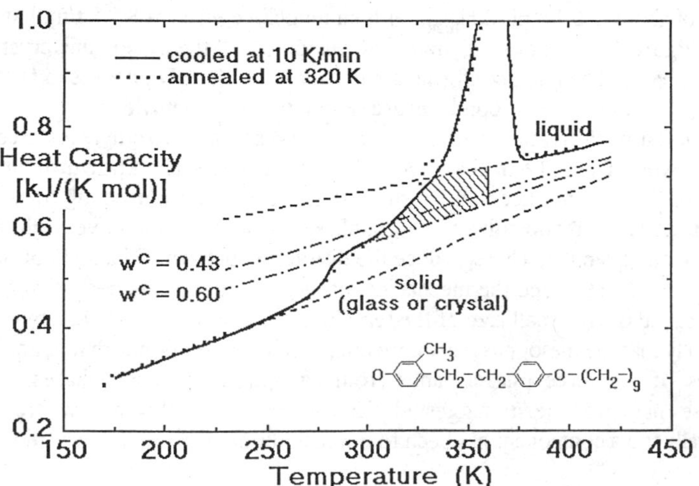

Figure 10: Thermal analysis of MBPE-9.

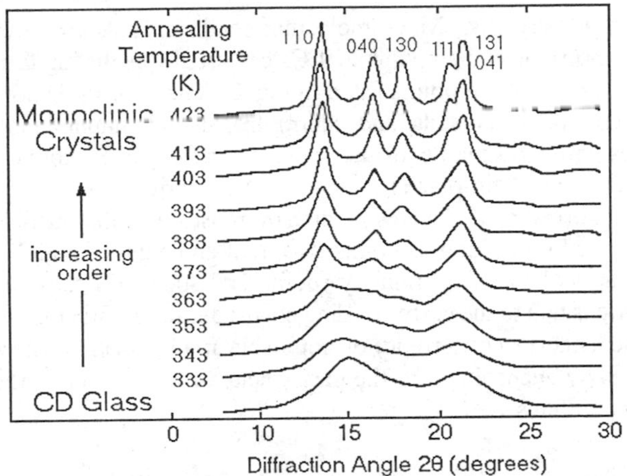

Figure 11: Polypropylene X-ray diffraction patterns after quenching from the melt to the CD glass (bottom curve) and subsequent annealing (as indicated in the ordinate). The peaks are labelled according to the monoclinic crystal structure.

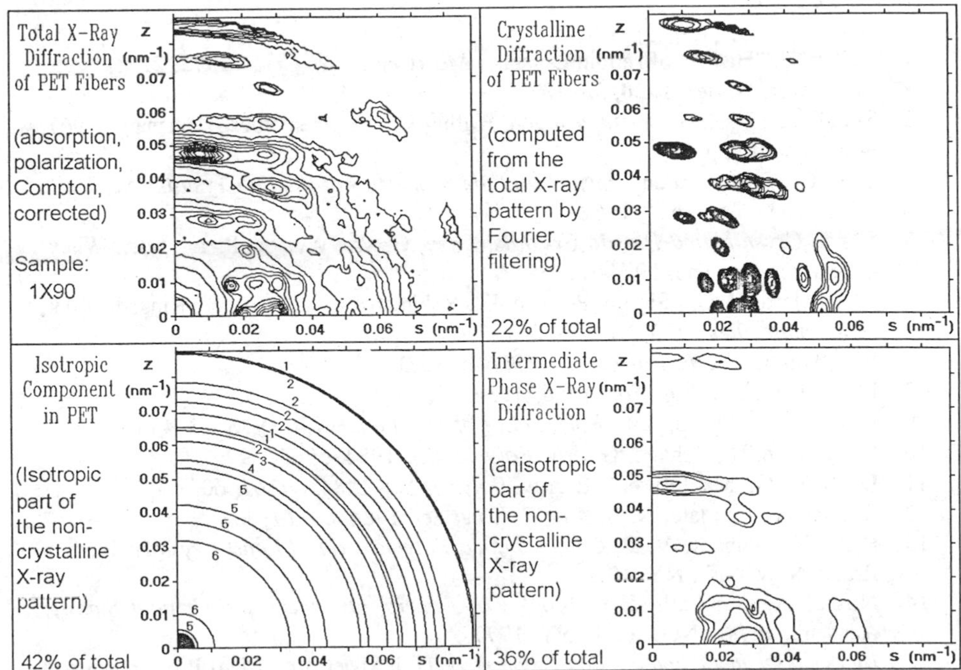

Figure 12: The X-ray diffraction pattern of drawn PET (upper left) and the three patterns it is composed of (three-phase model). The lower right pattern has the appearance of a mesophase.

facile than in "rigid" crystals. Macromolecular condis crystals are often only partially crystalline. Thermotropic and amphiphilic LCs can usually be distinguished from condis crystals by making use of the points listed in Fig. 8. There exists, however, a region of overlap between condis crystals and thermotropic and amphiphilic LCs in linear macromolecules with rigid segments interspersed in a flexible backbone. Thermotropic LC phases with main-chain-placed mesogens may derive most of their entropy of ordering from conformational changes in the nanophase-separated flexible spacer. Such phases may still be LCs as long as points two, four and eight of Fig. 8 are fulfilled and the mesogen shows liquid-like mobility (point 5). Such LCs show usually at lower temperatures a transition to the corresponding condis phase. In other cases the LC-phase may be only metastable (monotropic) or not stable at all. Unstable mesophases can be made metastable by quenching into the glassy state or by restraining the molecules, as in drawn fibers and films.

Acknowledgments

This work was supported by the Div. of Materials Res., NSF, Polymers Program, Grant # DMR 92-00520 and the Div. of Materials Sci., Office of Basic Energy Sciences, DOE, under Contract DE-AC05-84OR21400 with Lockheed Martin Energy Systems, Inc.

Literature Cited

1 Kelker, H. "History of Liquid Crystals" *Mol. Cryst. Liq. Cryst.* **1973**, *21*, 1.
2 Staudinger, H. *Ber.* **1920**, *53*, 1073.
3 Staudinger, H. *Arbeitserinnerungen*; Hüthig Verlag: Heidelberg, Germany, 1961; Nobel Lecture, p. 317.
4 Gray, G. W., *Molecular Structure and the Properties of Liquid Crystals;* Academic Press: New York, NY, 1962.
5 *Liquid Crystals and Plastic Crystals;* Gray, G. W.; Winsor, P. A., Eds.; Wiley: Chichester, England, 1974; Vol.1.
6 Voigt-Martin, I. G. ; Simon, P.; Yan, D.; Yakimansky, A.; Baur, S.; Ringsdorf, H.; *Macromolecules* **1995**, *28*, 236, 243.
7 Robinson, C. *Trans. Farad. Soc.* **1956**, *52*, 571.
8 Flory, P. J. *Proc. Roy. Soc.* **1956**, *A234*, 60, 73.
9 Roviello, A.; Siguru, A. *J. Poly. Sci., Polymer Letters Ed.* **1975**, *13*, 455.
10 Finkelmann, H.; Rehage, G. *Adv. Polymer Sci.* **1984**, *60-61*, 99.
11 Dobb, M. G;. MacIntyre, J. E. *Adv. Polymer Sci.* **1984**, *60-61*, 60.
12 Shibaev, V. P.; Platé, N. A. *Adv. Polymer Sci.* **1984**, *60-61*, 173
13 Platé, N.; Shibaev, V. P. *Comb-Shaped Polymers and Liquid Crystals*; Plenum Press: New York, NY, 1991.
14 Noshay A;. McGrath, J. E. *Block Copolymers, Overview and Critical Survey;* Academic Press: New York, NY, 1977.
15 *Block Copolymers: Science and Technology*; D. J. Meier, Ed., MMI Press-Harwood Acad. Publ.: New York, NY, 1983.
16 Wunderlich, B.; Grebowicz, J. *Adv. Polymer Sci.* **1984**, 60-61, 1.

17 Wunderlich, B.; Möller, M.; Grebowicz J.; Baur, H. *Conformational Motion and Disorder in Low and High Molecular Mass Crystals*; Adv. Polymer Sci.,Vol. 87; Springer Verlag: Berlin, Germany, 1988.

18 Smith, G. W. In *Advances in Liquid Crystals*; Brown, G H., Ed.; Academic Press: New York, NY, 1975; Vol. 1; p. 193.

19 Leadbetter, A. J. In *Thermotropic Liquid Crystals*; Gray, G. W., Ed.; J. Wiley, Chichester, England, 1987; p. 20.

20 Wunderlich, B. *Macromolecular Physics, I. Crystal Structure, Morphology, Defects;* Academic Press: New York, NY, 1973.

21 Wunderlich, B. *Macromolecular Physics, II. Nucleation, Crystallization, Annealing;* Academic Press: New York, NY, 1976.

22 Wunderlich, B. *Macromolecular Physics, III. Crystal Melting;* Academic Press: New York, NY, 1980.

23 Wunderlich, B. *Pure and Appl. Chem.* **1995**, *67*, 1919.

24 Becker, E. D. *High Resolution NMR;* second ed.; Academic Press: New York, NY, 1980.

25 Maier, W.; Saupe, A. Z. *Naturforsch.* **1958**, *13a*, 564.

26 Cheng, J.; Chen, W.; Jin, Y.; Wunderlich, B. *Mol. Cryst. Liq. Cryst.* **1994**, *241*, 299.

27 Annis, B. K.; Schwark, D. W.; Reffner, J. R.; Thomas, E. L.; Wunderlich, B. *Makromol. Chem.* **1992**, *193*, 2586.

28 Kelker, H.; Hatz. *Handbook of Liquid Crystals*; Verlag Chemie: Weinheim, Germany. 1980.

29 Lindau, J.; König, H-J.; Dörfler, H-D. *Colloid Polymer Sci.* **1983**, *261*, 236.

30 Cheng, J.; Jin, Y.; Liang, G.; Wunderlich, B.; Wiedemann, H. *Mol. Cryst. Liq. Cryst.* **1992**, *213*, 237.

31 Wunderlich, B.; Melillo, L. *Makromolekulare Chemie* **1968**, *118*, 250.

32 Hikosaka, M. *Polymer* **1987**, *28*, 1257.

33 Hikosaka, M. *Polymer* **1990**, *31*, 4584.

34 Wunderlich, B; Czornyj, G. *Macromolecules* **1977**, *10*, 906.

35 Noël, C. In *Polymeric Liquid Crystals;* Blumstein, A. Ed.; Plenum Press: New York, NY, 1983.

36 Wiedemann, H.-G.; Grebowicz, J.; Wunderlich, B. *Mol. Cryst. Liq. Cryst.* **1986**, *140*, 219.

37 Ungar, G. *Polymer* **1993**, *34*, 2050.

38 Sumpter, B. G.; Noid, D. W.; Liang, G. L., Wunderlich, B. *Adv. Polymer Sci.* **1994**, *116*, 27.

39 Jin, Y.; Xenopoulos, A.; Chen, W.; Wunderlich, B.; Diak, M.; Jin, C.; Hettich, R. L.; Compton, R. N.; Guiochon, G. *Mol. Cryst. Liq. Cryst.* **1994**, *257*, 235.

40 Yandrasits, M. A.; Cheng, S. Z. D.; Zhang, A.; Cheng, J.; Wunderlich, B.; Percec, V. *Macromolecules* **1992**, *25*, 2112.

41 Jin, Y.; Cheng, J.; Wunderlich, B.; Cheng, S. Z. D.; Yandrasits, M. A.; Zhang, A. *Polymers for Advanced Technology* **1994**, *5*, 785.

42 Cheng, J.; Jin, Y.; Wunderlich, B.; Cheng, S. Z. D.; Yandrasits, M. A.; Zhang, A.; Percec, V. *Macromolecules* 1992, *25*, 5991.

43 J. Cheng, Y. Jin, W. Chen, B. Wunderlich, H. Jonsson, A. Hult, and U. W. Gedde *J. Polymer Sci., Part B: Polymer Physics*, **32**, 721 (1994).

44 Natta, G.; Peraldo, M.; Corradini, P. *Accad. Naz. Lincei* **1959**, *24*, 14.

45 Grebowicz, J.; Lau, S.-f.; Wunderlich, B. *J. Polymer Sci., Symposia* **1984**, *71*, 19.
46 Bu, H. S.; Cheng, S. Z. D.; Wunderlich, B. *Makromolekulare Chemie, Rapid Commun.* **1988**, *9*, 75.
47 Bunn, A.; Cudby, M. E. A.; Harris, R. K;. Parker, K. J.; Say, B. J. *Polymer* **1982**, *23*, 694.
48 Fu, Y.; Busing, W. R.; Jin, Y.; Affholter, K. A.; Wunderlich, B. *Makromolekulare Chemie* **1994**, *195*, 803.
49 Fu, Y.; Annis, B.; Boller, A.; Jin, Y.; Wunderlich, B. *J. Polymer Sci., Part B: Polymer Physics* **1994**, *32*, 2289.
50 Fu, Y.; Chen, W.; Pyda, M.; Londono, B.; Annis, B.; Boller, A.; Habenschuss, A.; Wunderlich, B. *J. Macromol. Sci., Part B:* to be published.

Chapter 16

Orientation and Molecular Motion in an Aromatic Copolyester Studied by Proton NMR Spectroscopy

P. G. Klein, B. W. Evans, and I. M. Ward

Interdisciplinary Research Centre in Polymer Science and Technology, University of Leeds, Leeds LS2 9JT, United Kingdom

Proton NMR spectroscopy has been used to characterise the degree of chain orientation, and the variation in molecular motion with temperature, in a thermotropic liquid crystalline copolyester derived from 4-hydroxybenzoic acid (HBA) and 2-hydroxy-6-naphthoic acid (HNA). Analysis of the NMR lineshape of samples arranged parallel and perpendicular to the magnetic field indicates a high degree of chain alignment along the macroscopic sample axis. The variation of the $T_{1}\rho$ with temperature has been studied for samples containing 73 and 30% HBA. The $T_{1}\rho$ values display minima, coincident with the temperatures of the γ and β relaxations previously observed by dielectric and dynamic mechanical analysis. The $T_{1}\rho$ - temperature data have been fitted to the Cole-Davidson distribution function, which indicates a broad distribution of correlation times. The activation energies obtained from the fitting are higher than from dielectric data, but, in a qualitative sense support the contention that the γ relaxation is associated with the motion of HBA units, and the β relaxation with the motion of HNA units.

The random copolyester formed from 4-hydroxybenzoic acid (HBA) and 2-hydroxy-6-naphthoic acid (HNA), (produced by Hoechst-Celanese under the name Vectra), has been studied in some detail over the last decade, particularly with regard to its crystallinity *(1,2)*, crystal structure *(3,4)* and mechanical properties *(5)*. The material was originally developed as an alternative to polymers such as Kevlar, the idea being that the bulky HNA units reduce the melting temperature so that melt processing can be possible without chemical degradation. The copolymer exists as a nematic liquid crystalline phase above about 300 °C, so that significant molecular orientation tends to be present in products which are processed in elongational flow fields, optimising the mechanical properties.

Despite the random chain substitution of the phenyl and naphthyl moieties in

0097–6156/96/0632–0249$15.00/0

the copolymer, it has been established by Windle and co-workers *(4,6)* that crystallinity can arise as a result of sequence matching between units on adjacent chains, producing structures termed non-periodic layer (NPL) crystals. In a previous publication *(7)*, we have reported on the effect of sample processing and annealing on the development of NPL crystallinity, detected using principally proton NMR. The levels of crystallinity are relatively low; usually in the region of 10-20%, a maximum of about 25% having been recently reported *(3)*.

The viscoelastic behaviour of the majority amorphous phase is important in terms of the mechanical properties of the material. Dynamic mechanical *(5)* and dielectric studies *(8)* reveal three relaxations, labelled α, β and γ in order of decreasing temperature. Variations in the strength of these relaxations with systematic changes in the mole fraction of each component have led to the association of the γ relaxation with the HBA component and the β relaxation with the HNA component, with the α relaxation displaying features typical of a glass transition process. Support for these assignments has been obtained from analysis of the proton NMR second moments *(9,10)*.

The present work extends the NMR investigations by acquiring information in two areas. First we examine the influence of processing on chain orientation by studying the proton NMR lineshape. Secondly, the variation in the proton $T_{1\rho}$ with temperature, for different compositions of the copolymer is reported. In particular, we relate the temperatures of the $T_{1\rho}$ minima to the locations of the viscoelastic relaxations, and also attempt to quantify the distribution of molecular correlation times and the activation energies for the relaxations.

Experimental

Sample Details. All materials were supplied by Hoechst Celanese. Two compositions were examined, containing 73 and 30 mole % HBA. The samples were melt processed at 300 °C. Samples designated 73R and 30R were in the form of cylindrical rods of about 5 mm diameter, injection moulded into a mould maintained at 100 °C. The sample designated 73M is an extruded monofilament of about 1 mm diameter, subjected to an annealing treatment following processing, of 4 h at 250 °C, followed by a further 15 h at 280 °C.

NMR Measurements. Proton NMR spectra were recorded on a Chemagnetics CMX 200 system, at 200 MHz. For the $T_{1\rho}$ measurements, the excitation pulse width was 2 μs, the spin-locking field strength 62.5 kHz, and the spectral width 1 MHz. Each point is typically the result of 1000 accumulations, with a recycle delay of 30 s.

Results and Discussion

Molecular Orientation. In a previous publication *(7)*, we have established that the processing of the copolymer is important in terms of the development of NPL - type crystallinity. Based upon a combination of DSC and NMR proton $T_{1\rho}$ relaxation, the crystallinities were calculated as approximately 12% for the 73M and 6% for the 30R. The 73R sample is essentially amorphous. The proton NMR lineshape can also provide a good indication of the orientation produced by the processing in the liquid crystalline phase. Figure 1 shows the broad-line proton spectra from the 73M and 73R samples at room temperature, with the fibre or rod axis parallel or perpendicular to the magnetic field. Comparing the two samples in the perpendicular orientation, the 73M spectrum is slightly broader, and also shows a well defined doublet splitting, and we consider that both of these features are attributable to the presence of the small amount of rigid, crystalline fraction in this sample. The doublet splitting, or Pake pattern, is due to the dipolar interaction between the nearest neighbour protons on the phenyl ring of the HBA component. The expression for the splitting, Δf, in kHz is given *(11)* by Equation 1:

$$\Delta f = \left(\frac{\mu_0}{4\pi}\right)\frac{3\gamma^2\hbar}{4\pi r^3}\left(3\cos^2\theta - 1\right) \tag{1}$$

where γ is the gyromagnetic ratio, r the distance between the protons, and θ the angle between the inter-proton vector and the magnetic field. Molecular modelling on BIOSYM gives a value of 0.243 nm for r, and so therefore Δf can be calculated as $12.5(3\cos^2\theta-1)$ kHz. Inspection of Figure 1 shows that the doublet splittings for the 73M in the parallel and perpendicular orientations are approximately 25 and 13 kHz, respectively, consistent with the polymer chain axis being highly aligned with the fibre axis. The spectra for 73M are similar to those obtained by Allen and Ward (*10*), on oriented, annealed tapes of the same composition, which would be expected to contain a similar level of crystallinity. For the 73R sample, the splitting cannot be resolved at room temperature, but Figure 2 shows this sample at 120 °C, where the doublet can again be seen. The magnitude of the splitting, and its change with sample orientation is again consistent with a high degree of chain alignment. For the 30R sample, the spectrum is dominated by the protons on the HNA ring, and the various interactions do not produce a simple doublet pattern which can be analysed in the same way. However, the processing conditions for the rods were identical, and hence it is reasonable to assume a similar degree of chain orientation in the 30R sample.

The Proton $T_{1\rho}$. Variation with Temperature, and Comparison with Dielectric and Dynamic Mechanical Studies. The proton spin-lattice relaxation time in the rotating frame ($T_{1\rho}$) relaxation of these samples has been examined over a wide range of temperature, with the primary aim of relating the variation in relaxation time to the viscoelastic transitions which have been observed by dielectric and dynamic mechanical analysis. The technique involves spin-locking the magnetisation in the x-y plane following the initial 90° excitation pulse. The strength of the spin-locking fields used means that the $T_{1\rho}$ is sensitive to molecular motions in the kHz range, that is, similar to the frequencies accessible by, and therefore comparable to, dielectric relaxation studies.

For a homogeneous sample with a single $T_{1\rho}$ the variation of the integrated spectrum area with spinlock time t will be a single exponential, Equation 2.

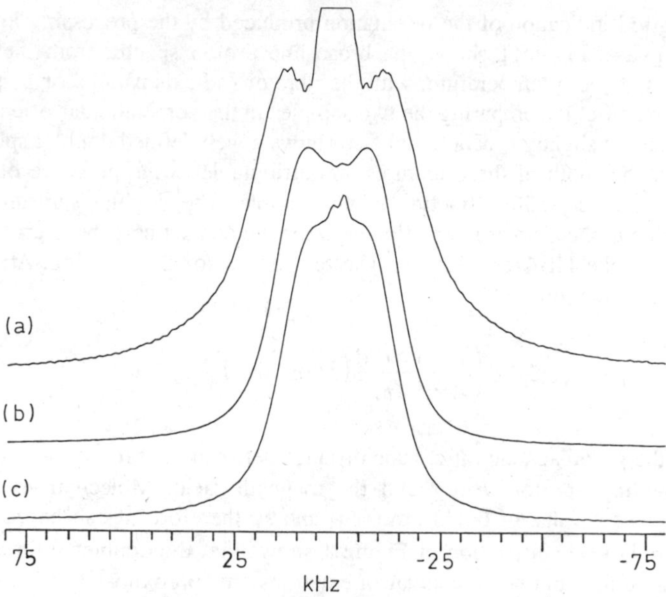

Figure 1. Broad-line proton NMR spectra at room temperature for 73M ((a) and (b)) and 73R (c). In (a), the fibre axis is parallel to the magnetic field, in (b) and (c) it is perpendicular. The signal in the centre of the spectrum in (a) is from the sample holder.

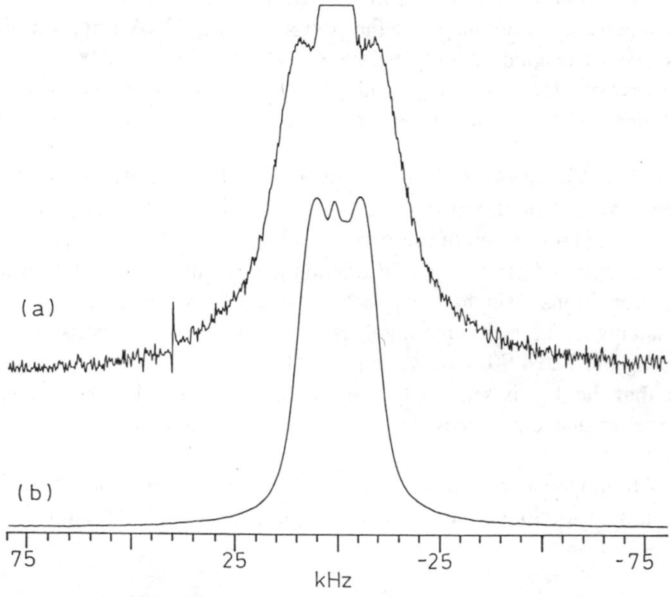

Figure 2. Broad-line proton NMR spectra at 120 °C for 73R. Sample axis parallel (a) and perpendicular (b) to the magnetic field.

$$I = I_0 \exp(-t/T_{1\rho}) \qquad (2)$$

In general, however, the decays for these samples are well described by a sum of two exponentials, as can be seen in Figure 3, with the slower relaxing, longer $T_{1\rho}$ component relating to the crystalline phase, as has been described in a separate publication *(7)*. Although two separate, and quite different, $T_{1\rho}$ values can be obtained from the bi-exponential fitting, the contribution of spin diffusion should be considered. In general, because these systems are heterogeneous, containing a small crystalline fraction, spin diffusion between protons in different regions will tend to partially average the intrinsic $T_{1\rho}$ values. In the present work, we are concentrating on the amorphous phase, represented by the faster relaxing early time, shorter $T_{1\rho}$ component, and its variation with temperature. Since the crystalline fraction is relatively small (12% maximum) *(7)*, and constant, the effect of spin diffusion on the measured $T_{1\rho}$ of the amorphous phase, over the temperature range studied, will not be very significant, but it should be noted that the values may differ from the intrinsic $T_{1\rho}$ values of the amorphous phase. The results are shown in Figure 4. All three samples display a minimum in the $T_{1\rho}$, which is indicative of the onset of a relaxation process in the material. In order to relate this feature to the transitions observed by dielectric and dynamic mechanical analysis, the locations of the $T_{1\rho}$ minima have been added to the relaxation map *(5,8)* for the copolymer, as shown in Figure 5. Here, the temperatures of the α, β and γ relaxations, observable as maxima in tan δ, are plotted as a function of the frequency of the oscillating strain or electric field applied. At first sight, it is tempting to associate the location of the minima in the $T_{1\rho}$ plots directly with the three relaxation processes observed by dielectric studies, as these points lie almost exactly on the log frequency - reciprocal temperature graphs, extrapolated to the NMR frequency of 62.5 kHz. However, it has to be recognised that the situation is undoubtedly more complex than this. In the dielectric and dynamic mechanical analyses, the γ and β peaks are separately resolvable, whereas in the NMR we see only one broad minimum over the same temperature range, presumably due to the onset of convergence of the two different activation energy processes at the NMR frequency, and the broad correlation time distribution, which will be addressed in due course. For the 30R sample, though, the NMR relaxation will be dominated by the protons on the HNA ring, and the fact that the $T_{1\rho}$ minimum occurs at this particular temperature is consistent with the β process being associated with the HNA units, as has been concluded from dielectric *(8)* and dynamic mechanical *(5)* studies. Similarly, the 73R sample, being HBA rich, will be dominated by the relaxation of protons on the phenyl ring, and the temperature of this $T_{1\rho}$ minimum supports the idea that the γ relaxation is associated with the HBA units.

The result for the 73M is more intriguing. The location of the minimum suggests we are observing the β relaxation, but since the sample is HBA rich, this is probably a coincidence. A more likely explanation is that this is a γ relaxation shifted to higher temperature, perhaps due to the influence of the crystalline phase.

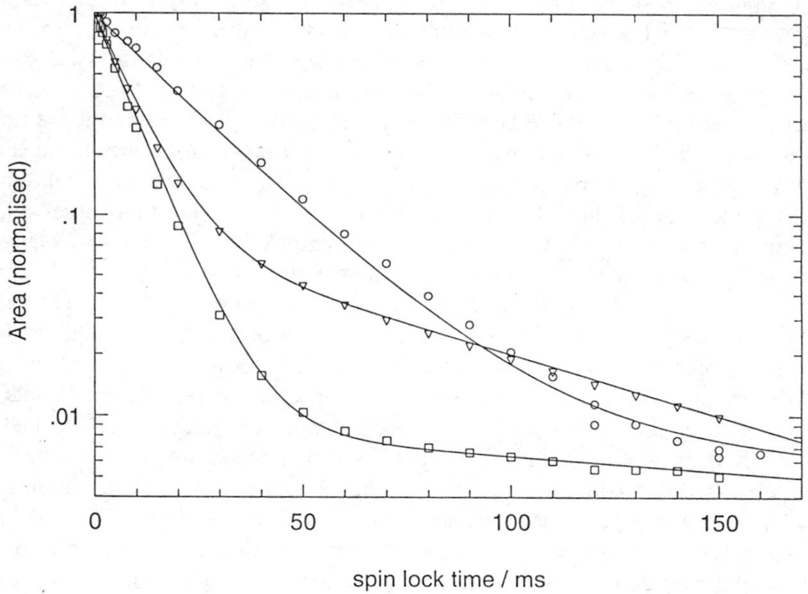

Figure 3. Area of spectrum versus spin-lock time for 73R (□), 73M (▽) and 30R (○). The lines through the data points are fits to a sum of two exponential decays.

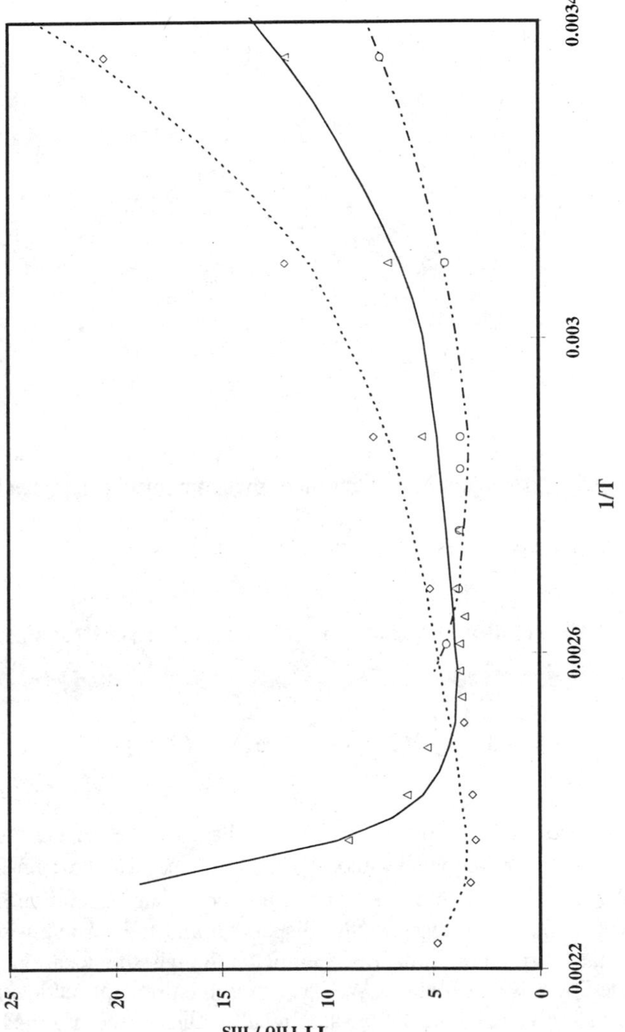

Figure 4. Proton $T_{1\rho}$ versus reciprocal temperature for 73R (O), 73M (Δ) and 30R (◊). The lines through the data are fits to the Cole-Davidson distribution function.

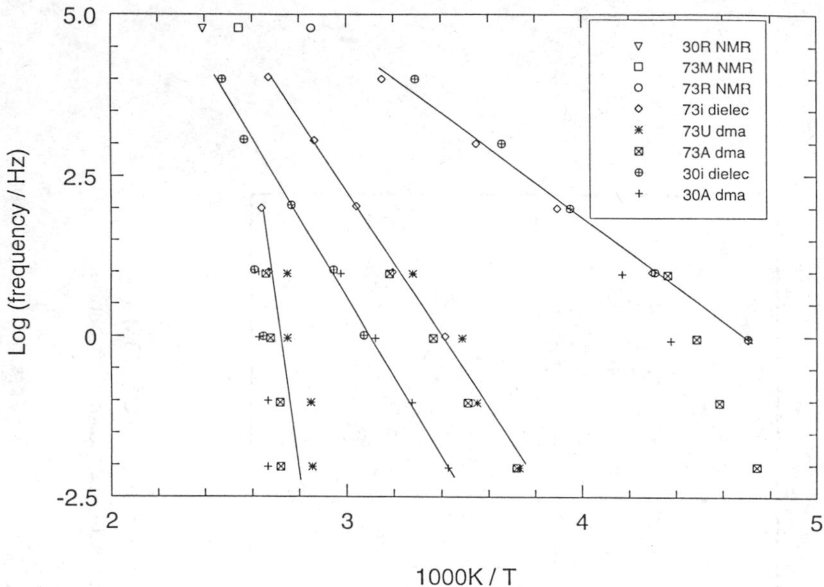

Figure 5. Log frequency versus reciprocal temperature, samples as indicated. Dielec and dma refer to dielectric studies and dynamic mechanical analysis, results taken from references 5 and 8. U, A and i refer to unannealed, annealed and isotropic samples, respectively. For further details, see text.

The Distribution of Correlation Times. The full expression *(12)* for the $T_{1\rho}$ is given by Equation 3.

$$\frac{1}{T_{1\rho}} = \frac{2}{3} M_2 \left\{ \frac{3}{2} J(2\omega_1) + \frac{5}{2} J(\omega_0) + J(2\omega_0) \right\} \tag{3}$$

Where M_2 is the second moment of the NMR lineshape, J the spectral density function, with ω_0 the Larmor frequency, and ω_1 the frequency of the spin-locking field. The spectral density can be written in terms of the molecular correlation time, τ, and the overall shape of the $T_{1\rho}$ - temperature dispersion and the relatively shallow minima are due to the correlation time distribution, although the location of the minimum is unaffected by this distribution. We have examined several models for the distribution, all of which give essentially the same results. One of the more simple is the Cole-Davidson function *(13)*, which has also been applied to the analysis of dielectric relaxations. The relevant expression for the spectral density in this case is given by Equation 4.

$$J(\omega) = \frac{2}{\omega} \left\{ \frac{\sin[\delta \tan^{-1}(\omega\tau)]}{(1+\omega^2\tau^2)^{\delta/2}} \right\} \tag{4}$$

where the parameter δ varies from 0 to 1 and indicates the width of the distribution, a value of 1 relating to a single correlation time. In the present work, the experimental variable is temperature, and since the data in Figure 5 show that the γ and β processes obey an Arrhenius relationship, we can relate the correlation time τ to the temperature via Equation 5.

$$\tau = \tau_0 \exp\left({E_a}/{RT} \right) \tag{5}$$

where E_a is the activation energy and R the gas constant. For each sample, then, the $T_{1\rho}$ - temperature data can be fitted using δ, τ_0 and E_a as free parameters, the second moment M_2 being measured directly from the spectrum at each temperature. In Figure 4, the curves through the data are fits to the Cole-Davidson distribution, and Table I gives the values of δ, τ_0 and E_a obtained from the fit for each sample.

The relatively small value of δ in each case indicates a fairly broad distribution of correlation times. The same value of 0.17 is obtained for 73R and 73M, which suggests that the small amount of crystallinity induced by the annealing of the 73M has no effect on the distribution. The value for the 30R is somewhat lower, at 0.12, which implies that increasing the number of naphthalene units in the chain broadens the correlation time distribution.

The derived NMR activation energies, E_a, can be compared to those obtained from the dielectric data. From the gradients in Figure 5, the γ and β processes observed by dielectric relaxation have activation energies of 50.2 and 105 kJ mol^{-1}, respectively.

Table I. Width Parameters, δ, Activation Energies, E_a and Pre-Exponential Factors, τ_0 Obtained From the Cole-Davidson Distribution Function

Sample	Width Parameter, δ	Activation Energy E_a / kJ mol^{-1}	Pre-exponential factor τ_0 / s
73R	0.17	93.9	4.04×10^{-19}
73M	0.17	115	3.70×10^{-21}
30R	0.12	178	2.41×10^{-27}

From Table I, it may be seen that the NMR E_a values are approximately the same for the 73% HBA samples, at about 100 kJ mol^{-1}, and 178 kJ mol^{-1} for the 30R. So, the absolute values of the activation energies obtained by the two methods do not agree, which, perhaps, is not entirely surprising since the processes involved are not the same. However, the ratio of the high to the low activation energy obtained is about 2:1 by both methods, which lends some additional support to the idea that the NMR $T_{1\rho}$ relaxation is sensitive to the γ process in the case of the HBA rich samples, and to the β process in the case of the HNA rich samples. The fact that the activation energies for both the 73% HBA samples are similar is further evidence for the $T_{1\rho}$ minimum in 73M being a shifted γ relaxation, and not a β relaxation.

Conclusions

The proton NMR spectrum for a sample rich in HBA is dominated by the dipolar interaction between the protons on the phenyl ring, and hence shows a well resolved doublet splitting, making the technique a good indicator of the polymer chain orientation. It is clear that processing the polymer in the nematic melt phase produces materials with a high degree of orientation, which is important with regard to the mechanical properties.

The $T_{1\rho}$ values for the amorphous phase display minima, which indicates that the NMR is sensitive to the viscoelastic relaxations in the material. The analysis is complicated by the fact that the NMR sees the contribution from all the protons in the sample, but it appears that samples containing 30 or 73% HBA are sufficiently rich in the HNA or HBA component for the $T_{1\rho}$ minimum to be located at the temperatures of the β and γ relaxations, respectively. This, combined with the activation energies obtained from fitting the data to the Cole-Davidson function, provides support for the association of the mechanical and dielectric data with specific molecular groups in the copolymer. An obvious extension of this work, which we are currently pursuing, is to use high-resolution carbon-13 NMR, which is capable of distinguishing directly between HBA and HNA groups, and may thus resolve some of the difficulties mentioned. In addition, it may help to quantify the extent to which the carbonyl group motion is co-operative with the motion of the aromatic groups, which is still an unresolved issue in these, and similar systems.

Literature Cited

1. Donald, A. M., Windle, A. H. *J. Mater. Sci. Lett.* **1985,** *4,* 58.
2. Cheng, S. Z. D. *Macromolecules* **1988**, *21*, 2475.
3. Wilson, D. J., Vonk, C. G., Windle, A. H. *Polymer* **1993**, *34*, 227.
4. Windle, A. H., Viney, C., Golombok, R., Donald, A. M., Mitchell, G. *Faraday Discuss. Chem. Soc.* **1985**, *79*, 55.
5. Troughton, M. J., Davies, G. R., Ward, I. M. *Polymer* **1989**, *30*, 58.
6. Hanna, S., Hurrell, B. L., Windle, A. H. *NATO ASI Series C, Mathematical and Physical Sciences* **1993**, *405*, Ch.67, 559.
7. Klein, P. G., Evans, B. W., Ward, I. M. *ACS PMSE Preprints* **1995**, *72*, 485.
8. Alhaj-Mohammed, M. H., Jamad, S. A., Davies, G. R., Ward, I. M. *Polymer* **1990**, *31*, 579.
9. Clements, J., Humphreys, J., Ward, I. M. *J. Pol. Sci., Pol. Phys. Ed.* **1986**, *24*, 2293.
10. Allen, R. A., Ward, I. M. *Polymer* **1991**, *32*, 202.
11. Harris, R. K. *Nuclear Magnetic Resonance Spectroscopy;* Longman: Avon, U.K. 1986 p. 145.
12. McBrierty, V. J., Packer, K. J. *NMR in Solid Polymers;* Cambridge University Press: Cambridge, U.K. 1993 p. 42.
13. Beckmann, P. A. *Phys. Reports* **1988**, *171*, 87.

Chapter 17

Temperature Evolution of the Structure of Liquid-Crystalline Main-Chain Copolyesters

E. M. Antipov[1], S. D. Artamonova[1], I. A. Volegova[2], Yu. K. Godovksy[2], M. Stamm[3], and E. W. Fischer[3]

[1]A. V. Topchiev Institute of Petrochemical Synthesis, Russian Academy of Sciences, Leninksy Prospekt 29, Moscow 117912, Russia
[2]Karpov Research Institute of Physical Chemistry, ul. Obukha 10, Moscow 103064, Russia
[3]Max-Planck-Institut für Polymerforschung, Postfach 3148, 55021 Mainz, Germany

Here we report the results of the X-ray scattering study in a wide temperature range of oriented samples of rigid-chain wholly aromatic random copolyesters of different chemical composition. It is common in these materials that segregated, layered structures already exist in the non-crystalline state of the original samples (as-spun fibers) and also coexist together with the crystalline phase developed during annealing, when the crystallization is completed. In other words, pre equilibrium structure of the material contains crystalline regions within either a periodic or aperiodic LC smectic matrix depending on, whether the monomer unit lengths are the same or different, respectively.

The increased interest in thermotropic wholly aromatic LC polyesters is related above all to their commercial applications. Their ability to exist as anisotropic melts, combined with high molding and extrusion characteristics, allows very easy processing of these materials into various pats and articles. The fundamental aspect of this interest is associated with the fact that, with these polymers, it is possible to examine the effect of chain stiffness and of the influence of regular chain structure on evolution of macromolecular order in polymers on the whole.

From our viewpoint, there are three most interesting aspect of those materials to study. First, the particular features of crystallization of these compounds, generally random copolymers, are of quite general interest. Two recent and essentially different models for crystallization of such irregular polymeric materials, that is, the model of nonperiodic layers *(1)* and the model of the paracrystalline lattice (2), are still under discussion.

0097–6156/96/0632–0259$21.25/0

Second, the ability of completely aromatic polyesters to form mesomorphic structures is interesting. Obviously, in such cases, the appearance of nematic LC states seems most probable. Furthermore, some polybenzoates *(3-8)* show mesophases, which cannot be described in terms of the conventional classification of liquid crystals, with respect to their arrangement, and which are similar to mesomorphic structures of flexible-chain polyorganosiloxanes *(9)* and polyphosphazenes *(10)*. Finally, studies of relaxations and phase transitions in rigid-chain LC polyesters, in particular, their molecular mobility in the solid state, i.e., below the melting temperature of crystalline phase, are of great interest.

The main aim of this paper is to increase the understanding of the mechanism of phase formation and relaxational transitions in two representative thermotropic LC copolyesters, namely, ternary copolyester (CPE-1) containing 10 mol % p-hydroxybenzoic acid (HBA), 45 mol % phenylhydroquinone (PHQ), and 45 mol% terephthalic acid (TPA), as well as copolyester (CPE-2) composed of monomer units of 4-hydroxybenzoic acid (HBA) and 6-hydroxy-2-naphthoic acid (HNA) (Schematics 1).

The choice of these copolyesters was not accidental. Concerning the first copolymer, one should remind that quite recently, a comprehensive analysis of the structure of CPE-1 fibers subjected to various thermal treatments at high temperatures was published *(11)*. Under certain annealing conditions, this copolyester was shown to experience crystallization, which was accompanied by the formation of two crystalline orthorhombic phases with slightly different unit cell parameters. The origin of such phase separation was not understood. It was shown in *(12, 13)*, that these two crystalline phases melt at 300 and 320°C. The melting is accompanied by a transition to the nematic LC state and then to the isotropic melt. In the temperature range between the melting temperature of the crystalline phases and the temperature of appearance of the nematic LC structure, the possible existence of an additional mesomorphic state was also reported *(11, 13)*.

With this material Johnson and Cheng *(14)* advanced a hypothesis concerning the coexistence of at least two more mesomorphic phases (in addition to the common nematic phase). The structure of one of these phases is assumed to be similar to pseudohexagonal, indicating that this phase belongs to mesophases of a new type *(9, 10)*. The above results show that the phase composition in CPE-1 is complicated, and the related problems require still further study.

In connection with this, one of the main objectives of this paper was to analyze in particular the noncrystalline states in the copolyesters under study at relatively low temperatures (up to 250°C). In this temperature region, either copolyester crystalline structures are not yet formed (in the as-spun samples) or, when such structures are formed (annealed samples), they are at temperatures much lower than that of their intensive disruption. The second copolymer under study is one of the popular copolymers of the class of thermotropic LC copolyesters, produced on a relatively large indu-

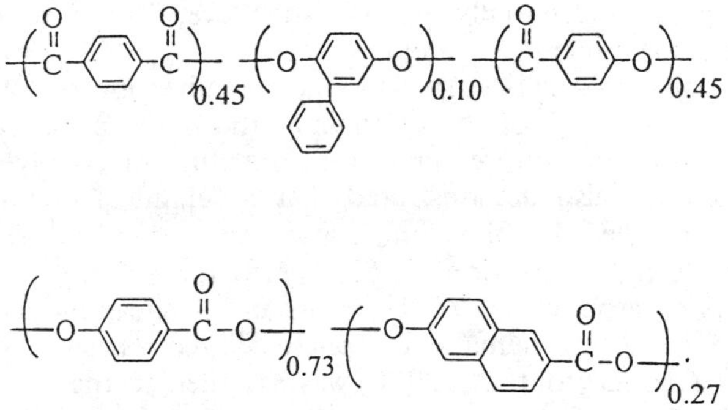

Schematics 1. Chemical composition of used copolyesters.

strial scale. The structure of CPE-2 was examined in many studies. In particular, it was found that this copolymer shows polymorphism, that is, depending on its thermal history, it may exist in different crystalline forms, viz., orthorhombic and hexagonal *(15, 16)*. Butzbach et al. *(17)* demonstrated that crystallites of the copolymer are rather imperfect. As a result, the enthalpy of melting is 10-20 times smaller than that expected for the ideal crystal and the change in volume that accompanies the melting transition is small. Nevertheless, according to *(17)*, the degree of crystallinity of CPE-2 may reach 60%, which is a rather high value for a statistical copolymer.

The morphology of CPE-2 was also extensively studied. It was found that the typical dimension of crystallites was 200 Å in cross-section and 400-800 Å in the longitudinal direction *(18)*. As a result of annealing, the longitudinal dimension of crystallites may reach 1500 Å or larger *(19)*. For annealed CPE-2 Hanna et al. *(19)* also observed a SAXS reflection that had never been reported for similar systems. The presence of this reflection suggests an essentially biphasic structure, for the copolymer with periodically arranged phase elements.

During study of high-temperature recrystallization of CPE-2 and of the rheological properties of the copolymer in its LC state, it was found that the optimum temperature range for annealing of copolyester is 280-290 °C *(20)*. It was also demonstrated that mechanical characteristics of CPE-2 are controlled by the temperature and duration of annealing. Moreover, in X-ray diffraction studies performed during a rheological experiment, Nicholson et al. *(21)* discovered a shear-induced structure formation of CPE-2 in a nematic LC phase. Three relaxational transitions were reported *(22, 23)*; one at 120 °C was assigned to the glass transition, whereas the other two low-temperature transitions were assigned to defreezing of the local mobilities of phenylene and naphthene monomer units.

Studying dynamic mechanical properties of various rigid-chain LC polyesters, including CPE-1 and CPE-2 *(24)*, we found that each structural unit of LC polyesters is characterized by its own characteristic temperature interval associated with the appearance and manifestation of its local mobility, which is virtually independent of the content and type of other structural units also present in the material. This result suggests that changes in molecular mobility in the thermotropic rigid-chain LC polyesters induced by an increase of the temperature are the consequence of the consecutive appearance of local mobility in polyester structural units. The temperature interval associated with the appearance of the mobility of the most rigid structural unit in LC polyester can be identified with the glass transition temperature of the polymer. Further examination of the validity of the above assumption is the second objective of this paper, which is related to the study of molecular mobility in the solid state by means of dynamic mechanical relaxation (DMR), differential scanning calorimetry (DSC), and X-ray diffraction analysis.

Although the structure, morphology, phase and relaxational transitions, as well as specific features of rheological and physicomechanical behavior of CPE-1 and 2 have been considered in many publications in the last decade, one still is far away from complete understanding of these phenomena. Analysis of the published data reveals much controversy in the identification of phase and relaxational transitions and in the interpretation of the evolution of structure in copolymers during heating and cooling.

We believe that the lack of a uniform conclusion is primarily related to methodological ambiguity of the publications cited. As a rule, the structure, morphology, and other characteristics of CPE were examined either at room temperature or after thermal treatments, rather than during heating at the elevated temperatures. It was assumed that, after cooling, the copolymer retains the structure evolved at elevated temperatures. However, when such procedures are used, polymorphic transformations and/or simple redistribution of phase components in the material may occur which may depend on cooling conditions. In addition, the extent of imperfection exhibited by a phase state at elevated temperature may be so different that a phase state that, at 20°C, could be described as a conventional three-dimensional crystal would show only, near the melting point, a two-dimensional order characteristic of a mesomorphic state. In this study we analyze the evolution of the structure of CPE using X-ray scattering directly during heating in a wide temperature range.

Finally, we would like to mention one more problem. According to Windle et. al. *(25)* the limited degree of crystallinity observed for these systems is thought to be attributed to a process of segregation. Random but similar sequences of the different monomer units on adjacent chains come into register to form a layered structure with long-range order periodicity perpendicular to the chain axes but sometimes only with aperiodic order parallel to the chains, giving rise to "non-periodic layer" crystals. On the other hand it is believed, that a semicrystalline polymer is a two phase system. The question on the structure of the non-crystalline phase in such a material is puzzling. To answer this question is the third objective of this paper.

Experimental

Materials. Completely aromatic random copolyester CPE-1 was synthesized by melt polycondensation, where TPA, PHQ, and HBA (45, 45, and 10 mol %, respectively) were used as initial components. To increase the molecular mass, the final stage of polycondensation was carried out in the solid state in vacuum at 260°C. Varying the duration of this final stage, one obtaines samples with different molecular weight. The molecular weight was estimated from the measurement of relative viscosity of copolyester solutions (0.5 g/dl) in a trifluoroacetic acid-chloroform (60:40%) mixed solvent. For the CPE-1 samples studied, the specific viscosity was 2.7, 3.4, 4.4, 4.7, and 6.4. We studied highly oriented CPE-1 fibers. The fibers were spun in the melt at 320°C, and then underwent thermal orientational drawing. The oriented fibers were rapidly cooled down to room temperature. We studied both as-spun and annealed samples. Thermal treatment of the as-spun fibers was performed in vacuum and involved several stages: (1) slow heating for 2 h from 20 to 290°C, (2) annealing at 290°C for 12 h, (3) annealing at 300°C for 9 h, and (4) annealing at 320°C for 6.5 h.

CPE-2 fibers were prepared by spinning of Vectra A900 (Hoechst Celanese Corp.) in the nematic LC state. According to the characteristics determined, the copolyester obtained had statistical structure and was composed of 73 mol % of HBA and 27 mol % of HNA. The samples were examined as oriented fibers. The diameter of monofilament was ~10 μm. Both freshly spun fibers and those subjected to prolonged annealing (280°C; 8-10 h) were examined.

Specimens for X-ray diffraction studies were prepared by stacking about 300 monofilaments aligned in parallel into a compact bundle. Temperature-dependent measurements were carried out under isometric conditions (i.e., the ends of a specimen were fixed). For DSC measurements, we used samples weighting 10-20mg and placed into standard pellets.

Methods. X-ray diffractioon studies were carried out with Siemens D 500T and DRON-3.0 diffractometers using copper radiation; the diffraction pattern was registered with a scintillation counter. Equatorial and meridional diffraction patterns were recorded in reflection and transmission modes, respectively. Slit collimation was used; the primary X-ray beam was monochromatized with a graphite crystal or a curved quartz crystal. In addition, a pulse discriminator sensitive to CuKα radiation (λ=1.54 Å) was used.
X-ray patterns in wide diffraction angles were registered with a 18-kW Rigaku X-ray generator. A rotating anode was used as a source of copper radiation. The primary beam, 0.5 mm in diameter, was monochromatized with two graphite crystals. A two-dimensional position-sensitive coordinate detector was used to record patterns. X-ray diffraction patterns were registered in transmission mode with a specimen placed in a special chamber under vacuum. In a typical case, exposure time did not exceed 10 h. Measurements were carried out at room temperature. Depending on the geometry, measurements in both small- and wide-angle X-ray scattering were carried out.

DSC measurements were performed with a DSC-7 (Perkin-Elmer) instrument in a temperature range extending from -50 to +380°C at a heating or cooling rate varied from 10 to 40 K/min. All experiments were carried out with CPE fibers in free state (nonisometric conditions).

Dynamic viscoelastic characteristics of CPE fibers (dynamic Young's modulus E′ and mechanical loss tangent tan δ) as a function of temperature were studied by examining low-frequency acoustic properties of the fibers with a resonance techniques that uses longitudinal vibrations of vertically suspended fiber. This method and its advantages, as well as formulas for calculation of E′ and tan δ using the resonance frequency, are described in detail in *(24)*.

Results and Discussion

Before presenting structural studies for both CPE-1 and CPE-2 copolymers, the DMR and DSC data for samples with different thermal history will be described.

Dynamic Mechanical Relaxation (DMR) and Differential Scanning Calorimetry (DSC) Measurements

CPE-1. Figure 1 presents the temperature dependences of modulus and loss tangent obtained during the first and second heating runs of the as-spun CPE-1 fibers with a specific viscosity of η_{sp}=4.4. During the first heating run, the loss modulus E′ of CPE-1 fibers first remains unchanged or even slightly increases (Figure 1, curve 1) while it strongly decreases when the glass transition temperature is reached (see below). Upon subsequent cooling to room temperature, CPE-1 fibers show an increased E′ at low temperatures. This value depends on the temperature up to which the samples were heated and on the cooling rate. This behavior is characteristic of both as-spun and annealed samples.

As shown in *(24)*, dynamic mechanical properties of thermotropic LC polyesters seem to be stabilized after their heating from room temperature up to the temperature corresponding to the high-temperature maximum of tan δ and subsequent slow cooling at a rate of 5 °C/min. For CPE-1 fibers preheated under such conditions, the temperature dependence of E′ (Figure 1, curve 1′) remains virtually unchanged under repeated heating and cooling runs. Therefore, we will consider the temperature dependences of E′ and tan δ of the fibers treated under such conditions.

Examination of the temperature dependences of E′ and tan δ of as-spun and annealed CPE-1 fibers (Figure 1, curves 1′ and 2′) reveals that solid CPE-1 shows relaxational transitions in temperature intervals from -60 to +20°C and from 110 to 230°C. In these temperature intervals, E′ tends to decrease. In the temperature dependence of tan δ, the above transitions are associated with maxima. At 250°C, as-spun CPE-1 fibers start to flow, which is accompanied by a sharp drop in E′ and increase in tan δ. Such behavior is characteristic of all the as-spun CPE-1 fibers, independent of their specific viscosity.

The resonance DMR method applied in this work does not allow one to estimate the activation energy of relaxational transitions. Naoki and Akira *(26)* studied thermotropic LC polyester, whose chemical structure was similar to that of CPE-1; however, the ratio between the components involved was somewhat different: 33.3 mol % NPA, 33.3 mol % HBA,and 33.3 mol % PHQ. The methods of DMR and dielectric relaxation (DR) were applied to investigate the two principal relaxational transitions that take place in the same temperature intervals as the relaxation transitions in

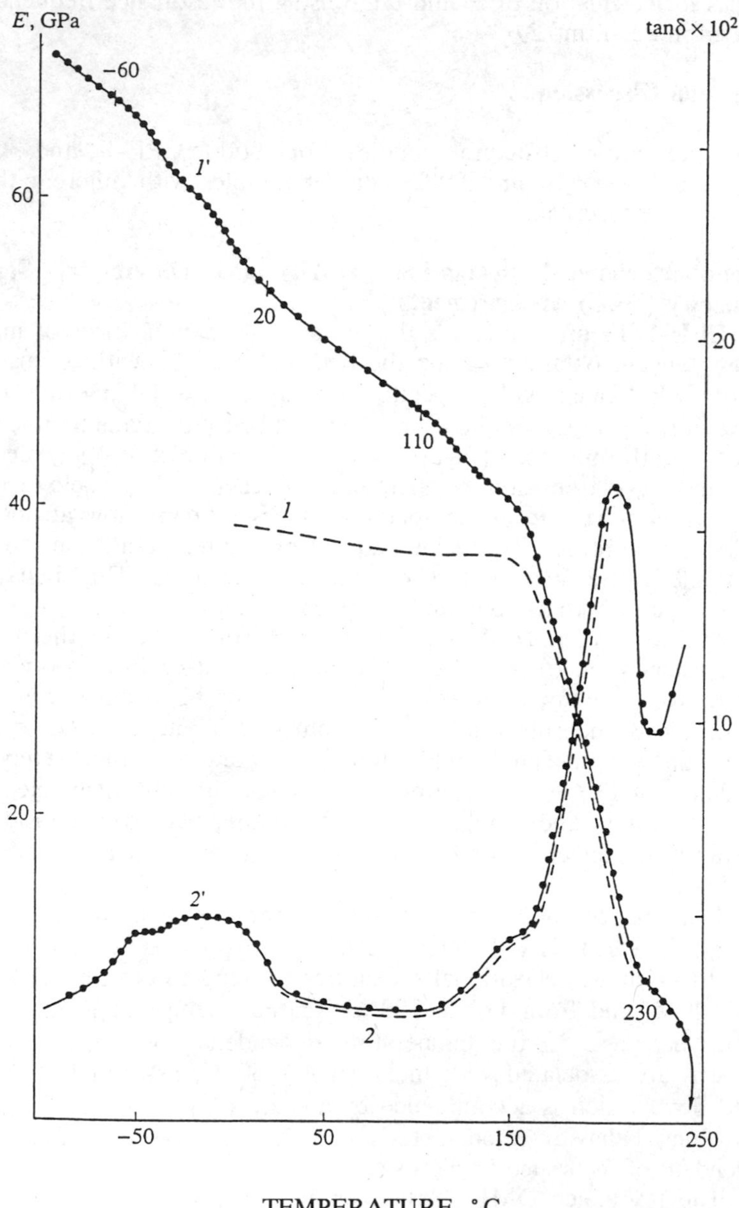

Figure 1. Temperature dependences of (1, 1′) Young modulus E′ and (2, 2′) loss tanδ of the as-spun CPE-1 fibers (η_{sp}=4.4) on (1, 2) the first and (1′, 2′) repeated heating runs.

CPE-1. The low-temperature transition with low activation energy (60 kJ/mol) is related to the local mobility of the aromatic ring and COO groups, and the high-temperature transition with higher activation energy (600 kJ/mol) is attributed to the glass transition.

Hence, the transition in the temperature range from 110 to 230°C, as revealed by the DMR method, can be identified as the glass transition. This conclusion is supported by DSC data, even though an adequate identification of this transition in the DSC heating curves of CPE-1 samples is difficult.

Curve 1 in Figure 2 presents a typical DSC curve corresponding to the first heating run of the as-spun CPE-1 fibers (η_{sp}=4.4). As in dynamic mechanical studies of CPE-1 fibers, in order to obtain reproducible DSC traces, one should anneal the test samples at temperatures above the glass-transition temperature range. Figure 2 (curve 2) shows the heating curve of the as-spun CPE-1 fibers (η_{sp}=4.4), which were preheated to 230°C at a heating rate of 10 °C/min. A comparison of curves 1 and 2 (Figure 2) allows one to conclude that, as a result of preheating of the CPE-1 samples, the corresponding DSC curves show a well-pronounced glass transition temperature. The temperature range of these glass transitions coincides with that of the as-spun CPE-1 fibers as revealed by the DMR method. The preheating of the as-spun fibers to 230°C has no marked effect on the development of crystalline structure in CPE-1.

According to *(11, 27)*, this crystallization process commences just above the glass transition, and, at 260-280°C, the ctystallization rate is maximal, as evidenced by the exothermal effect in the corresponding DSC heating curves of the as-spun CPE-1 fibers (Figure 2, curves 1 and 2). Two endothermic peaks at 300 and 320°C, the total heat effect of which does not exceed 2-3 J/g, were assigned *(11-13)* to the melting of two equilibrium crystalline structures that are formed above the glass transition temperature in the nonequilibrium CPE-1 fibers. The phase transitions in CPE-1 at temperatures above 250°C have been well studied *(28-32)*. Therefore, we focuse on the temperature interval below 250°C. In this temperature range, either no crystalline phase in CPE-1 is observed or only its nucleation commences. Hence, the flow temperature of the as-spun CPE-1 fibers, as determined by the DMR method (250°C), slightly exceeds the upper limit of the glass transition temperature interval.

Quite a different situation is observed for the annealed CPE-1 fibers. In this case, the corresponding DSC curve (Figure 2, curve 3) suggests that, as a result of slow heating of the test samples to temperatures exceeding the upper limit of the glass transition temperature interval, which involves prolonged annealing of the samples at each stage, crystallites with higher melting point and melting heat (7-8 J/g) are formed. According to the X-ray diffraction data, the crystallinity of the CPE-1 samples after prolonged annealing was as high as 35-40% *(29)*. Hence, the flow temperature of the annealed CPE-1 samples (320°C) is higher than that of the as-spun fibers (250°C). This conclusion is confirmed by the temperature dependences of E′ and tan δ of the annealed

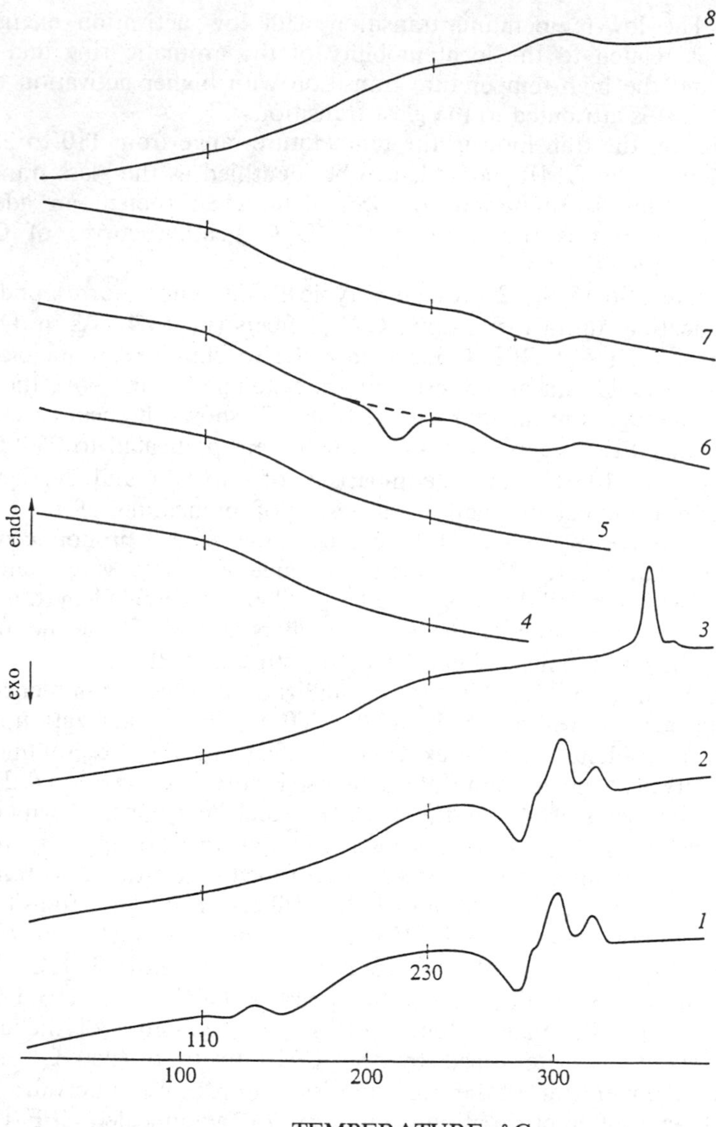

Figure 2. DSC curves of the as-spun CPE-1 fibers (η_{sp}=4.4) and annealed CPE-1 fibers recorded at a heating rate of 40 °C/min. DSC curves 1 and 2 correspond to the as-spun fibers on first heating and to the samples that were preliminarily heated to 230°C at a heating rate of 10 °C/min, respectively. Curve 3 corresponds to the annealed fibers that were preliminarily heated at a heating rate of 10 °C/min to 230°C. Curves 4-7 correspond to cooling of the (4, 6) as-spun and (5, 7) annealed fibers from (4) 280, (5) 320, and (6, 7) 380°C at a cooling rate of 40 °C/min. Curve 8 corresponds to the as-spun and annealed fibers that were preliminarily heated to 380°C.

CPE-1 fibers (η_{sp}=4.4) presented in Figure 3. Note that this is the only difference between the as-spun and annealed CPE-1 samples (Figures 1 and 3); in particular, the temperature intervals of the two relaxational transitions for both types of fibers coincide.

The experimental evidence obtained allowed us to conclude that, in the region studied, the molecular mobility of CPE-1 in the solid state is independent of molecular weight and the degree of crystallinity. This agrees well with the mechanism of development of molecular mobility of thermotropic LC polyesters in the solid state that was suggested in *(24)*. According to this mechanism, upon heating from low temperatures, changes in molecular mobility involve the consecutive appearance of local mobility in different structural units in CPE-1. In the case studied, this concerns TPA, HBA, and PHQ fragments. In the temperature range from -60 to 20°C, TPA fragments gain mobility, and, at 30-90°C, correspondingly HBA fragments *(24)*. However, in the case of CPE-1, the low content of HBA fragments does not allow identification of this transition in the temperature dependences of E′ and tan δ. The PHQ fragments are the last ones to achieve molecular mobility. This is only observed at temperatures varying from 110 to 230°C; this transition regime corresponds to the glass transition in CPE-1.

Figure 2 also shows the DSC curves on cooling of the as-spun (curve 4) and annealed CPE-1 samples (curve 5) at a cooling rate of 40 °C/min. The samples were coooled from a starting temperature of 280 and 320°C, respectively. These temperatures are substantially lower than the melting temperatures of crystalline phases in the CPE-1 fibers of both types. The thermograms show only the glass transition in the temperature range from 230 to 110°C. This temperature interval coincides with the temperature interval in which the glass transition in CPE-1 is observed upon heating. This good coincidence seems to be unusual for glassy and partially crystalline polymers.

Another specific feature of glass transition in CPE-1 is well pronounced upon cooling from the nematic melt. Figure 2 shows the DSC curves (curves 6 and 7) corresponding to cooling of the as-spun and annealed copolyester fibers from 380°C at a cooling rate of 40 °C/min. In the case of the annealed CPE-1 samples, the cooling curve (curve 7) shows glass transition at 230-110°C and a small exothermal peak is observed in a temperature interval from 310 to 250°C. The cooling curve of the as-spun CPE-1 fibers (curve 6) also shows an exothermal peak at 215°C, which coincides with the onset of the glass transition in CPE-1 upon cooling.

A comparison of curves 4-7 (Figure 2) suggests that the appearance of exothermal peaks in the DSC traces upon cooling of CPE-1 samples from 380°C is likely to be related to the crystallization of CPE-1 from the nematic melt. This assumption was advanced in *(11-13)*, where the existence of two exothermal peaks in the DSC cooling curves of the as-spun CPE-1 samples was correlated to the development of two different types of crystals.

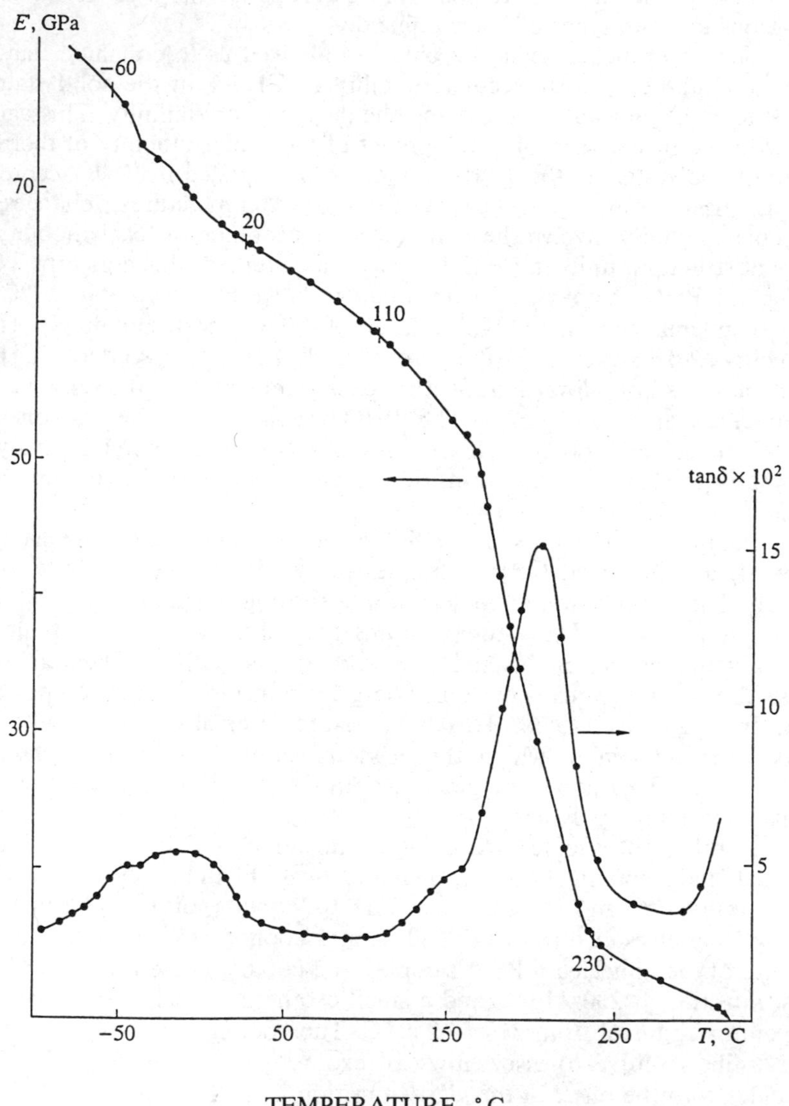

Figure 3. Temperature dependences of Young modulus E′ and loss tangent tanδ of the annealed CPE-1 fibers (η_{sp}=4.4).

There is a marked difference between the corresponding thermograms of cooling of the as-spun and annealed CPE-1 samples from 380 °C. However, upon repeated heating of the samples after cooling from the nematic melt (Figure 2, curve 8), the DSC curves appear to be identical for both types of CPE-1 fibers and show glass transition and a low endothermic peak within a temperature range of 270-320 °C. This evidence suggests that heating to the state of nematic melt results in a decrease in the ability of both types of CPE-1 fibers to crystallize. It could be related to a partial but marked disorientation of CPE-1 fibers. This conclusion is also supported by the corresponding X-ray patterns.

Furthermore, the second (low-temperature) exothermal peak observed upon cooling of the as-spun CPE-1 samples from 380°C seems not to be related to CPE-1 crystallization from the nematic melt, but to the completion of crystallization, which coincides with the onset of glass transition in CPE-1.

CPE-2. As mentioned above, dynamic mechanical properties of thermotropic LC polyesters become reproducable and stable after heating to the temperature of the high-temperature tan δ maximum and slow cooling back to room temperature. Therefore, the plots of E′ and tan δ versus temperature considered below were obtained with CPE-2 fibers heated to 200°C and cooled at 5 °C/min.

Analysis of E′ and tan δ versus temperature dependences for as-spun CPE-2 fibers (Figure 4, curves 1 and 2) reveals that solid CPE-2 shows a complex relaxational transition, in a wide temperature range between 30 and 170°C. In this interval, a significant nonmonotonic decrease in E′ is observed, which is accompanied by a broad maximum of complex shape on the tan δ versus temperature plot. At 270°C , CPE-2 starts flowing; this is accompanied by a sharp decrease in E′ and an increase in tan δ.

In dynamic mechanical and dielectric relaxation studies, Naoki and Akira *(26)* discovered that the principal relaxation transition in CPE-2 shows a bimodal pattern (Figure 4, curve 3), which, according to *(24)*, may be related to defreezing of local mobility of the two constituent structural units (i.e., constituents of HBA and HNA). Mobilities of the units of these aromatic hydroxyacids are not very different: at 30-90°C, the mobility of HBA units defreezes: then, the mobility of HNA units is set free at 100-170°C.

This pattern is confirmed by data of NMR studies. According to Allen and Ward *(22)* and others, at -60°C, all aromatic rings in CPE-2 are immobile; at 100°C, the aromatic rings of HBA acquire rotational mobility, whereas the rings of HNA are still immobile; above 150°C, all aromatic rings in CPE-2 are free to rotate.

In addition, “defreezing” of the mobility of each structural unit in thermotropic LC polyesters was found to proceed via two stages *(24)*. The physical origin of the two-stage pattern for defreezing of mobility of structural units in thermotropic LC polyesters remains unclear. Within each of the temperature intervals specified above, a decrease in E′ is accompanied by two steps on the corresponding plot. In the first temperature interval, steps can be identified at temperatures between 30 and 60°C and between 70 and 90°C, in the second interval, at temperatures between 100 and 120°C and between 130 and 170°C (Figure 4, curve 1). In plots of tan δ versus temperature, these steps are associated with four maxima (Figure 4, curve 2).

Unfortunately, the resonance technique used in dynamic mechanical studies in this work does not allow assessment of activation energies for relaxational transitions. However, activation energies for "defreezing" of the mobilities of both structural units in CPE-2 were reported in *(26)*. The low activation energy reported for the first transition (92 kJ/mol) suggests that this transition may be associated with local mobility in macromolecules of CPE-2. The second transition, which is characterized by a higher activation energy (644 kJ/mol), may be assigned to the glass transition. This conclusion agrees with the mechanism of molecular mobility in thermotropic LC polyesters suggested in *(24)*; according to this mechanism, "defreezing" of the stiffest structural units is identical to the glass transition.

In this case, DSC curves obtained on heating of CPE-2 from 100 to 170°C must show a jumpwise variation of heat capacity. As reported in *(28)*, examination of glass transitions in thermotropic LC polyesters by means of DSC is a difficult task. As in dynamic mechanical studies, preheating of the sample to the upper boundary of the glass transition region is required to obtain reproducible DSC curves in the glass transition region. Figure 5 (curve 1) shows a DSC pattern obtained on heating of as-spun CPE-2 fibers preheated to 200°C and cooled at 10 °C/min. The pattern shows that heat capacity significantly varies in a temperature region coinciding with the temperature range covering the relaxational transition (peak in tan δ) discovered in CPE-2 by means of dynamic mechanical measurements (30-170°C).

Apparently, defreezeng of the mobility of HBA units (their content in CPE-2 copolymers significantly exceeds the content of HNA units) contributes to the variation of heat capacity in CPE-2 during heating to an extent comparable with that due to the glass transition. As a result, DSC registers defreezing of local mobility and the successive glass transition as a single process extending over a wide temperature range. The observed rise of heat capacity, which is equal to 0.23 J/(gK), corresponds to a typical value of ΔC_p characteristic of the glass transition in polymers; however, it is slightly less than that characteristic of the glass transition in rubbers. The temperature corresponding to the middle of this broad temperature interval is 100°C; it is usually identified with T_g of CPE-2 *(33, 34)*. However, the fact that the glass transition temperature measured by NMR *(22)* lies at 120°C confirms the observation that, in CPE-2, the glass transition occurs in a broad temperature interval between 100 and 170°C.

CPE-2 melts at 250-270°C, as is evidenced by the endothermic peak observed at these temperatures on the DSC pattern of as-spun CPE-2 fiber

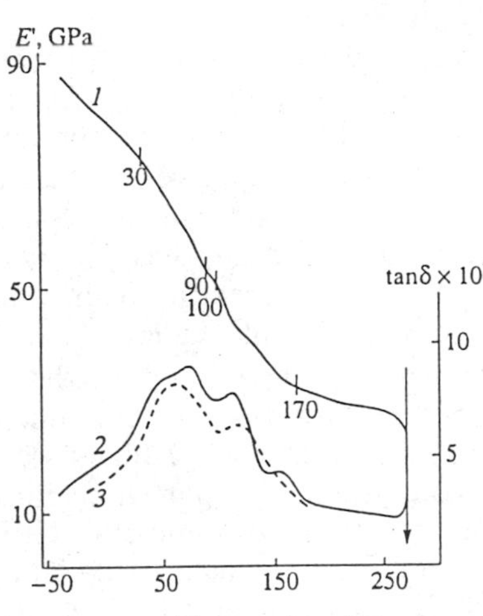

Figure 4. (1) Dynamic Young modulus E′ and (2, 3) mecanical loss tangent tanδ as-spun CPE-2 fibers as a function of temperature. Curve 3 was taken from (26).

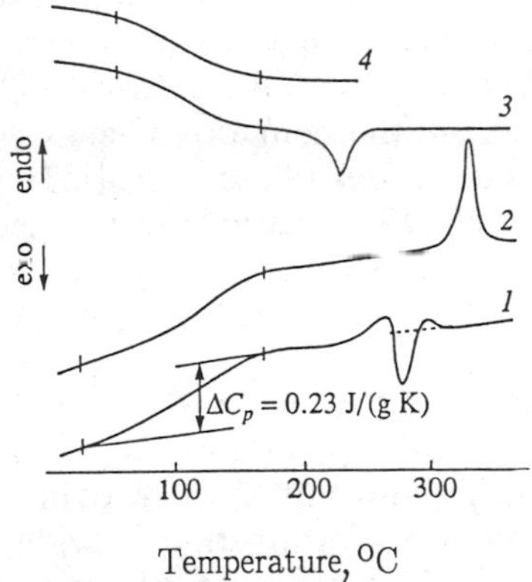

Figure 5. (1, 2) DSC traces obtained on heating and (3, 4) DSC traces obtained on cooling of CPE-2 fibers from, respectively, 350 and 250°C. Heating and cooling rates were 40 K/min. (1) As-spun fiber; (2) annealed fiber.

(Figure 5, curve 1) and the flow temperature obtained for this fiber by means of the dynamic mechanical technique (this was found to be 270°C). Recrystallization is superimposed with the onset of melting; as a result, crystallites more perfect than those in the as-spun fiber are formed. Therefore the DSC pattern of as-spun CPE-2 fiber shows a succession of endothermic, exothermic, and again exothermic effects in the interval between 250 and 300°C. Annealing for several hours at the temperature of exothermic maximum (275°C) results in the formation of crystallites showing higher melting temperature and heat of melting (Figure 5, curve 2). Because of this, for the annealed fiber, the flow temperature (determined by dynamic mechanical studies) is 320°C, which is much higher than the flow temperature determined for the as-spun fiber (270°C). No other differences in the patterns of E′ and tan δ versus temperature plots for the as-spun and annealed CPE-2 fibers were observed. Thus, an increase in the degree of crystallinity upon annealling does not affect the temperature intervals of relaxational transitions in CPE-2.

To conclude the discussion of data of dynamic mechanical studies and DSC, it is appropriate to consider the DSC patterns obtained by cooling. Figure 5 (curve 3) shows the DSC patterns of CPE-2 fibers of both types; the patterns were obtained by cooling nematic melts (350°C) at a cooling rate of 40 °C/min. As it is seen, an exothermic effect is observed at 230°C and glass transition occurs at 70-160°C. When cooling is started at a temperature below the melting region of the crystalline phase in CPE-2 fibers of both types (e.g., 250 °C), the DSC patterns reveal only the glass transition at 70-160 °C (Figure 5, curve 4). Comparison of curves 3 and 4 in Figure 5 demonstrates that the exothermic effect observed on the DSC pattern obtained upon cooling of CPE-2 from 350°C is associated with crystallization of CPE-2 from the nematic LC state.

The data on relaxation and phase transitions in CPE-1 and CPE-2 obtained by the DMR and DSC methods were used as a basis for X-ray diffraction studies of the structure of both copolymers performed at elevated temperatures.

X-ray Diffraction Analysis

CPE-1. X-ray diffraction analysis provides a direct means of studying the polymer structure. Examination of wide and small angle X-ray scattering (WAXS and SAXS) patterns of the as-spun CPE-1 fibers suggests that the structure is characterized by a well-defined orientational order of macromolecules, which are aligned parallel to the fiber axis.

Equatorial and meridional scans of the wide-angle X-ray scattering patterns of the as-spun CPE-1 fibers are shown in Figure 6. This X-ray pattern shows a well-pronounced broad halo at diffraction angles of 10°-35° and two weak diffuse maxima at 2Θ=5°-9° and 12°-16° (Figure 6a). On the meridian, one can distinguish three narrow and well-defined Bragg reflections (2Θ=14.20°, 28.67°, and 43.63°) that are associated with different orders of one reflection and which are located on the second,

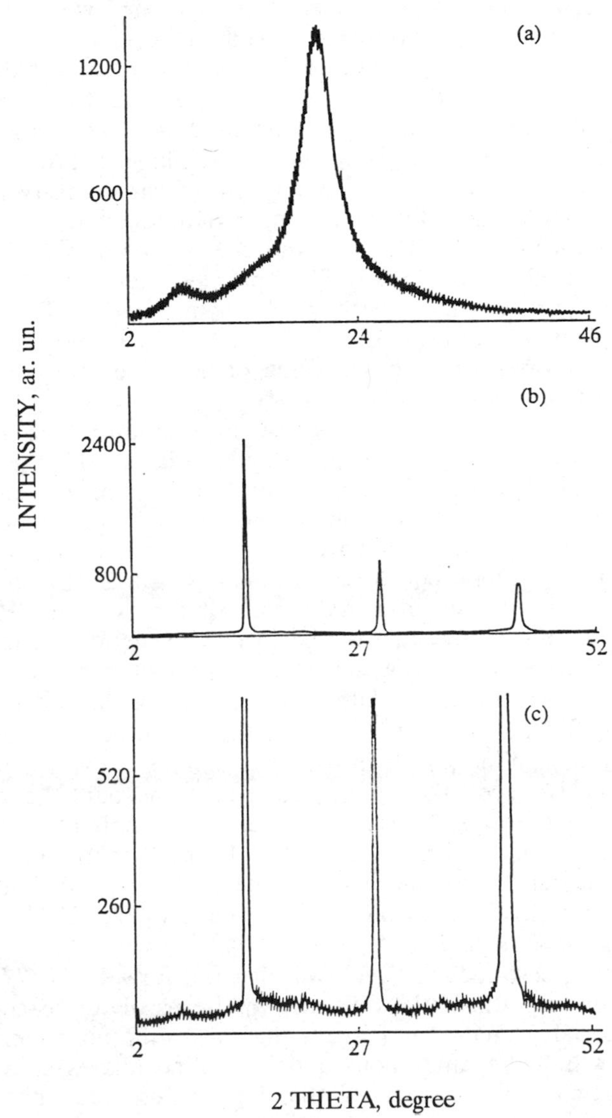

Figure 6. (a) Equatorial and (b, c) meridional diffractograms of the as-spun CPE-1 fibers obtained at 20°C at (b) conventional and (c) enlarged scales.

fourth, and sixth layer lines (Figure 6b). Meridional reflections are periodic, and interplanar spacings are d_{hkl}=6.22, 3.11, and 2.07 Å, respectively. Scattering on odd layer lines is rather weak, but reliably detected using an enlarged plotting scale (Figure 6c).

In studying the phase composition of the as-spun CPE-1 fibers, several possibilities should be taken into account. First, one should consider the possible existence of the oriented nematic phase, which at room temperature exists in a glassy state. This phase can show equatorial scattering observed as a diffusive halo and uniform intensity distribution along nonzero layer lines (Figure 6). In particular, on the meridian, the latter circumstance can manifest itself in the existence of weak maxima not only on odd, but also even, layers. Their profiles are clear in enlarged diffractograms (Figure 6c) below strong narrow reflections. The meridional pattern is typical of scattering on an individual macromolecule, and the equatorial halo corresponds to scattering on an ensemble of chains disordered in the lateral direction.

Second, it seems quite obvious that, in addition to the nematic LC phase (which is in a glassy state), one should consider the existence of at least one additional phase. The exact nature of this phase is still an open question. The existence of long-range periodicity along the chains (the Bragg meridional reflections) and weakly pronounced order in the lateral direction (several diffuse equatorial maxima) suggests the formation of three-dimensional structure in some regions of the as-spun CPE-1 fibers. Taking into account the fact that the half-width of meridional reflections is small (~0.1°), coherence along the chain is present even at distances as great as 800 Å. In contrast, the lateral dimensions of the ordered domains are rather small (less than 80 Å). Hence, this ordered phase can be identified as a crystalline phase, in which crystallites show a marked shape anisotropy, that is, needlelike crystals. Taking into account the fact that we are dealing with the crystallization of a random copolymer, small lateral dimensions of the crystallites are expected. Large longitudinal dimensions are likely to be related to the fact that the length of different structural units, which are involved in the CPE-1 chain (TPA, HBA, and PHQ fragments), is the same and equal to 6.2-6.4 Å.

This two-phase model, which was first put forward in *(11)*, provides an adequate description of the diffraction patterns of the as-spun CPE-1 fibers. One can reasonably anticipate, that, as with most semicrystalline polymers in which the amorphous and crystalline phases coexist, in the case of the two-phase system of the copolyester studied, small-angle scattering should reveal the existence of a well-pronounced periodicity.

However examination of the SAXS pattern, which is not shown here, reveals only a weak diffuse meridional scattering. This evidence suggests the absence of conventional long-range periodicity in the system studied. However, this conclusion does not mean that our speculations concerning its two-phase nature are necessarily wrong. It is known that discrete small-angle X-ray scattering is primarily associated with a marked difference between the densities of two phases which are periodically arranged. In the case studied, the densities may be close, as was observed, for example, for poly(4-methylpentene-1) *(35, 36)*, and the contrast may not be sufficient for discrete SAXS peaks. There is a reasonable alternative explanation. According to this explanation, there is no periodicity in the spatial arrangement of the phase components, for example, as a result of wide size distribution of the domains. The domains in CPE-1 can be assumed to be of quite different nature as compared to conventional semicrystalline polymer. There they are due to chain entanglaments and chain folding, which because of the rigidity of CPE-1 chains are not expected to be present in our case. Finally, one may suggest that the system studied is characterized by high values of long spacing (as was mentioned earlier, the longitudinal dimensions of the crystallites are close to 1000 Å), and hence the resolution of the instruments used (typically 800 Å) may be insufficient.

However, if we assume that there is no long spacing in the X-ray diffraction pattern of CPE-1 and that this behavior is not related to the above reasons, then we must reject the speculations on the two-phase model of the as-spun CPE-1 fibers. A more detailed consideration shows that this material may well be described by a single-phase model.

The existence of periodic meridional reflections and diffuse equatorial maxima can be explained by a single-phase LC structure corresponding to coherent regions of 1000 Å. The existence of weak, but reliably identified, scattering on odd layer lines may be considered as a consequence of a certain deviation of the chain conformation from an ideal $P2_1$, which was observed for CPE-1 *(11, 29)*. A set of diffuse equatorial maxima with a half-width (~3.5 degrees) less than that typical of amorphous polymers (for example, atactic PS or PMMA are characterized by a halo of half-width of 5-8 degrees) suggests a relatively well-defined two-dimensional short-range order within the smectic layer, i.e., in the direction perpendicular to the fiber axis.

Hence, this alternative hypothesis provides an adequate explanation of the experimental evidence obtained. However, the physical reasons responsible for the development of layer (smectic) structure in the absence of a typical mesogen in the copolyester chains are still unclear. Periodicity along the chain axis is 12.4 Å. According to *(11)*, this value corresponds to the *c*-translation or lateral dimensions of the unit cell of orthorhombic CPE-1 crystals. Along the fiber axis, the unit cell can accommodate only two of three structural units in CPE-1, which, with respect to the spatial $P2_1$ group, constitute a specific conformational mesogen of the macromolecules involved. One can assume that, upon rapid cooling of the fibers below the glass transition temperature, there is no time for the development of a three-dimensional long-range order in the CPE-1 macromolecules. As was discussed earlier, the occurence of the glass transition in CPE-1 involves inhibition in the mobility of PHQ fragments. This process is accompanied by ordering of the side phenyl substituents of PHQ in the lateral direction, resulting in the development of an extended smectic layer. However, there may be no time for the development of a three-dimensional structure. According to *(11, 29),* the development of this ordered structure is possible only under prolonged annealing of the as-spun CPE-1 fibers at temperatures above the glass transition temperature.

Hence, in the case of the initial as-spun CPE-1 fibers, three phase components may exist: nematic mesophase, needle-like crystals with a well-defined shape anisotropy, and a mesophase of smectic A type. The system could have two coexisting phases. In this case, the first (nematic) phase can coexist with the second or third phase component. One may, on the other hand, well assume that a single-phase system is formed. Then, a smectic mesophase structure seems to be most probable.

In this paper, an attempt is made to identify the structure of CPE-1 fibers more thoroughly by direct X-ray studies at elevated temperatures. However, let us first shortly recall the earlier data on the structure of the annealled CPE-1 fibers obtained in *(11).*

Annealing of the fibers results in the appearance of discrete Bragg reflections both on the equator and in quadrants of the corresponding X-ray patterns (Figure 7a). In general, this 2D-pattern is similar to that typical of a three-dimensional crystalline structure of polymer fibers. On the meridian, the X-ray pattern shows four well-defined layer lines. When the sample axis is inclined at a certain angle with respect to the primary beam, one can obtain meridional Bragg reflections on the fourth, and even on the sixth layer lines. As in the case of the as-spun CPE-1 fibers, discrete meridional reflections are located on the even layer lines, whereas, on the odd lines, only very weak diffuse maxima are observed. This character of meridional scattering suggests that the spatial symmetry group is close to $P2_1$.

As mentioned in *(11),* the annealed CPE-1 fibers show two coexisting crystalline phases: crystal 1 and crystal 2; the total crystallinity does not exceed 35-40% *(30-32).*

As in the case of the as-spun CPE-1 fibers, the existence of a strong diffuse equatorial halo and weak diffuse maxima on all layer meridional lines allows the conclusion that, as a result of annealing, the system is still to some extend “semicrystalline”. At room temperature, two crystalline phases appear to coexist with the glassy oriented mesophase *(11, 29).* However, an attempt to verify this assumption (via a search for long-range periodicity in small-angle X-ray scattering) was again unsuccessful (Figure 7b). Discrete reflections are absent in the SAXS pattern; however, in contrast to the X-ray diffraction pattern of the as-spun CPE-1 fibers, a strong diffuse halo is observed both in lateral and longitudinal directions. This SAXS pattern is typical of a microporous system *(37)* and suggests the existence of micropores in the system studied. The difference between the as-spun and annealed CPE-1 fibers is related only to the fact that, in the former case, microvoids have anisotropic shape, and the long axis of micropores is oriented along the fiber axis. In the latter case (Figure 7b), the cross-shaped SAXS pattern suggests that the shape of the microvoids is close to a parallelepiped.

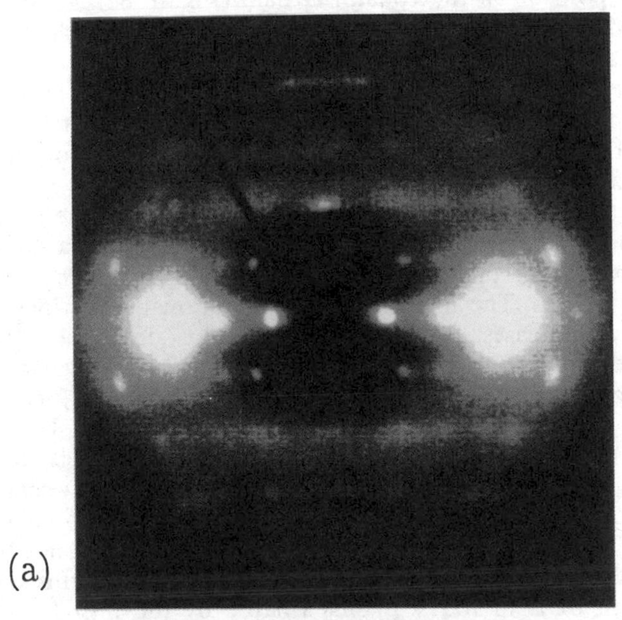

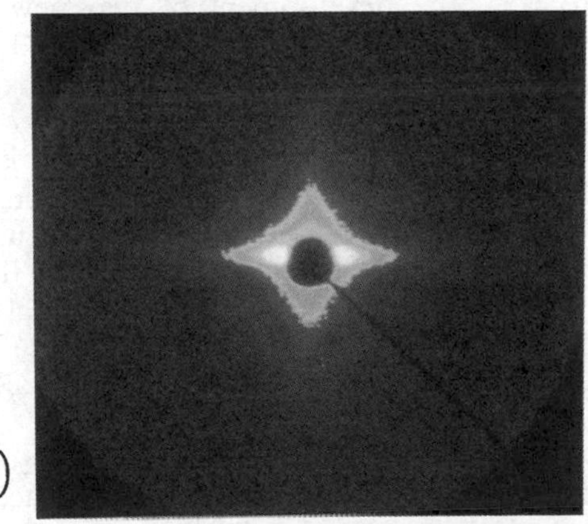

Figure 7. (a) WAXS and (b) SAXS patterns of the annealed CPE-1 fibers, ontained at 20°C. Angular ranges are approximately -35° to +35° and -2° to +2°, respectively.

Figure 8 shows experimental data obtained at elevated temperatures for the as-spun CPE-1 fibers (η_{sp}=4.4). The temperature dependence of the angular position of the equatorial halo (Figure 8a) shows a weak, but well-defined, inflection point at 180°C, which is reliably identified. According to *(38)*, this can be used to determine the glass transition temperature of the noncrystalline phase in the polymer. This value agrees well with the position of the maximum in the temperature dependence of loss tangent tan δ and with the middle of the glass transition interval in the DSC curves of CPE-1.

The temperature dependence of the intensity of the equatorial halo (Figure 8b) shows a maximum at 180°C, which suggests maximal ordering in the noncrystalline phase as formed upon the first heating run of the as-spun CPE-1 fibers. The intensity of the meridional reflections (Figure 8d) dramatically decreases (by approximately one order of magnitude) at temperatures above 180°C. This evidence suggests a significant conformational rearrangement of CPE-1 macromolecules, accompanied by destruction of the long-range order along the fiber axis. The longitudinal dimension of the coherence regions decreases from 1000 to 300-400 Å. This process is associated with a marked shrinkage of the macromolecules: the periodicity along the chain axis decreases from 6.24 to 6.09 Å. In other words, in the glass transition region, the conformation of nearly ideal rods ($P2_1$) is disrupted. The fraction of polymer chains with this conformation or the length of rigid rod segments (maybe by the introduction of kinks etc.) dramatically decreases at 180°C; then, as the temperature increases to 250°C, it gradually decreases to zero.

Parallel to this, the development of two new phases takes place (Figure 8d), which is most pronounced at 260-280°C, where an exothermal peak in the DSC heating curve of the as-spun CPE-1 fibers was observed (Figure 2, curve 2). According to *(11-13)*, the development of two crystalline phases takes place, each of which is associated with its own chain conformation and type of three-dimensional packing. Conformationally regular chains, which are typical of the structure of the as-spun fibers, are characterized by low packing density in the lateral direction. Above the glass transition temperature, conformational ordering of macromolecules is less favourable; however, formation of long-range order, which is characteristic of the three-dimensional crystal, results in substantial energy gain.

As follows from Figure 8d, conformational disordering commences at the glass transition temperature and is accomplished at 250°C. At this temperature, marked changes in the diffraction pattern are observed, that is, two-fold changes in both intensity of the equatorial halo (Figure 8b) and its half-width (Figure 8c). This behavior suggests a considerable disordering in the plane perpendicular to the chain axes. The lateral dimension of the coherence regions decreases from 80 Å at 180°C to 30 Å at 250°C. The latter value is typical for a polymer liquid *(39)*. Furthermore, at 250°C, as was mentioned earlier, the sample starts to flow.

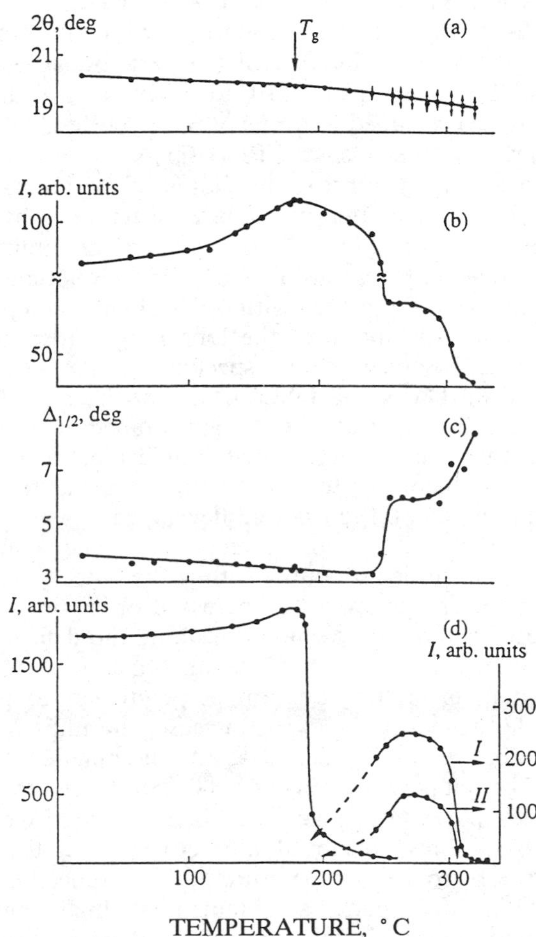

Figure 8. Temperature dependences of (a) angular position 2Θ, (b) intensity I, and (c) half-width $\Delta_{1/2}$ of the equatorial amorphous halo and the temperature dependence of (d) intensity I of the meridional (002) reflection of the as-spun CPE-1 fibers.

The experimental evidence led to the conclusion that the onset of the glass transition is accompanied by phase transformation in the structure of the as-spun CPE-1 fibers. This phase transformation is accomplished at 250 °C. Only at this temperature the development of crystalline phases in CPE-1 becomes well pronounced (Figure 8d). What is the origin of this transition at 180-250 °C, which is concurrent with the glass transition of copolyester (the temperature interval of the glass transition as detervined by DMR and DSC scans from 110 to 230 °C)? Let us analyze the experimental evidence obtained in terms of the hypothesis of a single-phase smectic LC structure in the as-spun CPE-1 fibers.

According to this hypothesis, the fading of meridional reflections at 180-250 °C implies the disruption of the layer structure, which is related to the appearance of mobility of the structural units in CPE-1 macromolecules that are responsible for the development of smectic layering, that is, PHQ fragments with side phenyl groups. Once these groups show mobility, disruption of the long-range order in the direction of the fiber axis is observed. These speculations are confirmed by the observation, based on DMR and DSC data, that the relaxation process of glass transition in CPE-1 is related to the appearance of the mobility of the most rigid structural unit of the macromolecule, that is, the PHQ fragment. Note that, according to *(24)*, at this moment, the local mobility of other structural units in CPE-1 is well developed.

Furthermore, disorder in the direction perpendicular to the chain axis suggests that the initial structure within the smectic layer in the as-spun CPE-1 fibers is characterized by increased order similar to the short-range order of a high-temperature liquid. Taking the data reported in *(14)* into account, one can reasonably assume that this two-dimensional packing of chain units in the direction perpendicular to the axis of the fiber is close to hexagonal packing. In this case, the initial structure of the as-spun CPE-1 fibers can be classified as LC mesophase of the smectic B type. Hence, in the temperature range of 180-250 °C, the transformation of smectic B to a new phase takes place. Let us try to identify this new phase.

Figures 9f-9k show equatorial diffractograms of the as-spun CPE fibers obtained at room temperature after annealing at elevated temperatures. Diffraction patterns obtained at high temperatures are presented in Figures 9a-9e. Examination of the data presented in Figure 9 shows that, as the temperature increases from room temperature to 250 °C, no principle changes in the shape and intensity of the equatorial halo are observed. The only difference is related to the shift of the angular position of the maximum in the diffraction curve corresponding to thermal expansion. A diffferent situation is evident during heating the system above 250 °C. In addition to the development of the crystalline phases, as indicated by the appearance of numerous Bragg reflections, the intensity of the halo (for example, at 275 °C) becomes almost two times less than that upon subsequent cooling to 20 °C. Important to note that this process is reversible. In other words, repeated heating from 20 to 275 °C followed by

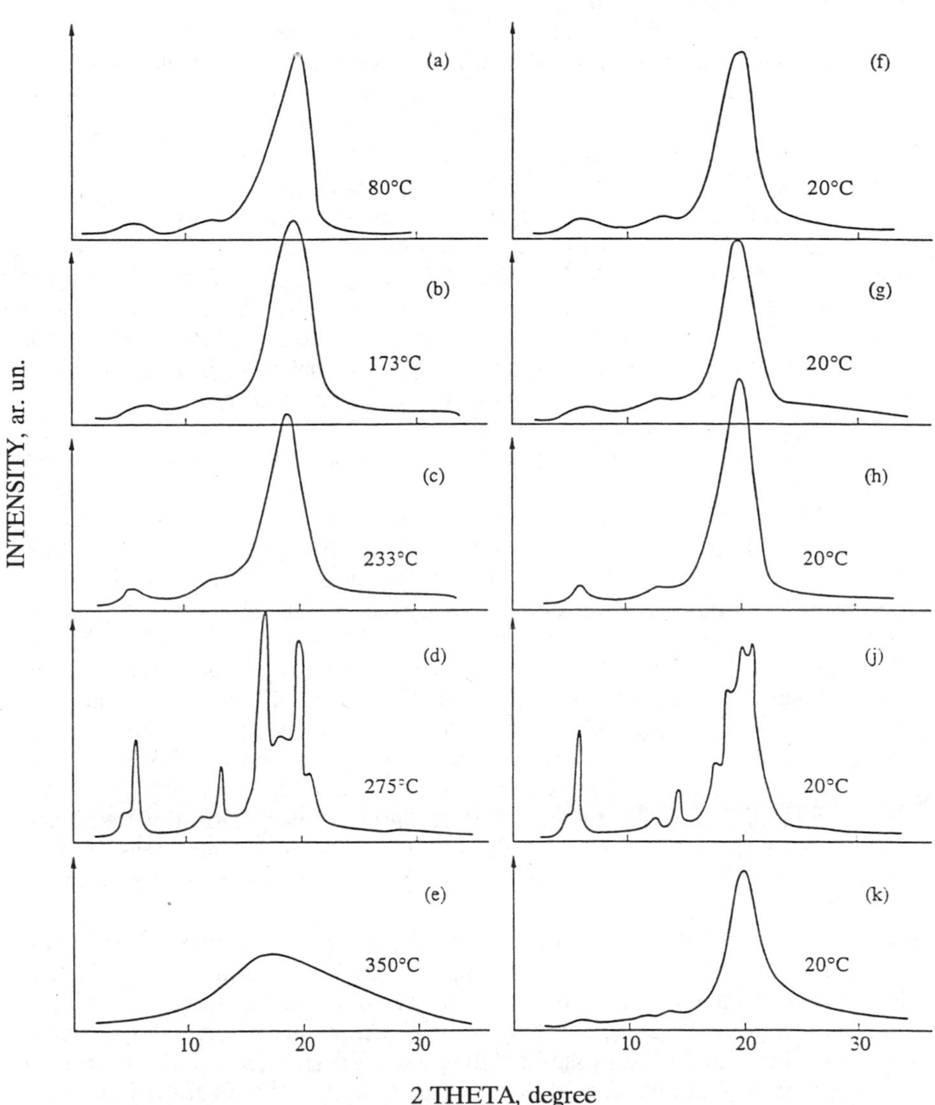

Figure 9. (a-e) Equatorial diffractograms of the as-spun CPE-1 fibers obtained at elevated temperatures and (f-k) after cooling to 20°C. Although the absolute intensity is in an arbitrary scale, the intensities in different plots are comparable.

cooling to 20°C results in similar changes in diffuse scattering, as shown in the different patterns of Figures 10a and 10c.

The annealed CPE-1 samples exhibit similar behavior (Figure 11). A characteristic feature is that, in both the as-spun and annealed samples, heating of the fibers above the melting point of the crystallites (350°C) and subsequent cooling to 20°C result in the disappearance of crystalline reflections in the corresponding X-ray diffraction patterns and in the appearance of quite different scattering patterns (Figures 9k and 9e).

Reversible changes associated with heating-cooling cycles are also observed on the meridian of the X-ray diffraction patterns from both as-spun and annealed CPE-1 fibers (Figure 12). In particular, strong reflections characteristic of the as-spun fibers vanish at temperatures above 250°C, and doublet reflections appear on each odd layer line. Two sets of "new" reflections correspond to the conformational state of chains in two crystalline lattices. If, as in the case of equatorial scattering, we subtract the scattering corresponding to the crystalline polymer fraction, the changes in the heating-cooling cycles are presented in Figures 10b and 10d.

The equatorial and meridional diffractograms presented in Figures 10c and 10d are typical of a nematic LC structure. Hence, one can conclude that, at 180-250°C, a phase transition (LC smectic B - LC nematic) takes place. This process is likely to be considered as a first-order transition. This is consistent, in particular, with the reversible character of this process. Moreover, the transformation should be accompanied by abrupt changes in density. Although, in this work, dilation measurements were not carried out, examination of the temperature dependences of interplanar distances which may be correlated to a density shows that, at least for meridional reflections, a jumpwise change is observed (Figure 8d). In the case of equatorial maxima, unfortunately, superposition of the Bragg reflections and the equatorial halo results in an increased measurement error (Figure 8a) and one cannot identify a jump at temperatures above 250°C

It is also difficult to identify the thermal effect corresponding to the smectic-nematic transition in the DSC heating and cooling curves. This is likely to be related to the complex character of DSC thermograms, which, in this temperature range, involve contributions from several processes. One cannot exclude the possibility that the thermal effect of the transition (mesophase 1 - mesophase 2) is so weak that its identification is beyond the sensitivity limit of the methods used. Despite the fact that the arguments in favor of the conclusion that this transition is a first-order transition are not complete, such a proposition seem to us most convincing.

Thus, the experimental evidence obtained and the results reported in *(11-13, 24, 27-32)* allow us to propose the folllowing series of phase and relaxation transitions of the ternary copolyester CPE-1.

The initial structure of the as-spun CPE-1 fibers may be assumed as LC mesophase of the smectic B type. This system is then single-phase. Any

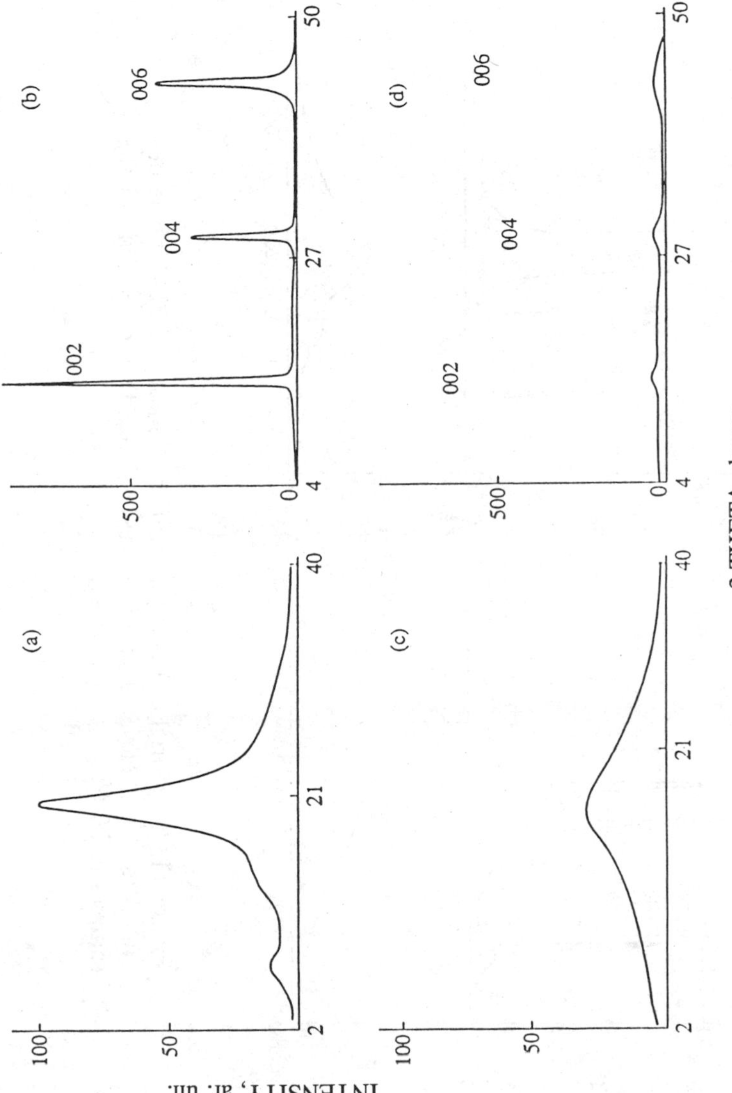

Figure 10. (a, c) Equatorial and (b, d) meridional diffractograms obtained as a result of the subtraction of X-ray scattering from the crystalline phases in the as-spun CPE-1 fibers. (c, d) correspond to heating up to 275°C, and (a, b) correspond to cooling to 20°C.

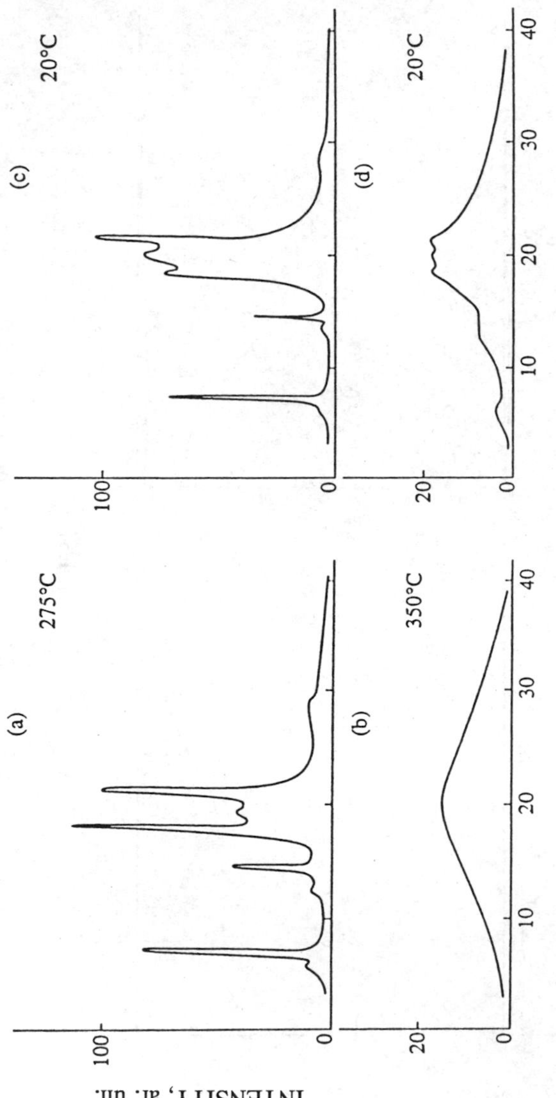

Figure 11. Equatorial diffractograms of the annealed CPE-1 fibers obtained (a, b) at elevated temperatures and (c, d) after cooling of the samples to 20°C.

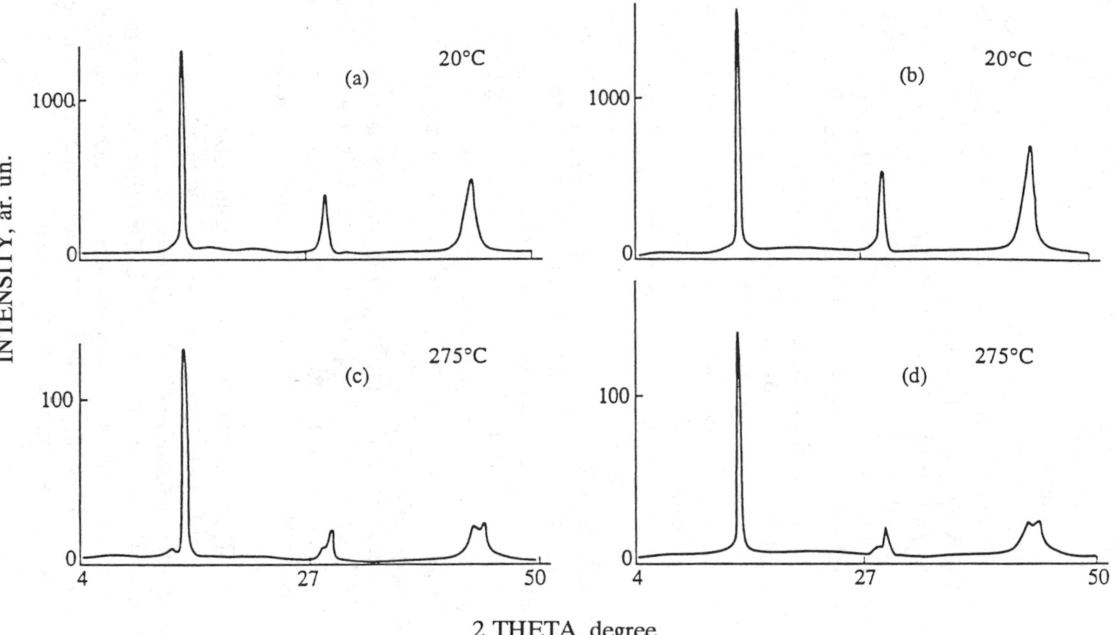

Figure 12. Meridional diffractograms of the (a, c) as-spun and (b, d) annealed CPE-1 fibers at (c, d) 275°C and (a, b) room temperature.

two of the three structural units involved in the copolyester, that is, HBA, PHQ, and TPA, which have similar dimensions, can serve as mesogenic groups, which are necessary for the development of layer packing. At room temperature (below the glass transition temperature), such a rigid-rod like conformational mesogen is characterized by a nearly ideal $P2_1$ structure The glass transition temperature, determined as the middle of the interval of the corresponding relaxational process (from 110 to 230 °C), is equal to 180 °C. The appearance of cooperative mobility in macromolecules in this temperature range is related to the manifestation of mobility of the most rigid structural unit in CPE-1, that is, the PHQ unit. This unit, because of its side substituent, is responsible for the development of the smectic layer. Note that, prior to the onset of the glass transition, the local mobilities of the two other structural units involved in CPE-1 were already obtained: the TPA unit, from-60 to20 °C, and the HBA unit, from 30 to 90 °C.

Upon heating, CPE-1 macromolecules lose their rigid conformation. Finally, the smectic phase transforms into the nematic LC phase at 180-250 °C. At such low temperatures, this state is not in equilibrium. As a result, starting from 250 °C, two thermodynamically stable crystalline phases are readily formed. In the annealed samples, the melting temperature of these phases is 340-350 °C, and net crystallinity is 35-40%. In the temperature range above 250 °C up to melting temperatures, two crystalline phases coexist with a nematic LC (60%) phase. The crystalline lattice serves as a physical network that reinforces the material and does not allow copolyester to flow at temperatures below the melting point. In the case of the as-spun CPE-1 fibers, the flow starts at 250 °C, i.e., at temperatures where devitrification is completed, and the smectic phase is completely transformed into the nematic, while the development of the crystalline phases is not yet completed.

At temperatures above the melting temperature of the crystalline phases, the system becomes single-phase again and is transformed into oriented nematic LC melt. The copolyester is isotropic (or melts) at temperatures of about 400 °C, at which temperatures its active chemical destruction takes place.

This scheme of phase and relaxational transitions in this representative thermotropic LC copolyester provides an adequate description of the experimental evidence obtained. The validity of this scheme and the universal character of the transitions described for other copolyester where all monomeric units have the same lengths, can be shown for the whole family of thermotropic LC copolyesters by comparing the temperature behavior of CPE-1 with other polymers of this type.

CPE-2. The X-ray patterns of as-spun CPE-2 fibers (which are not shown here) indicate the high extent of orientation of the copolymer chains. The corresponding equatorial and meridional diffraction scans are shown in Figure 13a, c.

The equatorial X-ray diffraction pattern shows a strong diffuse maximum; near this maximum, but on the first layer line, two other spots

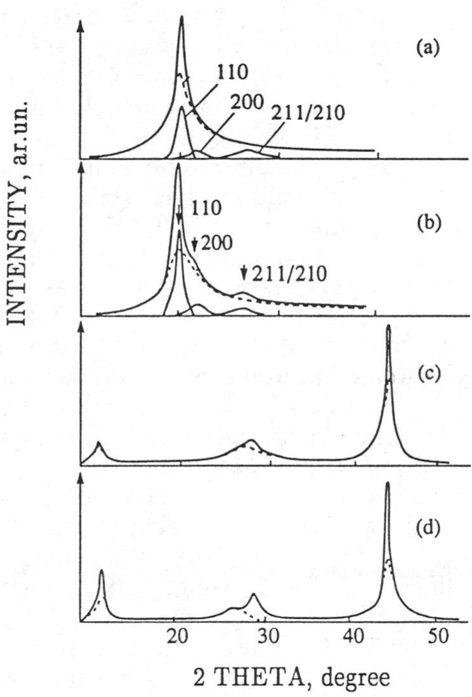

Figure 13. (a, b) Equatorial and (c, d) meridional X-ray diffraction patterns registered at room temperature: (a, c) as-spun CPE-2 fiber: (b, d) annealed fiber. The dashed curves indicate the diffuse maxima corresponding to scattering from non-crystalline part of the material.

are detectable, which suggest some three-dimensional order in the system. The meridional X-ray diffraction pattern shows distinct reflections on the second and fourth layer lines. Maxima of higher order cannot be registered in this mode, because they do not fit the reflecting condition.

Annealing of the samples gives rise to a set of Bragg's reflections (Figure 13b, d). In addition to the principal reflection on the equator, at least one more maximum becomes noticeable at wide scattering angles. In approximately the same range of angles, the first layer line reveals reflections, which are observed as a four-point pattern with respect to the center of the diffraction pattern. These features suggest that annealing of copolyester fiber is accompanied by an increase in crystallinity.

According to Kaito et al. *(5)*, macromolecules of the initial fiber (which is obtained by fast cooling from nematic LC state) fail to acquire a regular conformation. During annealing, aromatic planes of the structural units of the copolymer occupy random positions; as a result, the projection of the chain onto the crystallographic plane (001) is almost cylindrical. Consequently, the conformation of CPE-2 macromolecules is characterized by ternary symmetry, whereas the structure of the base plane is similar to a two-dimensional hexagonal pattern. Taking into account the conformational disorder frozen in by rapid cooling, the overall structure may be described as "condis" (conformationally disordered) crystalline *(40)*. During annealing, the macromolecules of the copolymer adopt a conformation that is closer to equilibrium; this conformation is no longer characterized by three-dimensional, but by two-dimensional symmetry. As a result, an orthorhombic structure is formed *(5)*.

In the case under consideration, analysis of equatorial diffraction patterns of as-spun (Figure 13a) and annealed (Figure 13b) CPE-2 fibers easily reveals (110), (200), and (210) reflections, assigned to an orthorhombic modification. The "wings" of azimuthally-distributed intensity of the (211) reflections lying on the first layer line are also registered on the equator of the X-ray diffraction pattern and vice versa. Hence, the data obtained in this study significantly differ from those reported in *(5)*.

Our assessment of the phase composition of CPE-2 before and after annealing is also different. Indeed, according to *(5, 17)*, the as-spun fiber shows single-phase structure, nearly pseudohexagonal. The copolymer becomes semicrystalline, that is, its structure becomes biphasic, only after annealing at elevated temperatures.

We believe that the structure of as-spun fiber is also biphasic. The first observation to support this assumption is the bimodal profile of the meridional reflection lying on the fourth layer line (Figure 13c). Careful analysis allows the asymmetry of the profile to be detected for the other two reflections on the meridian, which are positioned on the second and sixth layer lines. Distortion of the profiles of all three meridional reflections is due to the contribution of another phase component in addition to the major one. A similar situation is more explicitly observed for the annealed sample (Figure 13d).

The "wings" of the principal reflection on the equatorial diffraction curve of as-spun CPE-2 fiber are also markedly asymmetric (Figure 13a). At diffraction angles of 27°-28°, the presence of an additional maximum is easily discernible especially in the annealed sample. The entire reflection profile shows an "unusual" character, which could be explained by assuming that the scattering curves due to different phase components overlap. The dashed lines in Figure 13 illustrate how the diffraction patterns could be resolved into separate components. They are obtained by taking the mirror image of the reflections from the left side. Note also the relatively low intensity of the second and third reflections of the orthorhombic phase, whereas the first reflection (110) is, on the contrary, rather intense.

Such a rapid decline of intensity with increasing diffraction angle indicates that the crystalline phase is rather imperfect. Moreover, this phase is composed of small crystallites. The half-width of the principal reflection is rather large (~0.8°-1.0°) and corresponds to lateral crystallite dimensions of ~80-100 Å. On meridional X-ray diffraction patterns, this structure gives rise to Bragg's reflections localized on the second, fourth, and sixth layer lines, but at slightly different angular positions than the meridional maxima of the smectic phase discussed before for CPE-1. The possible deconvolution of the meridional pattern into reflections due to different phase components is illustrated in Figure 13. Like the lateral dimensions, the longitudinal dimensions of crystallites do not exceed approximately 100 Å.

Most of the material in as-spun CPE-2 fibers contributes to equatorial scattering as an amorphous halo. Quantitative evaluations reveal that the content of this phase in the system amounts to 90-95%. Aperiodic meridional reflections, shown in Figure 13c, d by the dashed lines, also correspond to the scattering on this phase component.

It was mentioned above that a similar phase in CPE-1 composed of structural units of equal lengths, gave rise to periodic meridional maxima on the second, fourth, and sixth layer lines of the X-ray diffraction pattern and, was explained by a smectic LC phase. The lengths of structural units (due to HBA and HNA) in CPE-2 are significantly different. Because of this and the random distribution of HBA and HNA fragments in the chain, the reflections on meridional diffraction patterns may be aperiodic. Taking into account the considerations discussed in a case of CPE-1, one may assume that, in the case under consideration, most of the material in as-spun CPE-2 fiber shows an aperiodic layer structure of LC smectic type. A characteristic feature of this structurc is the presence of aperiodic layers, which give rise to aperiodic reflections in the meridional X-ray diffraction pattern. The diffuse profile of the amorphous halo on the equatorial X-ray diffraction pattern could be explained by, short-range order (like in a liquid) in the packing of the centers of macromolecules; this order does not extend beyond 25-30 Å *(38)*. Along the fiber axis, the dimensions of coherence regions are greater (≤100 Å). This suggests the existence of long-range order in the arrangement of aperiodic smectic layers. Both types of

LC periodic and aperiodic smectic structures for CPE-1 and CPE-2 copolymers, respectively, are shown schematically in Figure 14.

Hence, the as-spun CPE-2 fiber shows an essentially biphasic structure. The content of the crystalline component is low and does not exceed 5-10%, whereas the crystallites are relatively small and imperfect. Annealing heals the defects and results in an increase of longitudinal dimensions of the crystallites to 150-200 Å (lateral dimensions are not affected by the annealing). During the annealing, the contents of phase components are redistributed. As a result, the degree of crystallinity of the annealed material may exceed 25%. The structure of the crystalline phase remains orthorhombic.

It is interesting to compare X-ray diffraction patterns registered at room temperature for as-spun and annealed CPE-2 fibers with X-ray diffraction patterns recorded at different temperatures. Figure 15 shows equatorial X-ray diffraction patterns recorded at different temperatures for the as-spun fiber. In Figure 16, the intensities and angular positions of the principal maximum are plotted as a function of temperature. It must be kept in mind that the plots in Figure 16 are not very rigorous. Indeed, because positions of the reflections due to both phase components are close, the deconvolution of the overall scattering curve into the two components may not be performed accurately and is not very reliable. Therefore, the data in Figure 16 allow only qualitative analysis and comparison with the data of DSC and dynamic mechanical studies.

As can be seen, the intensity at peak position of the principal reflection on the equatorial X-ray diffraction pattern starts growing almost from room temperature and grows monotonically as the temperature increases to 200 °C. However, the profile of the diffraction pattern does not sustain any significant changes (Figure 15), which indicates that the two phase components, crystalline orthorhombic phase and smectic LC phase, still coexist in the system. Regular drift of the angular position of the principal reflection (Figure 16) toward smaller angles (larger interplanar spacings) reflects the conventional thermal expansion occurring upon heating.

An increase in intensity of the principal equatorial maximum may be due to the higher crystallinity of the system. Corresponding assessments demonstrate that, in the temperature range extending from 20 to 200 °C, the content of crystalline phase almost does not vary and amounts to 5-10%. This means that such an integral characteristic as the area under the principal reflection remains approximately the same: the growth of intensity is compensated by a decrease in the half-width of the line (Figure 16).

Beginning from 200 °C (the upper boundary of the glass transition in CPE-2), the variation of X-ray diffraction parameters with temperature acquires a different pattern (Figure 16). The profile of the diffraction pattern also changes (Figure 15). The reflections due to the orthorhombic phase disappear; instead, in addition to the principal crystalline reflection, a new maximum appears on the equatorial X-ray diffraction pattern at

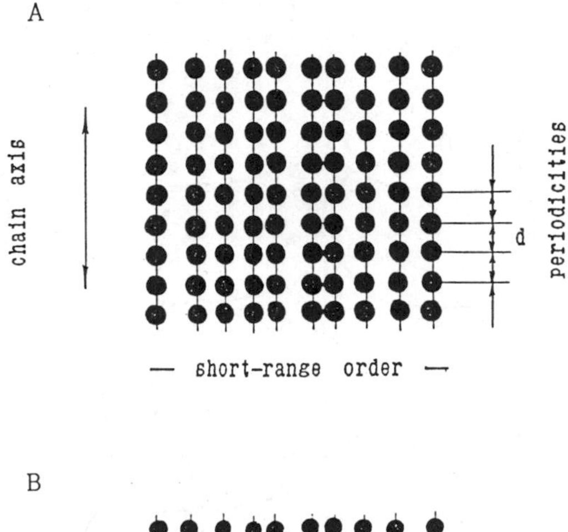

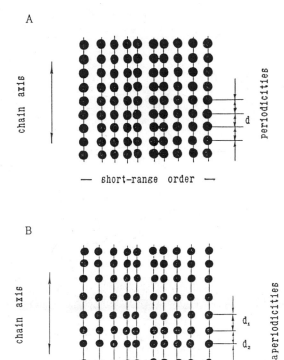

Figure 14. Schematic representation of (A) periodic and (B) aperiodic LC smectic structures.

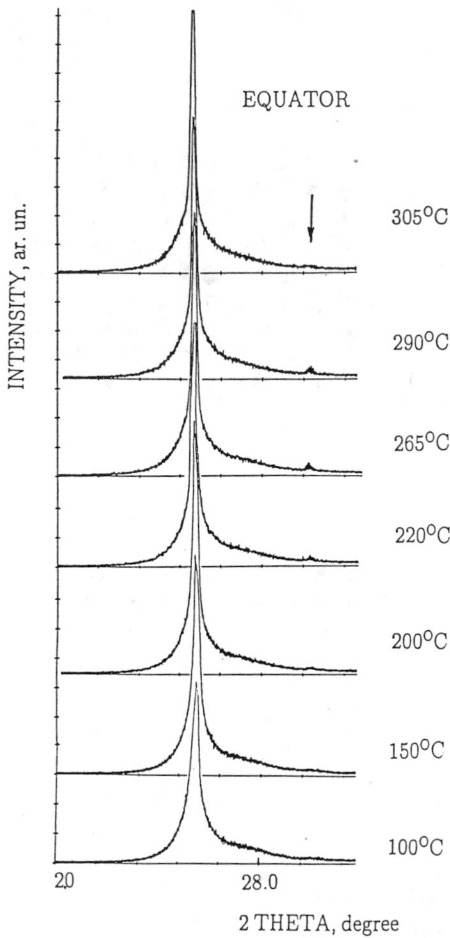

Figure 15. Equatorial diffraction scans of as-spun CPE-2 fibers obtained at elevated temperatures. The arrow indicates the position of the (110) reflection of hexagonal lattice.

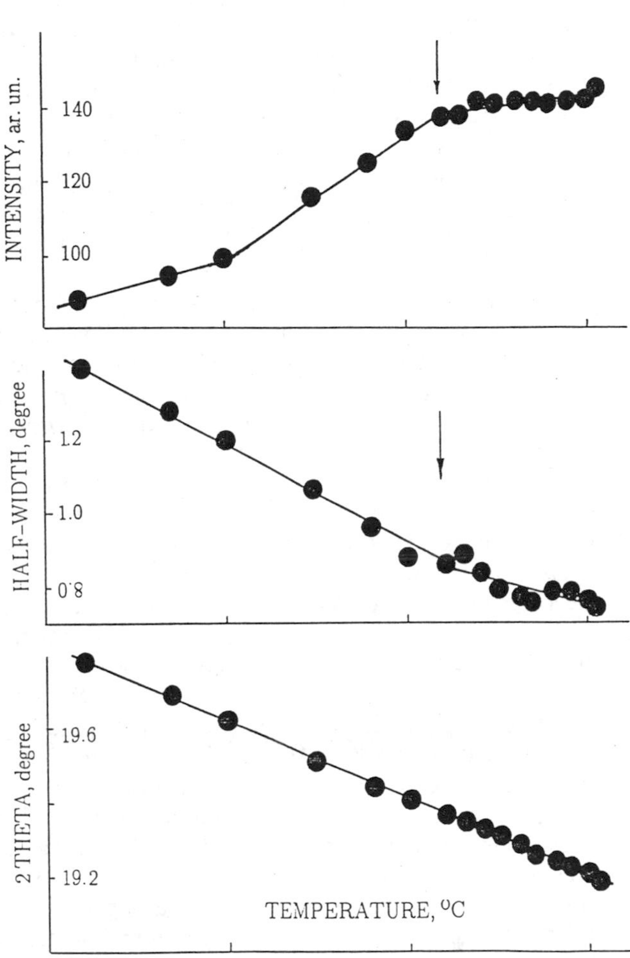

Figure 16. Temperature dependences of an intensity, half-width and angular position of the diffusive maximum on the equator of the X-ray pattern of as-spun CPE-2 fibers. The arrow indicates transition temperature from orthorhombic to hexagonal modification.

wide angles. At 200°C, interplanar spacings corresponding to the principal reflection and to the new maximum were evaluated to be 4.58 and 2.64 Å, respectively. The ratio of these two values is $\sqrt{3}$, which suggests hexagonal packing for the centers of chains in the base plane.

The ratio of (100) and (110) reflections of hexagonal lattice $d_{100}/d_{110}=\sqrt{3}$ remains invariable until CPE-2 undergoes transition to nematic LC state at 315-320°C (Figure 5, curve 1). Hence, in contrast to conclusions made in *(5)*, the orhombic structure of CPE-2 does not form during annealing near the melting temperature of the crystalline phase of the polymer. It is a hexagonal modification that is an equilibrium phase at high temperatures. Only on cooling of CPE-2 the orthorhombic structure forms as a result of polymorphic transition of (crystal 1)-to-(crystal 2) type.

The nature of this structural transition is unclear. Indeed, DSC patterns obtained both upon heating and upon cooling do not show endothermic and exothermic effects at 200°C. However, it is these thermal effects that identify this process as a first-order transition. Butzbach et al. *(17)* also reported that the unit cell volumes of orthorhombic and hexagonal phases are equal, and, therefore, the transition from one phase to the other must not be accompanied by a jumpwise variation of density. The fact that the angular position of the principal equatorial maximum (interplanar spacing) varies monotonically with temperature (see Figure 16) confirms the last conclusion.

Hence, the (crystal 1)-to-(crystal 2) transition is not a conventional phase transition of first order. This process coincides with sequential defreezing of the rotational mobility of aromatic rings of HBA and HNA in the temperature range from 30 to 170°C and, moreover, accompanies it. The hexagonal lattice, which is characteristic for the most compact packing of species showing cylindrical symmetry *(41, 42)*, is related to the occurrence of high-frequency rotational oscillations of macromolecular segments.

In polymers, such a transition is usually accompanied by the formation of a mesophase *(40)*, which is characterized by two-dimensional hexagonal packing of the centers of chains and the absence of long-range positional order (conformational disorder) along the axes of macromolecules, as is suggested by the diffuse meridional scattering. In the case considered, aperiodic reflections on the meridional X-ray diffraction pattern, which correspond to the scattering from the hexagonal phase, indicate an aperiodic layer structure. Thus, in the pseudocrystalline phase, long-range order exists along the axes of the chains. Chain segments between aperiodically arranged layers may be involved in rotational motion about their own axis; as a result, hexagonal structure is formed in the layer. According to *(17, 25)*, aperiodic reflections remain even in the nematic LC phase of the copolyester. Hence, in CPE-2, hexagonal structure must be identified with a mesomorphic state rather than with truly a crystalline state.

The meridional scattering pattern is less sensitive to heating of the sample (Figure 17). At 20-300°C (the structure of CPE is biphasic), each

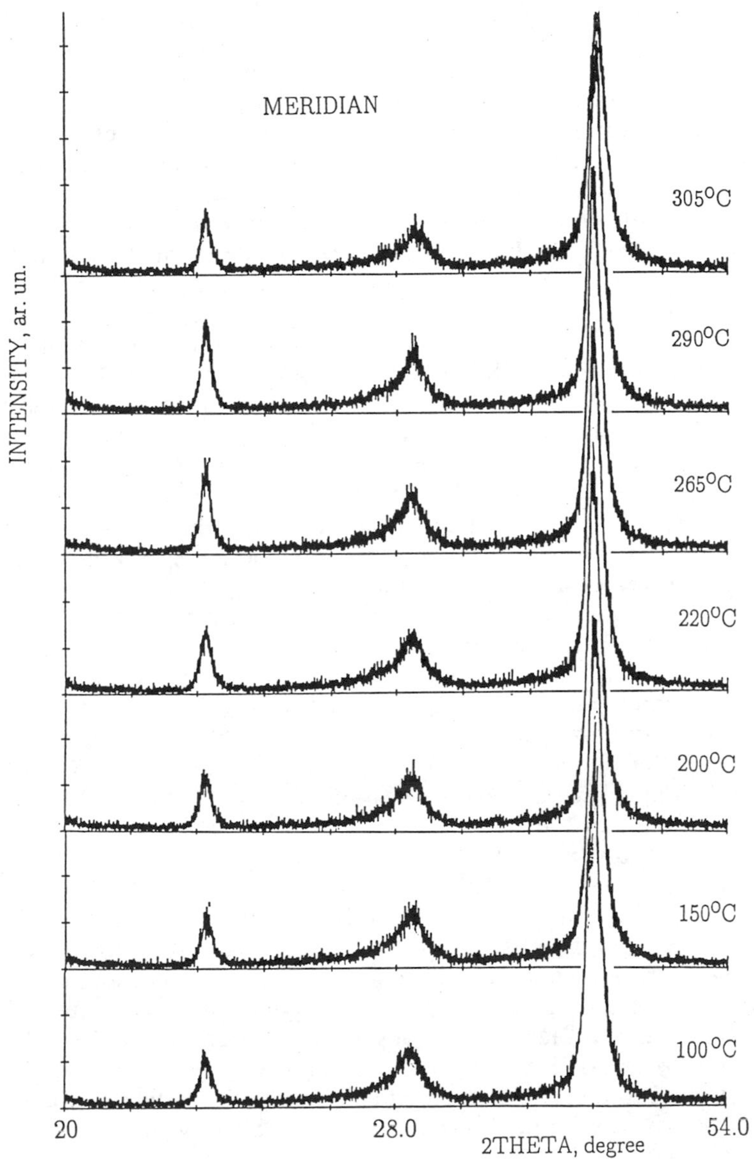

Figure 17. Meridional diffraction scans of as-spun CPE-2 fibers obtained at elevated temperatures.

of the reflections on the second, fourth, and sixth layer lines show a bimodal profile. We failed to deconvolute these maxima into separate components. However, beginning from 220 °C, redistribution of the intensity on the meridian is observed. This is manifested in changes of the asymmetry of the profile and indicates that the contents of phase components in the system are redistributed. The fraction of chains with a conformation corresponding to crystalline CPE-2 increases, whereas the content of macromolecules that are conformationally disordered and belong to the smectic LC phase decreases. For the annealed sample, the ratio between the contents of phase components was evaluated to be 30:70. Taking into account that the accuracy of the determination is not better than 20%, this assessment agrees with an evaluation based on the analysis of equatorial data.

Plots of intensity at peak position (Figure 18) of meridional reflections versus temperature are only illustrative (as in the other case discussed above, Figure 16), because the reflections were not resolved into components. Nevertheless, in the most important temperature intervals, corresponding to the onset of the glass transition in CPE-2 (~100°C) and to the transition of an orthorhombic crystal to a hexagonal crystal (~200°C), the slopes of the temperature dependences of X-ray diffraction parameters are subject to variation. This corresponds to transitions detected in dynamic mechanical and DSC studies.

The transformation of the structure of CPE-2 becomes understandable when one takes into account that the copolymer contains 73 mol% of HBA monomer units. For HBA homopolymers, two major crystal modifications were reported *(15, 16)*. The low-temperature modification was assigned to an orthorhombic lattice. If one assumes that it is mainly the crystallization of the HBA-rich fraction of CPE-2 that is detected, an obvious analogy comes to mind.

Naturally, in the case of CPE-2, the crystallizing chain fragments are not pure HBA, as would have been the case in a block copolymer or homopolymer. The fact that aperiodic reflections are observed at all temperatures (Figure 17) unambiguously proves that the structure of the HBA-rich fraction is statistical *(1, 2)*. This conclusion is supported by the large values of half-widths of the reflections in the orthorhombic phase as compared to similar values for the homopolymer and by a steeper decline in the intensity of Bragg reflections with an increase in the diffraction angle *(15-16)* for CPE-2 than for the homopolymer. These facts suggest that in CPE-2 the crystallites are smaller and the crystalline phase is less perfect than in the homopolymer (which is of course quite natural for a statistical copolymer).

Thus, during heating of as-spun CPE-2 fibers from room temperature to 200°C, a minor fraction of relatively small orthorhombic crystals that were formed on quenching of the copolymer is involved in a permanent (concomitantly with the defreezing of mobility of the structural units of CPE-2), rearrangement resulting in the formation of domains with hexagonal structure. Starting from 220°C, melting and recrystallization of

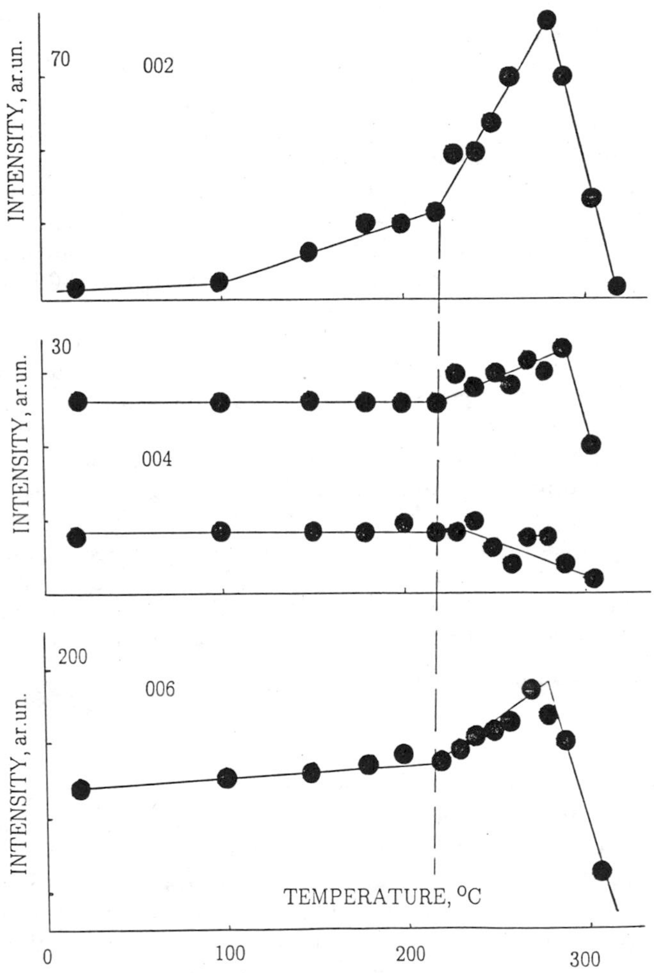

Figure 18. Temperature dependences of peak intensity of meridional maxima in the X-ray pattern of as-spun CPE-2 fibers.

orthorhombic crystallites into more perfect domains with hexagonal structure begins; this process reaches a maximum at 280°C. Sufficiently long annealing increases the content of the hexagonal phase to 25-30%. On heating above 320°C, the hexagonal phase melts. Apparently, the domains of the hexagonal phase act as physical junctions preventing the flow of material at elevated temperatures. When the domains have melted, a single-phase LC structure is formed in CPE-2. Upon cooling of the annealed CPE-2 (that was not molten) to room temperature, hexagonal packing transforms into an orthorhombic modification, the fractional content of which corresponds to the degree of crystallinity of the copolyester annealed at elevated temperature.

The structure of CPE-2 at any temperature up to the temperature of transition to the nematic LC state is always biphasic: the crystalline phase component of orthorhombic or hexagonal type coexists with the smectic LC component, the content of which may vary from 70 to 90% depending on the sample history. Aperiodic reflections on the meridional X-ray diffraction pattern, which correspond to the layer structure, and the diffusive halo on the equatorial pattern, which indicates the absence of long-range order in the layer, allowed this mesophase to be identified as an aperiodic smectic A structure, similar to the model of nonperiodic layers suggested by Hanna and Windle *(1)* for crystals of statistical copolyesters. Recently those authors reported about their observations of an aperiodic LC smectic structure in CPE-2 even above the melting point of the crystalline modification *(25)*. It may be that the aperiodic smectic first discovered in Vectra copolymer is also an inherent phase component in other thermotropic LC copolyesters of statistical nature, where the monomeric units constituting the main chain have substancially different lengths.

Conclusion

The detailed analysis of the structure, temperature behavior and phase as welll as relaxational transitions of two representatives of thermotropic LC copolyesters with a statistical sequence of monomers in chains shows the strong dependence of their phase composition on the chemical structure of monomeric units. One of the most important factors appears to be the geometrical dimensions of the monomeric groups, from which the chain of copolymer is constructed from, and the presence (or absence) of relatively large side substituents.

If the different monomers have the same lengths, as it happens, for example, in a case of CPE-1, the X-ray pattern of such an oriented material reveals periodic Bragg maxima on the meridian and numerous spots on the equator and in quadrants, indicating the developement of a three-dimensional arrangment of irregular rigid chains. Moreover, the crystallization of CPE-1 leads to the formation of two different crystalline modifications with orthorhombic latteces but slightly different unit cell parameters. The driving force of such an unusual microphase separation process is not clear yet, but the most probable reason seems to be related to configurational isomerism due to the possibility of "head to head" or "head to tail" connections along the chain during the synthesis of the copolymer.

If the monomeric units have different lengths, as it is the case for another copolyester under study CPE-2, aperiodic reflections in the meridian and several Bragg spots on the equator and in quadrants are observed in X-ray patterns of oriented fibers. That means that the crystalline phase of the material may be described by one of two models proposed by Windle *(1)* or Blackwell *(2)* for such unusual structures, namely, the model of non-periodic layer crystals and paracrystalline model, respectively.

The highest values of crystallinity achieved after annealing of fibers of CPE-1 and CPE-2 are 35 and 25%, respectively. Both copolymers under study appear to be semicrystalline materials. In contrast to the majority of flexible-chain polymers, the non-crystalline phase of both stiff-backbone copolyesters is not an amorphous but a mesomorphous one. It was shown that this structure may be identified as an ordered LC smectic state. However, the main difference between the non-crystalline structures of CPE-1 and CPE-2 consists in periodic or aperiodic packing of layers within the LC smectic phase, respectively.

In a case of CPE-1, PHQ groups containing large side phenyl rings play the role of distortions incorporated into the regular crystalline lattice. As a consequence, crystallites with relatively small lateral (<100 Å) and relatively large longitudinal (>1000 Å) dimensions are formed, indicating that we deal with the crystallization of a random copolymer. On the other hand, the PHQ group is the most thermostable unit in macromolecules of CPE-1 end responsible for the formation of layers of LC smectic phase. The onset of their local mobility on heating of the material results in the start of the cooperative mobility of the chains on the whole leading to the glass transition process and as a consequence to the destruction of layers linked by the parallel packing of side-chain phenyl rings before (on cooling of the as-spun fibers).

In a case of CPE-2, the most thermostable monomer is HNA. Only when the local mobility of HBA units is setting in during heating of the material the glass transition process takes place giving rise to a phase transition of the low temperature form of the orthorhombic crystalline modification into the high temperature one with pseudohexagonal structure. The non-crystalline phase of CPE-2 with aperiodic structure does not change substantially at this moment and coexists with the crystalline phase up to its melting point.

We believe that the mechanism proposed for the description of phase and relaxation phenomena in the two copolymers under study might have universal character for any other representatives of thermotropic LC main-chain copolyesters and might reveal their temperature behavior and phase composition. We also believe that taking into account the data presented here and based only on the knowledge of the chemical formula of the copolymer, one may predict its glass transition temperature, ability to crystallize and depending on, whether its monomeric units have the same or different lengths, its phases, and type of structure - periodic or aperiodic - which will be observed in both crystalline and non-crystalline phases of the material examined. We hope that further investigations (which are in progress now) will test and prove the validity of this statement.

Acknowledgments. This study was supported by Russian Fundamental Science Foundation (Grant No 94-03-08161) and INTAS Foundation (Grant No 94-4469). Their financial support is gratefully acknowledge. We would like to express also our sincere gratitude to Prof. A. Isayev for his considerable assistance concerning our participance and scientific presentation at the 209 Spring ACS Meeting 1995, in Anaheim, California.

Literature Cited

1. Hanna, S.; Windle, A. *Polymer.* **1988**, *vol.29,* no.2, p.207.
2. Biswas, A.; Blackwell, J. *Macromolecules.* **1988**, *vol.21,* no.12, pp.3146, 3152, 3158.
3. Li, L.S; Lieser, G.; Rosenau-Eichin, R.; Fischer, E.W. *Makromol. Chem., Rapid Commun.* **1987**, *vol.8,* p.159.
4. Bechtold, H.;Wendorff, J.H.;Zimmermann, H.J. *Macromol. Chem.* **1987**, *vol.188,* no.4, p.651.
5. Kaito, A.; Kyotani, M.; Nakayama, K. *Macromolecules.* **1990**, *vol.23,* no.4, p.1035.
6. Schwarz, G.; Kricheldorf, H.R. *Macromolecules.* **1991**, *vol.24,* no.10, p.2829.
7. Allen, R.A.; Ward, I.M. *Polymer.* **1991**, *vol.32,* no.2, p.202.
8. Percec, V.; Zuber, M.; Ungar, G.; Alvarez-Castillo. *Macromolecules.* **1992**, *vol.25,* no.5, p.1195.
9. Tsvankin, D.Ya.; Papkov, V.S.; Zhukov, V.B.; Godovsky, Yu. K.; Svistunov, V.S.; Zhdanov, A.A. *Polym. Sci., Part A: Polym. Chem.* **1985**, *vol.23,* no.7, p.2043.
10. Antipov, E.; Kuptsov, S.; Kulichikhin, V.; Plate′, N. *Makromol. Chem., Macromol. Symp.* **1989**, *vol.26,* p.69.
11. Antipov, E.; Stamm, M.; Fischer, E.W. *J. Mater. Sci.* **1994**, *vol.29,* p.328.
12. Antipov, E.; Stamm, M.; Fischer, E.W. *Am. Chem. Soc., Polym. Prepr. Am. Chem. Soc., Div. Polym. Chem.* **1992**, *vol.33,* no.1, p.300.
13. Antipov, E.; Stamm, M.; Fischer, E.W. *Mater. NATO Workshop on Crystallization of Polymer.* Kluwer Acad., Amsterdam, Ho, 1993.
14. Johnson, R.L.; Cheng, S.Z.D. *Macromolecules.* **1993**, *vol.26,* no.1, p.94.
15. Blackwell, J.; Lieser, G.; Gutierrez, G.A. *Macromolecules.* **1983**, *vol.16,* p.1418.
16. Hanna, S.; Windle, A.H. *Polym. Commun.* **1988**, *vol.29,* p.236.

17. Butzbach, G.D.; Wendorff, J.H.; Zimmermann, H.J. *Polymer.* **1986**, *vol.27*, no.9, p.1337.
18. Spontak, R.J.; Windle, A.H. *J. Polym. Sci., Part B: Polym. Phys.* **1992**, *vol.30*, no.1, p.61.
19. Hanna, S.; Lemmon, T.J.; Spontak, R.J.; Windle, A.H. *Polymer.* **1992**, *vol.3*, no.1, p.3.
20. Lin, Y.G.; Winter, H.H. *Macromolecules.* **1991**, *vol.24*, no.10, p.2877.
21. Nicholson, T.M.; Mackley, M.R.; Windle, A.H. *Polymer.* **1992**, *vol.33*, no.2, p.434.
22. Allen, R.A.; Ward, I.M. *Polymer.* **1991**, *vol.32*, no.2, p.202.
23. Allen, R.A.; Ward, I.M. *Polymer.* **1992**, *vol.33*, no.24, p.5191.
24. Godovsky, Yu.K.; Volegova, I.A. *Polymer Science, Ser.A.* **1994**, *vol.36*, no.3, p.419.
25. Hanna, S.; Romo-Uribe, A.; Windle, A.H. *Nature.* **1993**, *vol.366*, p.546.
26. Naoki, S.; Akira, N. *J. Toso Kenkyu Hokoku.* **1992**, *vol.36*, no.1, p.87.
27. Plate', N.A.; Kulichikhin, V.G.; Antipov, E.M. *Polymer Science, Ser.A.* **1993**, *vol.35*, no.11, p.1743.
28. Antipov, E.M.; Volegova, I.A.; Artamonova, S.D.; Godovsky, Yu.K. *Polymer Science, Ser.A.* **1994**, *vol.36*, no.9, p.1258.
29. Antipov, E.M.; Stamm, M.; Abetz, V.; Fischer, E.W. *Polymer Science, Ser.A.* **1994,** *vol.36*, no.11, p.1516.
30. Antipov, E.M.; Stamm, M.; Abetz, V.; Fischer, E.W. *Polymer Science, Ser.A.* **1995,** *vol.37*, no.1, p.60.
31. Antipov, E.M.; Stamm, M.; Abetz, V.; Fischer, E.W. *Acta Polymerica.* **1994,** *vol.45*, p.196.
32. Antipov, E.M.; Stamm, M.; Abetz, V.; Fischer, E.W. *Colloid. Polym. Sci.* **1995,** *vol.273*, p.23.
33. Yonetake, K.; Sagiya, T.; Koyama, K.; Masuko, T. *Macromolecules.* **1992,** *vol.25*, no.2, p.1009.
34. Cao, M.Y.; Wunderlich, B.J. *J. Polym. Sci., Polym. Phys. Ed.* **1985,** *vol.23*, no.3, p.521.
35. Kuz'min, N.; Matukhina, E.; Antipov, E.; Plate', N. *Makromol. Chem., Rapid Commun.* **1992,** *vol.13*, p.35.
36. Kuz'min, N.N.; Matukhina, E.V.; Polikarpov, V.M.; Antipov, E.M. *Polymer Science.* **1992,** *vol.34*, no.2, p.63.
37. Fischer, E.W.; Herchenroder, P.; Manley, R.St.; Stamm, M. *Macromolecules.* **1978,** *vol.11*, no.1, p.213.
38. Kuz'min, N.; Matukhina, E.; Makarova, N.; Polikarpov, V.; Antipov, E.M. *Makromol. Chem., Macromol. Symp.* **1991,** *vol.44*, p.155.
39. Ovchinnikov, Yu.; Antipov, E.; Markova, G.; Bakeev, N. *Makromol. Chem.* **1976,** *vol.177*, no.2, p.1567.
40. Wunderlich, B.; Moller, M.; Grebowicz, J.; Baur, H. *Adv. Polym. Sci.* **1988,** *vol.87*, p.1.
41. Antipov, E.M.; Kulichikhin, V.G.; Plate', N.A. *Polym. Eng. Sci.* **1992,** *vol.32*, no.17, p.1188.
42. Plate', N.A.; Antipov, E.M.; Kulichikhin, V.G.; Zadorin, A.N. *Polymer Science.* **1992,** *vol.34*, no.6, p.498.

Chapter 18

New Generation of Mesophases

Hydrogen Bonds and Microphase Separation in the Formation of Liquid-Crystal Structure in Nonmesogenic Species

R. V. Talroze, S. A. Kuptsov, T. I. Sycheva, G. A. Shandryuk, and N. A. Platé

A. V. Topchiev Institute of Petrochemical Synthesis, Russian Academy of Sciences, Leninsky Prospekt 29, Moscow 117912, Russia

A new approach to create mesomorphic ordered systems is discussed. Polyacids form H-bonded complexes with tertiary amines and their hydrochlorides having both mesogenic and nonmesogenic structure. The main feature of the order is the combination of the small angle maximum and amorphous halo in the X-ray patterns indicating the formation of lamellar smectic-like phase in solid complexes. The analysis of the structure in both crystalline and lamellar phases is given. On the basis of textured X-ray patterns the model for the layered structure is proposed. The change of the length and the structure of amine residue results in the change of the interplanar spacing in accordance with the model proposed. Low molecular weight models based on malonic and glutaric acids show the same tendency to mesophase formation. The mesomorphic behavior of complexes is interpreted in terms of microphase separated systems.

In recent years the LC world strongly increased due to the creation of mesomorphic structures of different types including LC and mesophase polymers. Together with well known synthetic routes which result in mesogen-containing liquid crystals, main and side chain LC polymers there appeared quite different approach. Molecular recognition principle so productive in biochemistry and molecular biology turned out to be very efficient in synthetic organic chemistry. It relies on steric effects, electrostatic interactions, and, ultimately, hydrogen bonding and results in many systems where non-covalent binding of different but on certain sense complementary to each other molecules or moieties neither one of which taken individually shows LC behavior but being sticked one to another create a LC ordering. For example, dimerization of p-alkoxybenzoic (*1,2*) or trans-alkyl-cyclohexanecarboxylic acids (*3*) was shown to result in mesogenic calomitic-like fragments composed of hydrogen-bonded molecules. A similar phenomenon was observed in binary systems comprising

0097–6156/96/0632–0304$16.75/0

bipyridines and alkoxybenzoic acids (*4,5*) and cis-s-cis aminoketones (*6*). Coordination with metals was demonstrated as a reasonable way to produce LC structures in metallo-complex compounds (*7*).

One of examples of controlled molecular design was reported by Brienne et al. (*8*) who prepared so-called supramolecular discotic like LC phases by associating complementary heterocyclic components. Discotic phases were also observed in systems of associated diols (*9*).

Kato, Frechet et al. (*10-12*) and Bazuin (*13*) applied a similar principle to prepare side chain LC polymers whereas Griffin (*14*) show the possibility to design main chain systems. In all cases intermolecular hydrogen bonds are involved in the formation of mesogenic-like fragments either in the side chains or in backbones of macromolecules.

The similar approach seems to be fruitful to create thermotropic mesophases based on amphiphilic systems where neither mesogenic structure nor strong shape anisotropy but the microphase separation process is responsible for the mesomorphic behavior as such. The main objective of this paper is to give some examples which show the correctness of this idea.

The design of an amphiphilic system due to non covalent binding strongly depends on the right choice of "complementary" components which is based on the well known ideas of the lyotropic mesomorphism. That is why we discuss here the structure and phase behavior of poly-and dicarboxylic acids complexes with tertiary amines (β-N-dimethylamino-4-alkyl-and alkyloxypropiophenones) and amine hydrochlorides having the following structure:

$(-CH_2-\underset{COOH}{CH}-)_n$ - (PAA), $(-CH_2-\underset{COOH}{C(CH_3)}-)_n$ - (PMAA),

$HOOH-CH_2-COOH$ - (MA) $HOOH-(CH_2)_3-COOH$ - (GA)

$(CH_3)_2-N-(CH_2)_2-CO-$ ⟨◯⟩ -R,

where **R = $C_{12}H_{25}$ (AC12), -⟨ ⟩- C_5H_{11} (AC5)**

$HCl{*}(CH_3)_2-N-(CH_2)_2-CO-$ ⟨◯⟩ -R

where **R=OC_8H_{17} -(AC8HCl), -$OC_{10}H_{21}$ (AC10HCl),**

-$OC_{12}H_{25}$ (AC12HCl), -⟨ ⟩- C_5H_{11} (AC5HCl)

This permits to compare the mesomorphic tendency in water solutions of corresponding salts with that in bulky complexes. At the same time one can determine how the length and chemical structure of the lyophobic fragment of the molecule as well as hydrogen bond interaction influences the general regularities for the formation of the ordered structure.

Lyotropic Phases in Amine Hydrochlorides Water Solutions.

Amines under study are insoluble in water whereas amine hydrochlorides exhibit typical properties of cationic amphiphiles which result from hydrophobic interactions between the non-polar "tails" and the solvent. The measurements of the conductivity (χ) of aqueous solutions (Figure 1) show the sharp rise of χ within the narrow concentration range indicating the spontaneous aggregation of individual molecules to form micelles at the critical micelle concentration (c.m.c.$_1$). The increase of c.m.c. with the lengthening of the alkoxy group (Table I) is in a good agreement with well known regularities described for ionogenic surfactants (*15*).

Table I. Critical micell concentration

Amphiphil	*c.m.c.$_1$*10^4, mol/l*	*c.m.c.$_2$*10^2 mol/l*
AC5HCl	7.53	2.05
AC8HCl	6.59	1.49
AC10HCl	6.09	1.02
AC12HCl	4.40	-

At the same time the chemical structure of the non-polar fragment strongly influences the value of the c.m.c.$_1$ which is one order of magnitude lower for aqueous solutions of β-N-dimethylamino-4-alkyloxypropiophenones than that measured for simple alkyl chains surfactants having the same length. But the stiffness of this hydrophobic fragment (pentylcyclohexyl derivative) increases the c.m.c.$_1$ even in the case the "tail" is shorter.

With the increase of the amine hydrochloride content one can observe the c.m.c.$_2$ when opalescence appears. It is 1.5-2 orders higher than c.m.c.$_1$ (Table I) and in accordance with (*15*) should correspond to the formation of anisotropic micelles.

If the concentration of the surfactant attains 10 wt.% wt. birefrigent confocal texture appears indicating the formation of the LC lyotropic phase which exists in broad temperature and concentration ranges (*16*).

At wide angles, X-ray diffraction patterns of the LC solutions containing up to 80 wt.%. of the amine hydrochloride show a single diffuse maximum with a Bragg's distance equal to 4.9 nm, indicating that there is only a short range order in the arrangement of the long molecular axes in the mesophase (Figure 2). In addition to a diffuse scattering, solutions containing more than 40-55 wt.% of surfactants give two diffraction maxima $\mathbf{d_1}$ and $\mathbf{d_2}$ (Table II) at small angles. With increasing concentration their intensity enhances and the position shifts to wider scattering angles (Figure 2, Table II). Comparing $\mathbf{d_1}$ with the length of the molecule of the corresponding salt leads one to suggest that, in solution, the molecules of the amine salts are arranged in a two-layer structure with the partial overlapping of hydrophobic "tails" (Figure 3). For the packing, observed within the entire range of the conditions of its existence, the $\mathbf{d_1/d_2}$ ratio is equal to 2, confirming the formation of a lamellar structure. A

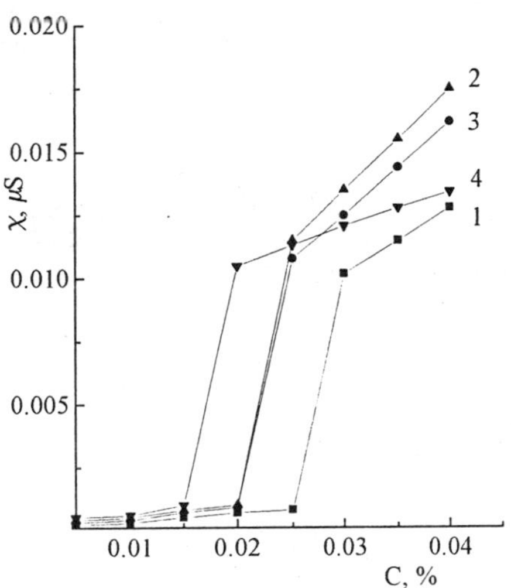

Figure 1. Conductivity as a function of concentration for water solutions of AC5HCl (1), AC8HCl (2), AC10HCl (3) and AC12HCl (4).

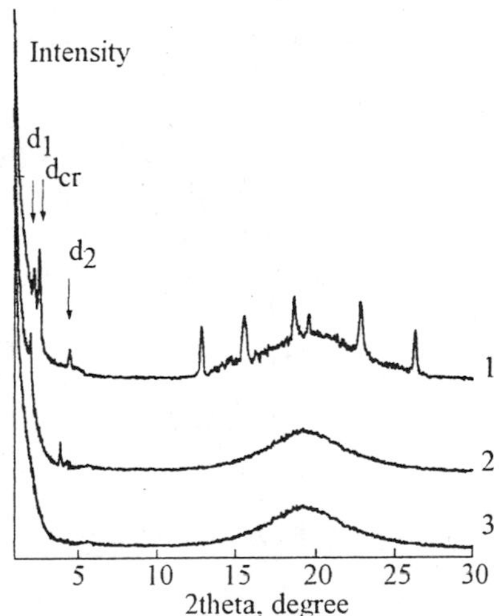

Figure 2. X-ray diagrams of AC5HCl water solutions at (1) 85, (2) 65 and (3) 35 wt. % concentration.

decrease in interlayer distance, observed when the concentration of the salt is increased, may be caused by thinner aqueous interlayers and compaction of the structure. At a concentration above 80 wt.%., the lamellar and crystalline phases coexist, as indicated by appearance of additional reflections (Figure 2).

Table II. X-ray data - interlayer distances for ACHCl water solutions

		AC5HCl			AC12HCl		
	c_{wt}, %	d_1, nm	d_2, nm	d_1/d_2	d_1, nm	d_2, nm	d_1/d_2
1	40	6.59	3.09	2.1	-	-	-
2	55	-	-	-	5.89	2.89	2.0
3	60	4.67	2.34	2.0	5.68	-	-
4	65	4.41	2.27	2.0	5.52	2.76	2.0
5	70	4.54	2.22	2.0	5.25	-	-
6	75	4.04	2.01	2.0	5.2	2.57	2.0
7	80	3.97	1.95	2.0	4.99	-	-
8	85	3.91	1.96	2.0	5.29	3.79	1.4
9	90	3.88	1.97	1.9	5.29	3.79	1.4

Thus, the examined low molecular mass amine salts show no tendency to thermotropic mesomorphism, whereas their concentrated aqueous solutions conform the main regularities valid for the lyotropic mesomorphism of ionogenic surfactants.

Solid Complexes: Type of Binding

It is well-known that reaction between carboxylic acids and amines may proceed via several pathways involving the formation of H-bond complexes (Type 1), charge transfer complexes (type 2) and, finally, complexes (salts) with complete proton transfer (type 3):

$$\mathrm{RCOOH+NR_3} \rightarrow \begin{cases} \mathrm{R\text{-}C(=O)\text{-}OH \cdots NR_3} & 1 \\ \mathrm{R\text{-}C(=O)\text{-}O^{\delta -} \cdots H^{\delta +}NR_3} & 2 \\ \mathrm{R\text{-}C(=O)\text{-}O^{-} \cdots H\overset{+}{N}R_3} & 3 \end{cases}$$

If examined amines are not soluble in water their solubility in ethanol is very high and the occurrence of a reaction with polyacids is easily detectable because mixing the initially transparent solutions causes a precipitate to form. In contrast to polymer acids, the products of reactions involving dicarboxylic acids remain in solutions.

However, regardless of whether a carboxylic acid is polymeric or a low molecular mass compound the reaction products in a solid state show the same spectral features: in the IR spectra of the reaction products, the bands due to ν(C-H) vibrations in methyl groups linked to the nitrogen atom in initial amines (2820 cm^{-1}) and the ν(C-N) vibrations (1200 cm^{-1}) as well as the ν(O-H) in the acid carboxylic group (2600-3600 cm^{-1}) disappear indicating proton transfer (17). The extent of proton transfer was analyzed in detail in (18); the lack of spectral features associated with the carboxylate anion and the appearance of two relatively weak bands at 2450 and 1900 cm^{-1} were attributed to the formation of a strong hydrogen bond with marked proton polarization (type 2). Formation the type 2 complex is supported by the similarity of the electronic density distribution in the molecules of examined complexes and to the specially synthesized organometallic complex A_2ZnCl_2 (*18*).

The characteristic complex composition corresponds to the ratio 2 COOH-groups :1 amine molecule which does not depend on the structure of the amine residue. As it is shown in (*18*) with low molecular dicarboxylic acids - malonic (MA) and glutaric (GA) the main reason for non stoichiometrical complex composition is the formation of intramolecular H-bonds. MA forms stable intramolecular H-bonds whereas GA is not able to form them but at the same time it is the best structure model for the section of the polyacrylic acid chain between two carboxylic groups. The analysis of IR spectra shows that MA forms characteristic complexes of the same composition as PAA does whereas both COOH groups in GA can react giving either or 2:1 and 1:1 complexes.

When hydrochloric salts (instead of amines) react with polyacids in ethanol solution, the precipitate obtained after evaporating the solvent manifests some spectral features which makes it different from the simple physical blend of initial reagents. The fact that the spectral features of the initial components are retained, whereas the ν (OH) band at 2600-3600 cm^{-1} partially disappears (Figure 4), suggests that the structure formed is stabilized by hydrogen bonds, and the reaction could be described by the following scheme:

$$AH^{+} \overset{1}{\cdots} Cl^{-} + HO{-}C({=}O){-} \quad = \quad AH^{+} \overset{1}{\cdots} Cl^{-} \overset{2}{\cdots} H \cdots O{-}C({=}O){-}$$

The cation NH^{+} preserves its ionic structure but the complexation weakens the H-bond 1. This results in the shift of $\nu(NH^{+})$ into the high frequency region.

Contrary to the acid-amine complexes the characteristic complex composition in this case corresponds to the ratio 1COOH-group : 1 salt molecule. This difference seems to be explained by the change of the proton affinity when one changes from neutral nitrogen atom (in the tertiary amine) /963.7 kJ/mol (*19*) to a chlorine anion (1395.3 kJ/mol (*20*)/. The latter being the stronger proton acceptor is able to distort intramolecular H-bonds in the polyacid, which may be responsible for the formation of 2:1 PAA-amine characteristic complex.

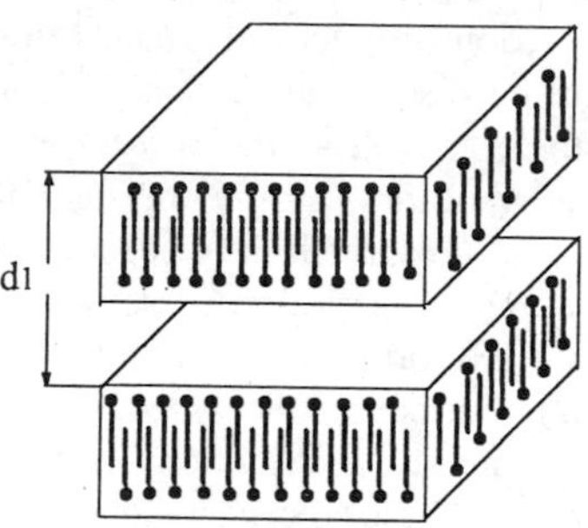

Figure 3. Model of lamellar structure of AHCl water solution.

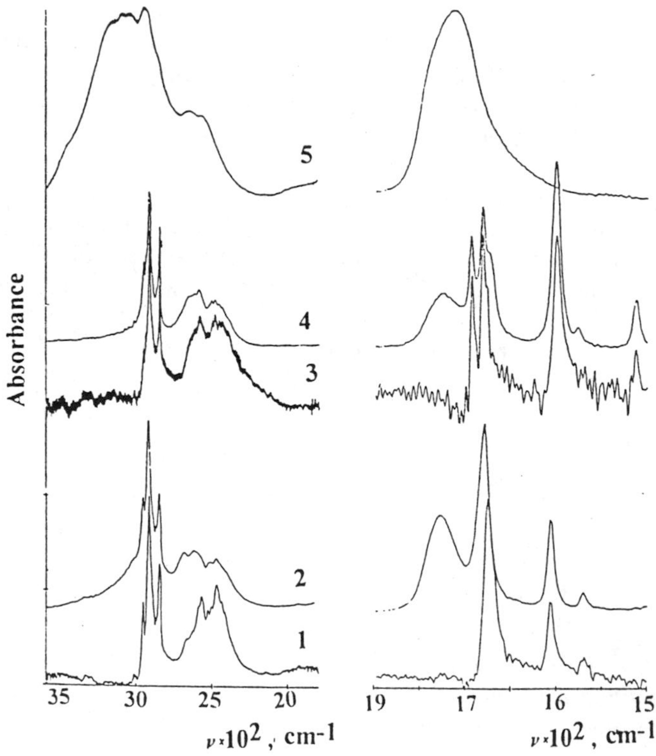

Figure 4. FTIR spectra of PAA (5), AC5HCl (1), AC12HCl (3) and 1:1 complexes (2,4).

LC Order in Polymer Complexes

Polyacid-amine systems. 2:1 complexes of polyacids with amines precipitating out of the reaction mixture are solid substances with very low crystallinity (about 2-3%) which is seen in X-ray diagrams (Figure 5, curve1) as diffraction maxima of low intensity at both wide and small angles combined with an intensive amorphous halo. Low crystallinity means that the hydrogen bonds prevent the crystallization of amine when complexes are formed at the conditions mentioned above.

The melting and crystallization of complexes proceeds in the rigid polymer matrix without any change in its aggregative state. Above the melting point the structure of complexes is characterized by small angle maximum and diffused halo at wide angles of X-ray scattering (Figure 5, curve 2) and optically anisotropic texture. These data indicate the order of the complex structure above the melting point. Interlayer distance **d** corresponding to the small angle maximum does not change at Tm. It means that below Tm complex contains two phases whereas above Tm the only one ordered phase exists in a very broad temperature range and disappears at the clearing point Tc (Figure 5, curve 3). As an example the DSC curve for the 2:1 complex PAA:AC12 is given in Figure 6 (curve 1). Endothermic peak at 150 ^{o}C corresponds to the disappearance of both the small angle maximum (Figure 5, curve 3) and the birefrigent texture. This transition is reversible although the structure is strongly sensitive to the cooling conditions.

At the same time **d**-spacing increases with the heating of the sample within the temperature range corresponding to the ordered structure of PAA complex (Figure 7, curve 1). One can suppose that the change of the slope in the curve **d**(T) relates to the complex glass transition, and the step in the DSC curve (Figure 6, curve 1) in the same temperature range confirms this idea. Tg of the 2:1 complex (90 ^{o}C) is lower than that for a pure PAA (106 ^{o}C). Contrary to PAA complexes **d**-spacing in complexes of PMAA does not change practically till the clearing point (Figure 7, curve 3) indicating much higher glass transition temperature.

The interplanar spacing depends on the length of the amine molecule (Figure 8): it increases with the lengthening of the amine fragment.

The orientation of the complex sample above Tg results in the textured X-ray pattern (Figure 9a) indicating the normal orientation of amine fragments to the deformation axes. Comparison of **d**-values with the certain length of amine molecules and polymer backbone thickness shows the formation of the smectic-like structure with overlapped alkyl groups independently on the structure of the amine molecule - mesogenic-like or non-mesogenic (Figure 9b).

With the decrease in the amine content the tendency to form crystalline phase becomes weaker. At the same time the change of the complex composition influences the structure of the smectic-like phase only slightly. Even small content of amine residues of about 10-20 % is enough to induce this kind of layer order in the polymer H-bond complexes. With the decrease of the content of amine in the complex the small angle maximum shifts to the wide angle region (Figure 10), and it corresponds to the decrease in the interlayer distance (Figure 8). The broadening of the maximum observed seems to result from the decrease of the correlation regions size.

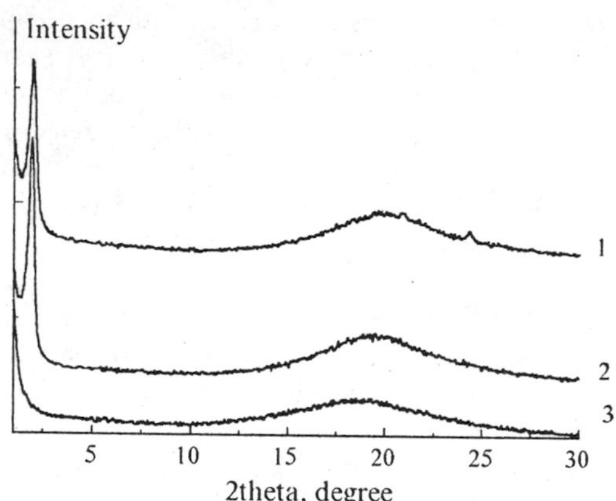

Figure 5. X-ray diffraction patterns of 2:1 complex of PAA:AC12 at 20 (1), 100 (2) and 150 ^{o}C(3). (Reproduced with permission from ref. *22*. Copyright 1996 A Publication of the American Chemical Society)

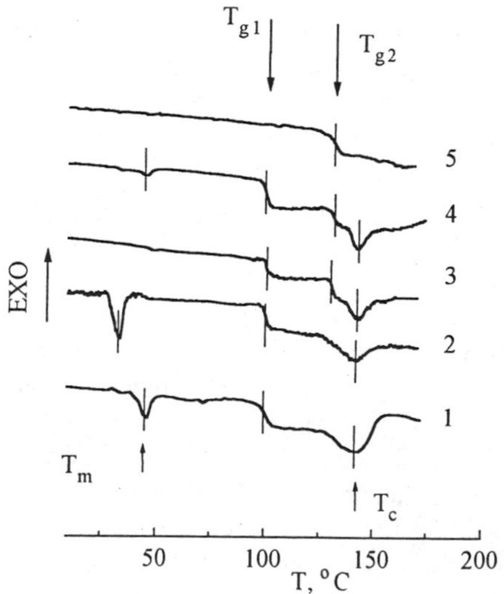

Figure 6. DSC curves for complexes PAA:AC12 2:1 (1, 2), 5:1 (3, 4) and PAA (5) after solvent evaporation (1, 3). Curves (2, 4, 5) - second heating. (Reproduced with permission from ref. *22*. Copyright 1996 A Publication of the American Chemical Society)

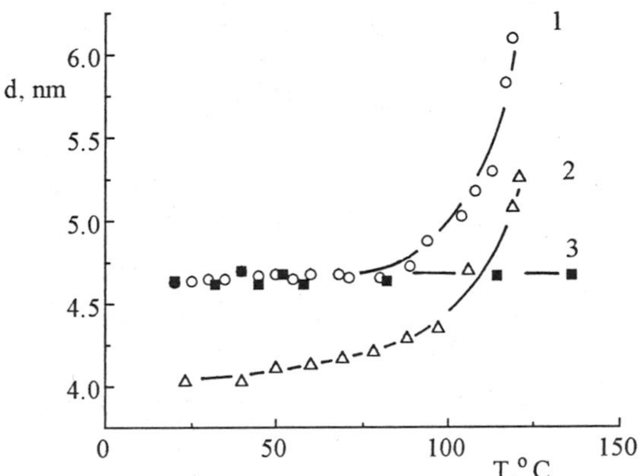

Figure7. Temperature dependence of interlayer distance **d** for complexes PAA-AC12 2:1(1), 5:1 (2) and PMAA:AC12 2:1 (3). (Reproduced with permission from ref. *22*. Copyright 1996 A Publication of the American Chemical Society)

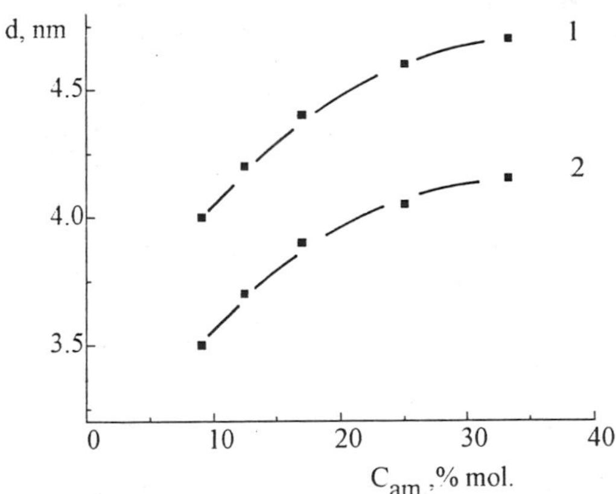

Figure 8. Interlayer distance **d** as a function of PAA:AC12 (1) and PAA:AC5 (2) complex composition.

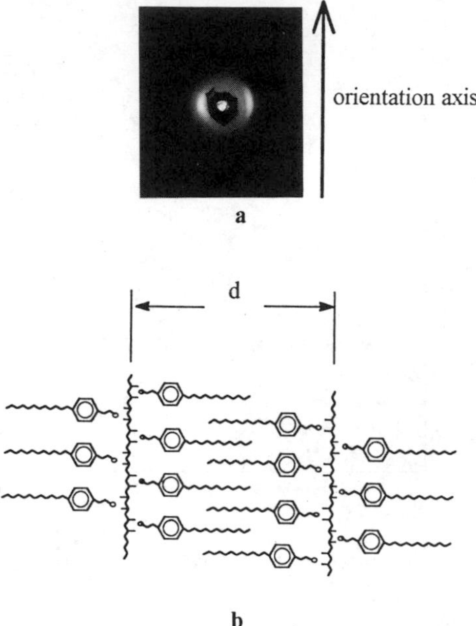

Figure 9. Small-angle X-ray diffraction pattern (a) and model (b) of PAA:ACn 2:1 complex. (Reproduced from reference 22. Copyright 1996 American Chemical Society.)

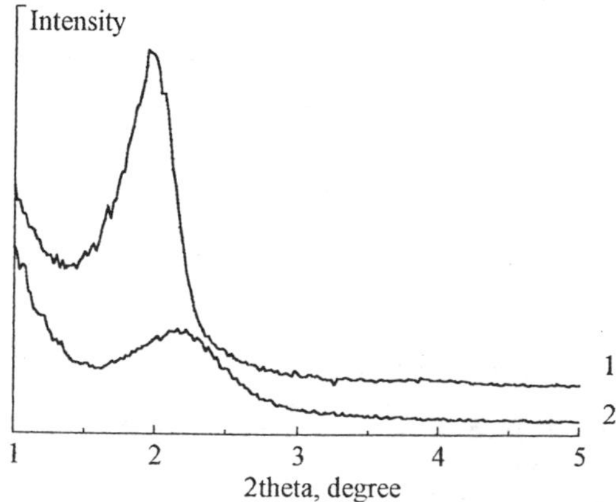

Figure 10. X-ray diffraction patterns of PAA:AC12 complexes of different composition: 2:1 (1) and 5:1 (2).

At the same time the temperature curve for the **d**-spacing in 5:1 complex (Figure 7, curve 2) is similar to that for 2:1 complex discussed above (Figure 7, curve 1) indicating practically the same glass transition temperature although the change of the copolymer composition should contribute in the glass transition temperature.

To explain such unexpected behavior one has to discuss the structure of non-characteristic ("diluted") complexes as such. There exist two possibilities for the distribution of "substituted" units of PAA in the chain. The first one corresponds to the statistical arrangement of complex units along the chain. In such a case the value of the resultant Tg should be between two glass transition temperatures corresponding to 2:1 complex and pure PAA.

On the other hand one can suppose the block distribution of different units along the chain. The latter could result in the microphase separation and the existence of two different Tg. However taking into account the small difference between the Tg corresponding to 2:1 complex prepared from solution (90 ^{o}C) and pure PAA (106 ^{o}C) it is not possible to make a choice between these structures for "diluted" complexes.

The situation becomes more clear if one analyses the DSC data (Figure 6) obtained for complexes which are cooled after the heating above the clearing point. In this case the crystallinity degree for 2:1 (Figure 6, curve 2) complex strongly increases whereas the glass transition temperature and the clearing point do not change.

At the same time one can observe two different Tg in "diluted" complexes (Figure 6, curves 3,4). The first one Tg_1 correlates with Tg indicated from X-ray data and coincides with that for 2:1 complex. The second Tg_2 corresponds to Tg of the waterless PAA (after annealing at 140 ^{o}C) (Figure 6, curve 5). Thus the idea on the possible microphase separation (PAA and complex blocks) in diluted polyacid complexes seems quite reasonable.

Coming back to the main reason for the formation of smectic-like supramolecular structure in complexes studied one can compare their structure with that of lamellar phase resulted from the packing of amphiphilic molecules of amine hydrochlorides in water.

We can suppose that weak amphiphilic properties of amine molecules become stronger due to hydrogen bonding with polyacids. As a result microheterogeneous system appears which contains regions where the interaction of alkylaromatic residues takes place and regions where polar groups are localized. In this case polyacid performs like "aqueous" interlayer.

Polyacid-amine hydrochloride systems. To discuss the structure and phase behavior of polyacid-amine hydrochloride complexes one has to mention that the salts under study crystallize in to different crystal modifications which convert into each other with the change of temperature. This transition at T_1 (Figure 11, Table III) for AC12HCl corresponds to the increase of the small angle maximum with the interlayer spacing equal to 3.54 nm and to the strong change of the set of diffraction maxima at wide angles. The temperature T_2 corresponds to the melting point.

X-ray data show the high crystallinity of 1:1 and 2:1 complexes (Figures 12 and 13, curves 1). Degree of crystallinity is about 80 and 40 % for 1:1 and 2:1 complexes of AC12HCl and 90 and 60% for AC5HCl complexes respectively (Table III). There is a great number of small- and wide angle diffraction maxima and diffused halo of

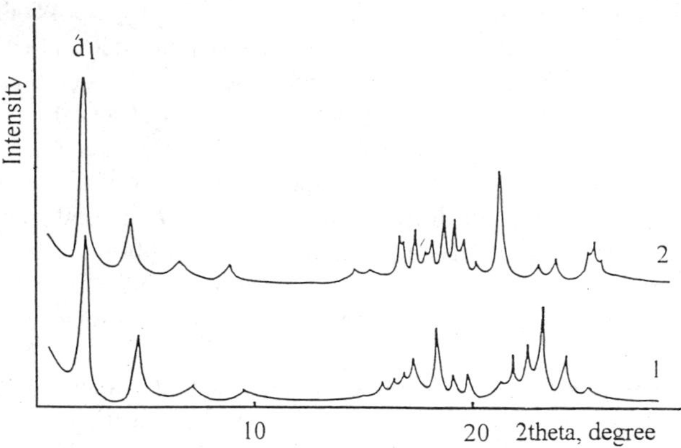

Figure 11. X-ray diffraction patterns of AC12HCl at 20 (1) and 130 °C (2).

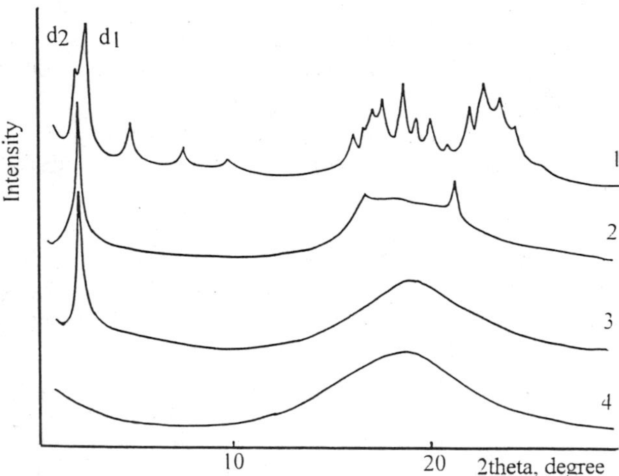

Figure 12. X-ray diffraction patterns of 1:1 complex PAA:AC12HCl at 20 (1), 105 (2), 130 (3) and 200 °C (4).

low intensity in X-ray diagrams of 2:1 complexes (Figure 13, curve 1) whereas there is no sign of the amorphous scattering in X-ray diagrams of 1:1 complexes (Figure 12, curve 1).

Table III Crystallinity degree and temperatures of phase transitions for PAA:ACnHCl complexes

Sample	*Composition*	*Crystallinity degree,%*	*Tm, ^{o}C (T_1, T_2)*	*Tc,^{o}C*
AC12HCl			110, 153	
Complex	1:1	80	95, 115	190
PAA: AC12HCl				
	2:1	40	60, 75	185
	5:1			
AC5HCl			146, 189	
Complex	1:1	90	183	196
PAA:AC5HCl				
	2:1	60	142	185
	5:1			

When compare the angle position of diffraction maxima for complexes and corresponding salts (Figures 14, curves 1-3) one can see some broadening and slight shift into the wide angles region although the general features of the X-ray pattern are still preserved. It means that the crystallization of hydrogen-bonded amine salts proceeds in the crystalline lattice with parameters closed to that for pure salts but slightly deformed. Polyacid molecules which do not hinder the salt crystallization has to be displaced not randomly but in crystallographical planes and the latter should strongly hinder the conformation set of macromolecules in such systems.

The decrease of the chlorine salts content in complexes results in the lowering of the crystallinity degree and corresponding X-ray maxima completely disappear in diagrams of 5:1 complexes (polyacid:amine hydrochloride) (Figure 14, curve 4).

Together with the maximum $\mathbf{d_1}$ equal to 3.54 nm (AC12HCl) and (AC5HCl) there is one more maximum $\mathbf{d_2}$ in X-ray patterns of 1:1 complexes (Figures 12) which is equal to 5.0 and 4.6 nm respectively. When complexes contain smaller amount of amine hydrochloride the intensity of $\mathbf{d_1}$ decreases (Figure 14), and the only one small-angle maximum $\mathbf{d_2}$ is still preserved. It prevails also in 2:1 complexes (Figure 13).

When compare DSC and X-ray data (Figures 12,13 curves 2, Table III) one can see that the heating of 1:1 and 2:1 complexes of AC12HCl above T_1 (95 and 60 ^{o}C respectively) results in the change of the crystalline structure whereas at T_2 (115 and 75 ^{o}C) crystals melt. X-ray maxima corresponding to the crystal disappear but $\mathbf{d_2}$ maximum coexisting with diffused halo is still preserved indicating the presence of the ordered structure above the melting point T_2.

This ordered system is optically anisotropic which is fixed as fine confocal texture within the broad temperature range up to Tc 190 (1:1) and 185 ^{o}C (2:1). Above the

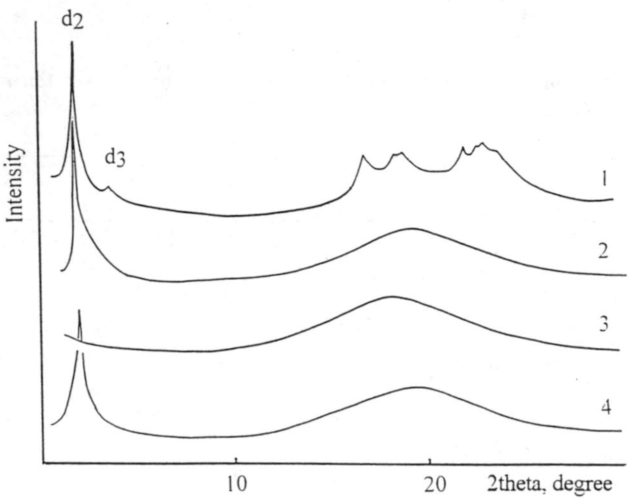

Figure 13. X-ray diffraction patterns of 2:1 complex of PAA:AC12HCl at 20 (1), 100 (2) and 200 ^{o}C; (2 - after annealing at 200 ^{o}C).

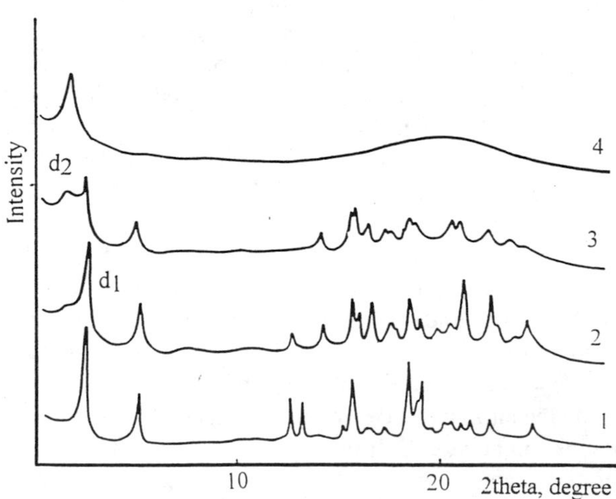

Figure 14. X-ray diffraction patterns of AC5HCl (1) and 1:1 (2), 2:1 (3), 5:1 (4) complexes of PAA:AC5HCl.

clearing point Tc optical texture becomes isotropic whereas small angle X-ray maximum disappears (Figure 12, curve 4 and Figure 13, curve 3). This process is completely reversible and with cooling the anisotropic texture and small-angle maximum appear again.

The orientation of the film prepared of 2:1 (PAA-AC12HCl) complex at temperatures above T_2 results in textured X-ray patterns at room temperature both at small ($\mathbf{d_2}$=5.0 and $\mathbf{d_3}$=2.51 nm) (Figure 15a) and wide angles (Figure 15b). The splitting of $\mathbf{d_2}$ and $\mathbf{d_3}$ in two arcs and the location of the latter in the equator of the X-ray pattern indicate the orientation of the backbone and the alignment of side fragments normally to it. The ratio $\mathbf{d_2/d_3}$ is equal to 2.

The experimental data given permit to suppose that in complexes based on polyacid and amine hydrochlorides the layered structure is formed. The main structural feature of the latter is the coexistence of the small angle maximum and amorphous halo in X-ray patterns which exist above T_2. When compare the length of amine hydrochlorides molecules with corresponding $\mathbf{d_2}$ values one can suggest the formation of the layered lamellar-like structure with the overlapping "tails" (Figure 15c) similar to that supposed for complexes based on pure amines (Figure 9b). The difference in schemes given results from the composition of the characteristic complex when it changes from polyacid:amine (2:1) to polyacid:amine hydrochloride complexes (1:1).

The presence of $\mathbf{d_2}$ maximum in X-ray patterns below T_1 and T_2 means that the layered (LC) phase coexists with the crystalline one. The slight increase in the $\mathbf{d_2}$ intensity simultaneously with the decrease in the $\mathbf{d_2}$ spacing (Figure 16) shows some perfection of the order in the lamellar phase. The strong enhancing of $\mathbf{d_2}$ intensity is accompanied by the decrease of the crystallinity degree (Figure 16a, curve 1) corresponding to the melting of the crystal phase whereas the angle position of $\mathbf{d_2}$ does not change. It means that at the melting point the crystalline phase undergoes transformation into the LC lamellar phase, and the latter dominate in the broad temperature range.

In order to understand the main regularities for the change of X-ray structure parameters with the temperature one has to come back to DSC data. Together with phase transitions mentioned in Table III one can observe the glass transition at 170 ^{o}C (Tg) which is much more evident for samples prepared by cooling than that prepared from the solution. It means that the melting and crystallization of the H-bond complexes containing amine hydrochlorides proceeds in the rigid matrix whereas the isotropic transition starts above the glass transition temperature. The sharp decrease in the small angle maximum intensity (Figure 16) and the increase in the $\mathbf{d_2}$ spacing (Figure 16) at 150 ^{o}C confirms this idea.

Thus tertiary amines or amine hydrochlorides having similar amine moieties being attached to polymer chains due to hydrogen bond interaction form stable ordered lamellar phases similar to that observed in water solutions of amphiphilic molecules.

Complexes Based on Dicarboxylic Acids.

The question arises whether the phenomena observed result from the ordering effect induced by polymer chains or it is related to the formation of the specific mesogenic

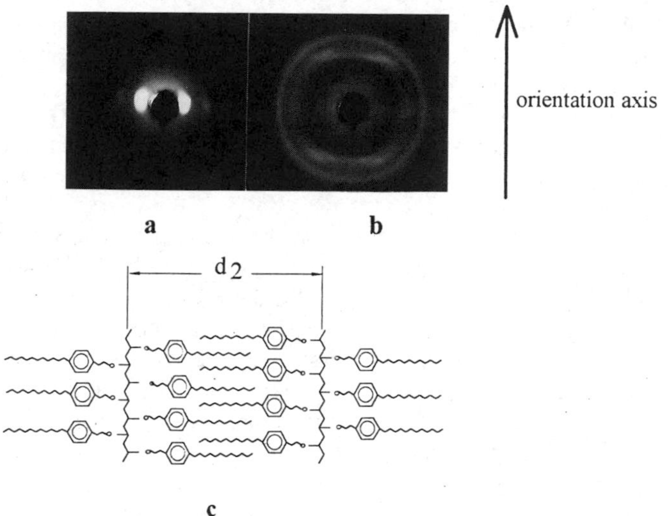

Figure 15. X-ray diffraction patterns of oriented PAA:AC12HCl (2:1) complex at small (a) and wide angles (b); c- model of the layered structure.

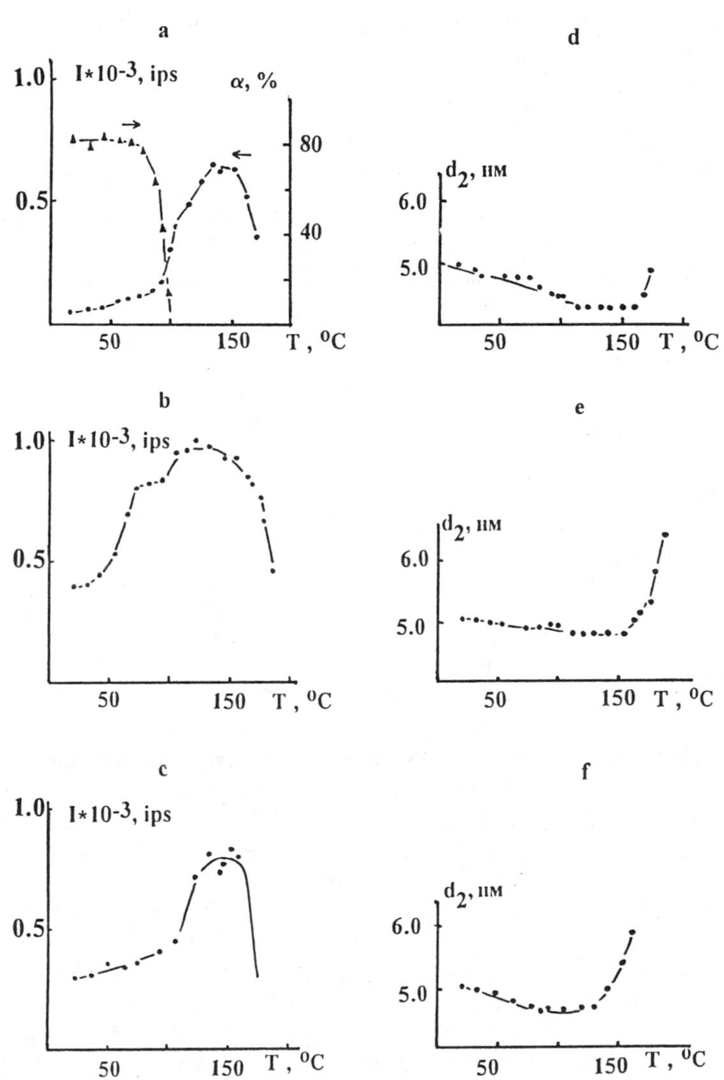

Figure 16. Temperature curves of the crystallinity degree α (a), reflection intensity **I** (a, b, c) and interlayer spacing $\mathbf{d_2}$ (d-f) PAA:AC12HCl complexes of 1:1 (a, d), 2:1 (b, e) and 5:1 (c, f) compositions.

structure due to the H-bond interaction between acids and tertiary amines (or amine hydrochlorides) under investigation. The study of model systems based on low molecular weight mono- and dicarboxylic acids helps to answer this question (*21*)

Complexes of monocarboxylic acids with amines are crystalline substances with melting points given in the Table IV.

Table IV. Temperatures of phase transitions for AC12 complexes with carboxylic acids

Acid	*Complex composition (COOH-group : amine)*	*Tm, ºC*	*Tc, ºC*
stearic	1:1	53	-
methylpropionic	1:1	42	-
benzoic	1:1	57	-
	2:1	42	-
glutaric	1:1	50	51
	2:1	56	57
malonic	2:1	82	110

(Reproduced with permission from ref. *21*. Copyright 1995 Gordon and Breach Science Publisher S.A.)

The melting of the crystalline 2:1 complexes of both MA and GA proceeds in several steps: DSC curves (Figure 17) show endothermic peaks indicating high enthalpies of melting (42 and 48 kJ/mol). At the same time one can see (shown with arrows) additional extremely small peaks ($\Delta H < 0.5$ kJ/mol). Optical microscopy data show that birefringent texture exists in the temperature range between the two peaks. It can be considered as fan- (Figure 18a) and oiled strips (Figure 18b). High mobility and the tendency to form homeotropic structure under shearing indicate the formation of the smectic-like structure which melts with the low enthalpy.

X-ray scattering measurements strictly confirm the conclusion given above. As an example X-ray diffraction patterns for 2:1 complexes of GA are shown in Figure 19. At room temperature X-ray diagrams present a great number of small- and wide angle reflections indicating the crystallinity of complexes. Most of the reflections disappear at melting points corresponding to first peaks at DSC curves (Figure 17). Above melting points complexes are characterized by small angle maximum and diffuse halo at wide angles. The latter is preserved above the second transition temperature when the small angle maximum disappears.

The set of experimental data leads to the conclusion that complexes of MA and GA with the amine studied form layered smectic-like liquid crystalline phases within the certain temperature and concentration ranges. As concerned for the structure of the smectic layer it is necessary to analyze the value of the interplanar spacing **d** corresponding to the small angle X-ray reflection for both complexes of different composition.

As it is seen from the data given in Table V one can conclude that **d** value practically does not depend on the acid structure as well as on the complex

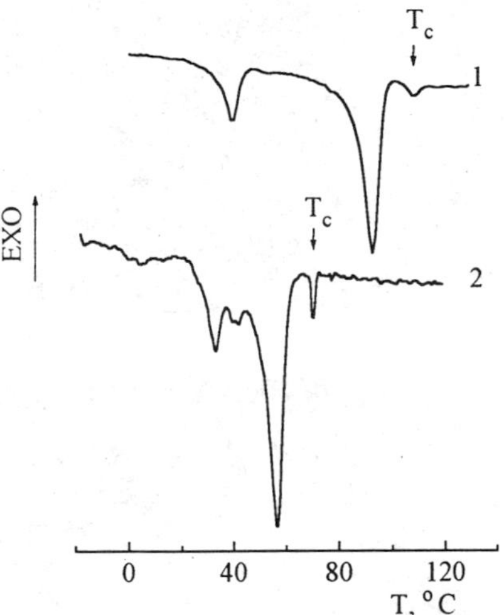

Figure 17. DSC curves for 2:1 complexes of MA (1) and GA (2) with AC12. (Adapted from ref. *21*)

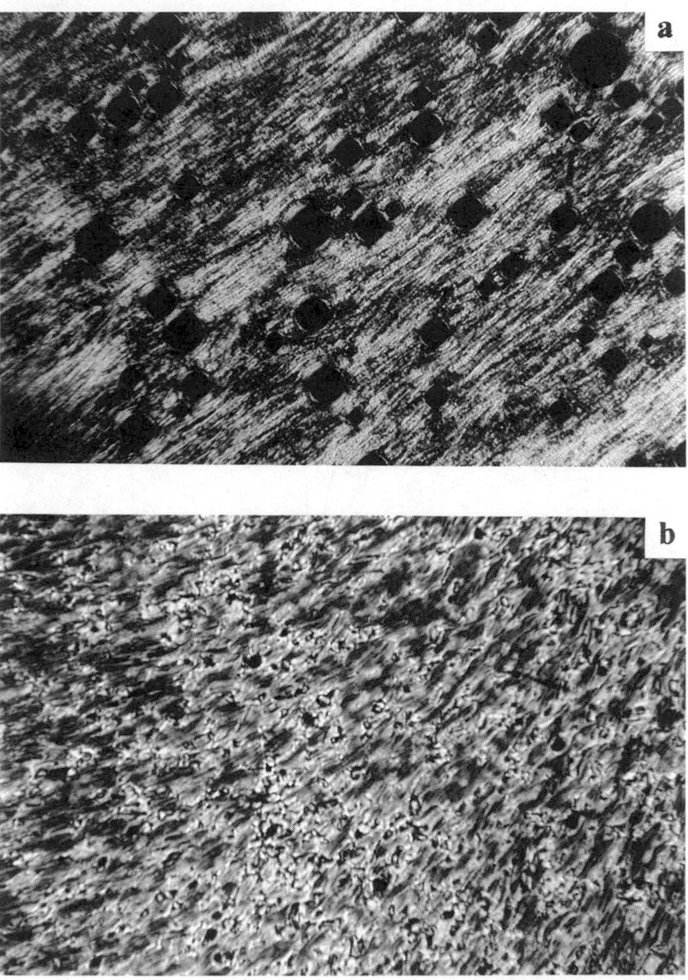

Figure 18. Optical textures of 2:1 complexes of MA:AC12 (a) and GA:AC12 (b) at 100 and 52 °C /250×/.(Reproduced with permission from ref. *21*. Copyright 1995 Gordon and Breach Science Publisher S.A.)

composition and is equal to 4.2-4.4 nm. In Figure 20 three main types of the layered structure for complexes of dicarboxylic acids are given.

Table V. Interplanar distances for lamellar phases of dicarboxylic acids complexes

Complex	*Model*	*d, nm (calculated)*	*d, nm (experimental)*
MA:AC12			
2:1	a	3.0-3.3	
	b	5.3-5.8	4.2-(at 100 ^{o}C)
	c	4.0-4.4	
GA:AC12			
2:1	a	3.2-3.5	
	b	5.5-6.0	4.4 (at 60 ^{o}C)
	c	4.1-4.6	
1:1	d	4.1-4.6	4.3 (at 48 ^{o}C)

Taking into account the length of the amine molecule (2.35-2.65 nm) and the size of dicarboxylic acids molecules equal to 0.8-1.0 nm (GA) and 0.7-0.9 nm (MA) as well as calculated values for the interplanar distance d shown in Figure one can conclude that one layered structure (a) as well as double layered structure (b) are not realized. The most realistic model corresponds to double layered ordering with overlapped aliphatic tails and acids residues (c). It shows also that both 2:1 (c) and 1:1 complexes (d) of GA have to form the identical layered structure with the same interplanar distance. This model permits to suppose that in complexes containing an excess acid its molecules are displaced between complex molecules within the domains enriched by acid residues. In this case the layered structure with the similar interplanar distance has to be preserved.

The very similar situation is observed in complexes with AHClC12. and the existence of the lamellar smectic like structure is proved by both X-ray (Figure 21) and DSC data (Figure 22).

In Figure 23 phase diagrams for GA-AC12 (a) and GA-AC12HCl (b) systems are given. One can see that in the first system the smectic phase appears at the point corresponding to 1:1 complex composition when there is no free amine in the system. The 2:1 complex has the highest temperature for smectic ⇔ isotropic transition as well as the highest melting point. Nevertheless the whole range of the smectic phase is broadened to the left side of the phase diagram where excess GA exists. It shows that the presence of the excess acid is not only possible but even necessary to stabilize smectic phase which provides the formation of the ordered LC structure being interacted with 2:1 complex due to the presence of free carboxylic groups in acid molecules. The second phase diagram differs from that mentioned above. The L.C. phase here is located at high amounts of amine hydrochloride and strongly reminds the phase behavior of isomorphous solutions.

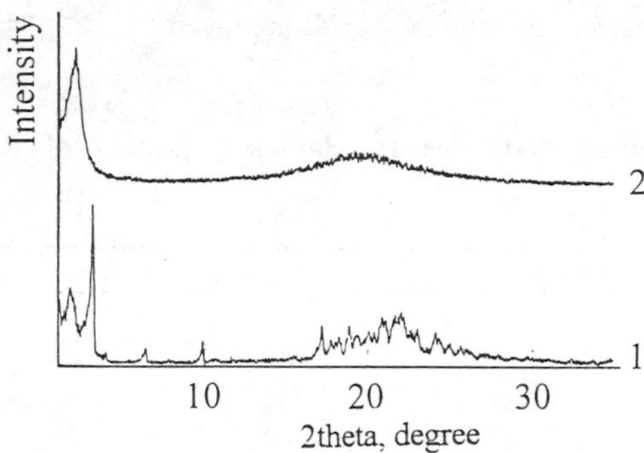

Figure 19. X-ray diffraction patterns of 2:1 complexes of GA:AC12 at 18 (1) and 60 ^{o}C (2). (Reproduced with permission from ref. *21*. Copyright 1995 Gordon and Breach Science Publisher S.A.)

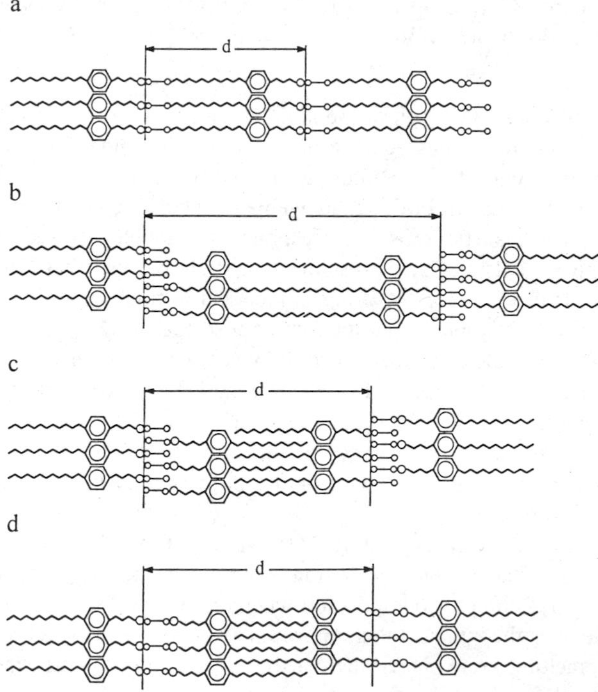

Figure 20. Models of smectic layer ordering in dicarboxylic acid:amine complexes. (Reproduced with permission from ref. *21*. Copyright 1995 Gordon and Breach Science Publisher S.A.)

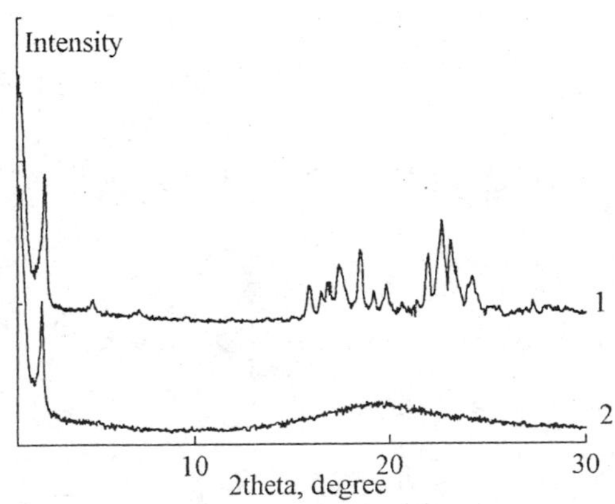

Figure 21. X-ray patterns of GA:AC12HCl (2:1) complex at 20 and 130 °C.

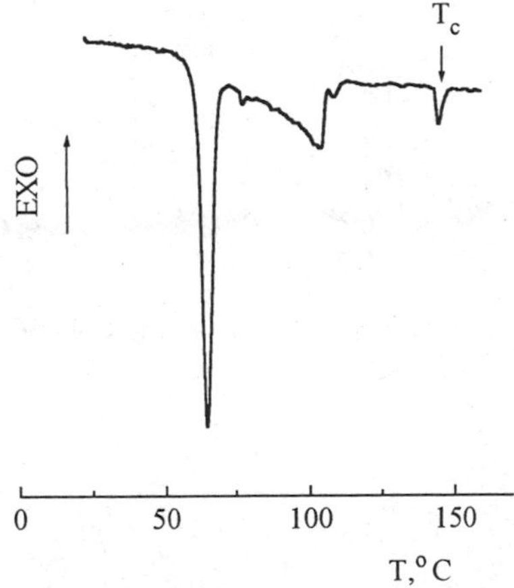

Figure 22. DSC curve for 2:1 GA:AC12HCl complex.

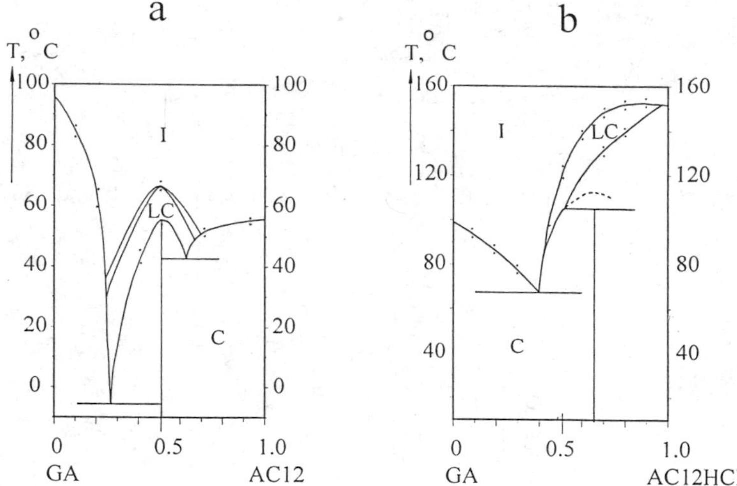

Figure 23. Phase diagrams for GA-AC12 (a) and GA-AC12HCl (b) systems.

The whole set of experimental data on the phase behavior of complexes based on dicarboxylic acids proves the fact of the stabilization of lamellar phases in hydrogen bond complexes independently of the structure of the carboxylic acid - polymer or low molecular mass one.

Conclusions

To explain the reasons for the mesomorphic behavior of complexes investigated one has to conclude that typical mesogenic cores as well as essentially high geometrical shape anisotropy can not be responsible for the creation of the LC ordering in these systems. On the other hand, the similarity in the ordering behavior of bulky complexes and water solutions of cationic amphiphiles means that hydrogen bond complexation results in the amphiphilic structure inducing micro segregation processes which have to be taken into account. In our case H-bond complexes represent micro heterogeneous systems with the layered structure resulting from the arrangement of amine residues. The latter are strongly binded with either dicarboxylic or polyacid microphase which performs as an aqueous interlayer. Such complexes could be considered as a particular group of liquid crystals which is displaced at the boundary between the thermotropic and lyotropic systems.

Acknowledgments

Authors express gratitude to Dr. V.S.Bezborodov for the synthesis of amines and T.L.Lebedeva for IR spectroscopy measurements and fruitful discussions. The research described in this publication was made possible in part by Grant No MR-300 from the International Science Foundation, Russian Foundation for Fundamental Research (Grant No 94-03-09535) and the Program "Universities of Russia".

Literature Cited

1. Bennett, M.G.; Jones, B. *J.Chem. Soc.* **1939,** N 2, p. 420.
2. Weygand, C.; Gabler, R. *Z. Phys. Chem.Abt. B.* **1940,** *vol. 46,* p.270.
3. Schuhbert, H.; Dehne, R.; Uhlig, V. *Z.Chem.* **1972,** *vol. 12,* p.219.
4. Kato, T.; Frechet, J.M. *J.Am.Chem.Soc.* **1989,** *vol. 1111,* p.8533.
5. Kresse, H.; Szulzewsky, I.; Diele, S.; Pashke, R. *Mol.Cryst.Liq.Cryst.* **1994,** *vol.238,* p.13.
6. Pyzuk, W.; Krowczynski, A.; Gorecka, E. *Liq.Cryst.* **1991,** *vol.10,* p. 593.
7. Giroud-Godquin, A.M.; Billard, J. *Mol.Cryst.Liq.Cryst.,* **1981,** *vol. 66,* p.247.
8. Brienne, M.-J.; Gabard, J.; Lehn, J.-M.; Stibor, I. *J.Chem.Soc.,Chem.Commun.* **1989,** p. 1868.
9. Ebert, M.; Kleppinger, R.; Soliman, M.; Wolf, M.; Wendorff, J.; Lattermann, G.; Staufer, G. *Liq.Cryst.* **1990,** *vol. 7,* p. 553.
10. Kato, T.; Frechet, M.J. *Macromolecules* **1990,** *vol. 23,* p. 360.
11. Kumar, U.; Frechet, M.J.; Kato, T.; Ijiie Seiji; Timura, K. *Angew.Chem.,* **1992,** *vol.31,* p. 1531.

12. Kato, T.;Kihara, H.; Kumar, U.; Fujihima, A.; Uryu, T.; Frechet, M.J. *Polym.Prepr. (Am.Chem.Soc.,Div.Polym.Chem.),* **1993,** *vol. 34,* p. 722
13. Bazuin, C.G.; Brandys, F.A. *Chem.Mater.,* **1992,** *vol. 4,* p. 970.
*14.*Lee, C.-M.; Jariwala, C.P.; Griffin, A.C. *Polymer,***1994,** *vol. 35,* p. 4550.
15. McBain, J.W.; Sierichs, W.C. *J.Amer.Oil Chem.Soc.,***1948,** *vol. 25,* p. 221.
16. Kornienko, E.V.; Sycheva, T.I.; Kuptsov, S.A.; Bezborodov, V.S.; Lebedeva, T.L.; Talroze, R.V.; Plate, N.A. *Polymer Science,Ser.A,* **1992,** *vol. 34,* p. 997.
17. Talroze, R.V.; Plate, N.A. *Polymer Science,* **1994,** *vol. 36,* p. 1479.
18. Lebedeva, T.L.; Shandryuk, G.A.; Sycheva, T.I.; Kuptsov, S.A.; Bezborodov, V.S.; Talroze, R.V.; Plate, N.A. *J.Mol.Struct.,1995,* (in press).
19. Bell, R.P. *The Proton in Chemistry;* Chapman and Hall: London, UK, **1959.**
20. Lebedeva, T.L. *J.Struct.Chim.* ***1982,*** *vol. 23,* p. 47.
21. Talroze, R.V.; Shandryuk, G.A.; Kuptsov, S.A.; Bezborodov,V.S.; Plate',N.A. *Mol.Cryst.&Liq.Cryst.,* **1995**, (in press).
22. Talroze, R.V.; Shandryuk, G.A.; Kuptsov, S.A.; Bezborodov,V.S.; Plate',N.A. Macromolecules, **1995,** (in press)

Synthesis and Properties

Chapter 19

Hybrid Liquid-Crystalline Block Copolymers

M. Laus[1], M. C. Bignozzi[1], A. S. Angeloni[1], G. Galli[2], E. Chiellini[2], and O. Francescangeli[3]

[1]Dipartimento di Chimica Industriale e dei Materiali, Università di Bologna, 40136 Bologna, Italy
[2]Dipartimento di Chimica e Chimica Industriale, Università di Pisa, 56126 Pisa, Italy
[3]Dipartimento di Scienze dei Materiali e della Terra, Università di Ancona, 60131 Ancona, Italy

The synthesis and some thermal properties of three series of block copolymers comprising both main-chain and side-chain liquid-crystalline (LC) blocks in the same macromolecular structure are described. The former block is a semiflexible LC polyester (block B), and the latter is an LC polymethacrylate (block A) containing a variously substituted mesogenic unit. The two structurally different blocks were partly phase-separated within the glassy and LC states and underwent distinct phase transitions. Significant deviations of the transition enthalpies relative to those of the corresponding homopolymers suggest the occurrence of a more or less diffuse interphase which may depend on the nature of the mesophase formed.

Block copolymers exhibit unique characteristics in that they are able to self-order to multiphase domain structures of submicron scale with various morphologies because of the relative incompatibility of the different blocks. On the other hand, the liquid crystalline (LC) mesophases provide an additional example of a state of matter characterized by non-crystalline order. The combination of these two different aspects into one single macromolecular architecture leads to block copolymers containing LC blocks (*1*). These materials can be valuable in elucidating specific aspects of polymer physics and, in addition, may be employed as highly versatile interfacially active additives and viscosity improvers potentially capable of providing enhanced optimization of material processing and performance.

By a synthetic procedure involving the use of azo macroinitiators (*2*), we have started to prepare and study new block copolymers consisting of semicrystalline/side-chain LC blocks (*3*) and amorphous/main-chain LC blocks (*4-6*).

0097–6156/96/0632–0332$15.00/0

Very recently, the first examples of hybrid block copolymers comprising in the same macromolecule both main-chain and side-chain LC blocks (*7*) have also been described. To gain a better insight into the effects of the macromolecular architecture on the phase transition behavior of this last class of block copolymers, in the present contribution we report on the synthesis and properties of three series of main-chain and side-chain LC block copolymers **1-3**, characterized by the following general structure:

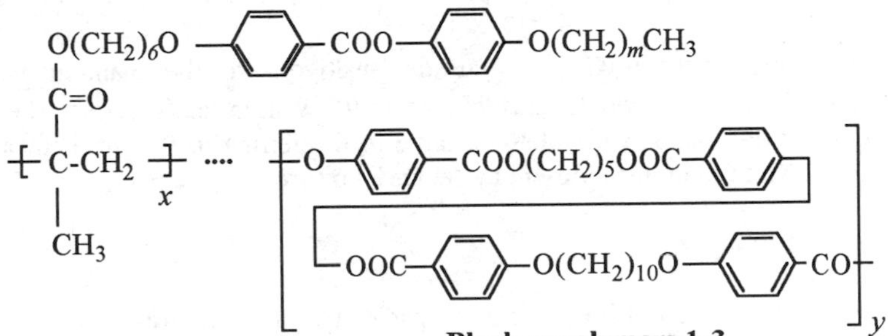

$O(CH_2)_6O$ — C_6H_4 — COO — C_6H_4 — $O(CH_2)_mCH_3$

C=O

$-[C\text{-}CH_2]_x-$ ···· $-[O-C_6H_4-COO(CH_2)_5OOC-C_6H_4-OOC-C_6H_4-O(CH_2)_{10}O-C_6H_4-CO]_y-$

CH_3

Block copolymers 1-3

m	0	3	5
series	**1**	**2**	**3**

Block copolymers **1-3** are constituted by a side-chain LC polymethacrylate block, containing a variously substituted mesogenic unit, and a main-chain LC polyester block made up by two mesogenic *p*-oxybenzoyl diads alternatively interspaced by aliphatic chains of five and ten methylene groups connected to the mesogenic cores by two ester and ether linking groups, respectively. The side-chain polymethacrylate homopolymers **4**, **5** and **6**, structurally analogous to the side-chain block, exhibited a nematic mesophase (**4**), smectic A (Sa) and nematic mesophases (**5**) and a Sa (**6**) mesophase.

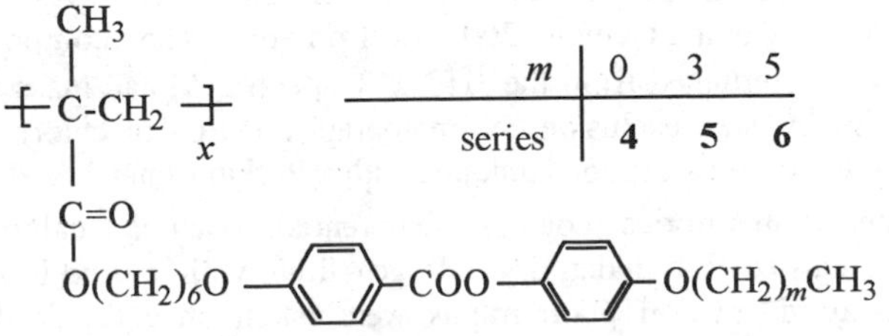

CH_3

$-[C\text{-}CH_2]_x-$

C=O

$O(CH_2)_6O$ — C_6H_4 — COO — C_6H_4 — $O(CH_2)_mCH_3$

m	0	3	5
series	**4**	**5**	**6**

Polymethacrylates 4-6

Homopolyester C_5C_{10}

Polyester homopolymer **C_5C_{10}**, structurally analogous to the main-chain block, formed nematic and smectic C mesophases (*8,9*). Within each series, the block copolymers are designated with a letter from **a** to **d** referring to the different amount of the methacrylate monomer present in the feed mixture.

Experimental Part

Methacrylate monomers **10-12** were prepared by a literature procedure (*10*). Macroinitiator **M-C_5C_{10}** was prepared according to the procedure reported in (*7*). Block copolymer series **1-3** were prepared following the synthetic route illustrated in Scheme 1.

Synthesis of Block Copolymers. In a typical copolymerization reaction, the required amount of the methacrylate monomer and 0.4 g of the macroinitiator **M-C_5C_{10}** were dissolved in 10 mL of anhydrous THF. The reaction mixture was introduced into a Pyrex glass ampoule, freeze-thaw degassed and then sealed under vacuum. After reacting for 20 h at 70 °C, the copolymer was recovered by addition of a ten-fold excess of methanol and purified from oligomers by extraction with boiling methanol in a Kumagawa extractor. The copolymer was then dried in vacuum for 24 h. The copolymerization yield was in the 60-70% range.

Four copolymer samples were synthesized according to the above procedure by keeping constant the amount of the macroinitiator (0.4 g) and using different quantities of methacrylate in the feed mixture: 0.3 g, **a**; 0.7 g, **b**; 1.0 g, **c**; 2.0 g, **d**.

Physicochemical Characterization. ^1H NMR and ^{13}C NMR spectra were recorded with a Varian Gemini 200 spectrometer. The composition of the copolymers was determined from the ^1H NMR spectra. Molar mass characteristics were determined by size exclusion chromatography (SEC) of chloroform solutions with a 590 Waters chromatograph equipped with a Perkin Elmer UV detector using a 10^4 Å Polymer Laboratories column. Differential scanning calorimetry (DSC) analyses were carried out under dry nitrogen flow with a Perkin-Elmer DSC 7 apparatus. X-ray diffraction photographs were taken on a Rigaku-Denki RU300 rotating anod generator equipped with a pin hole flat camera. Ni-filtered CuK_α radiation was used.

Results and Discussion

Synthesis. Block copolymer series **1-3** were synthesized by sequential polycondensation and free-radical polymerizations as illustrated in Scheme 1. In the first step, the polyester macroinitiator **M-C_5C_{10}**, containing a 10 mol-% of **8** with respect to the total content of diacid chlorides **7** and **8**, is prepared which is then used to initiate the polymerization of methacrylates **10**, **11** and **12** through the thermal decomposition of the azo group at 70 °C. Three block copolymer series (Table I), consisting each of four samples, were prepared by reacting the macroinitiator with the three methacrylate monomers, and using within each series, different amounts of methacrylate in the feed mixture. The content of the main-chain block, as evaluated by ^{1}H NMR, ranges from 19 to 72 wt.-%. In addition, a sample of **M-C_5C_{10}** was thermally decomposed at 70 °C in the presence of a large amount

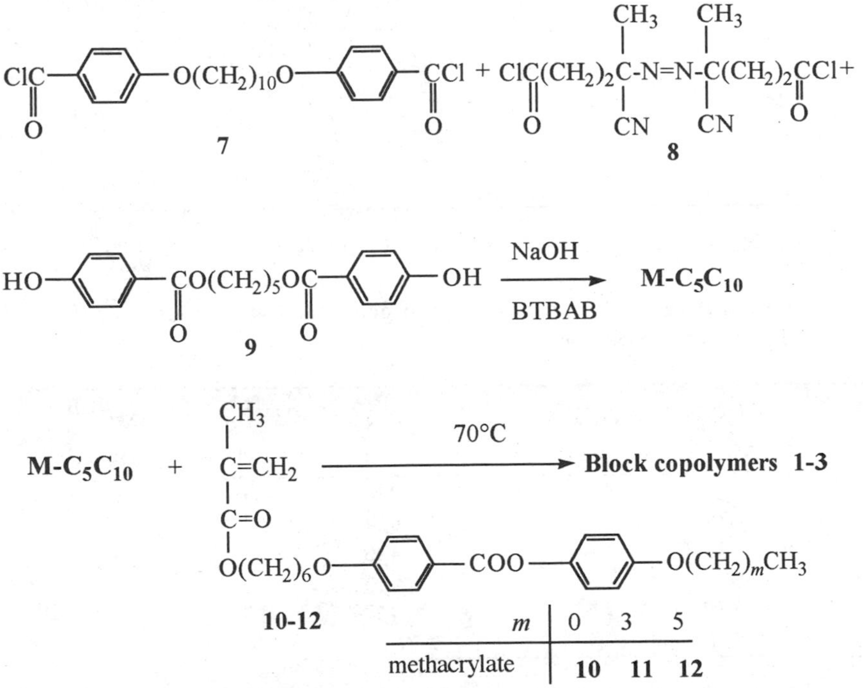

ClC(=O)–C₆H₄–O(CH₂)₁₀O–C₆H₄–C(=O)Cl (**7**) + ClC(=O)(CH₂)₂C(CH₃)(CN)–N=N–C(CH₃)(CN)(CH₂)₂C(=O)Cl (**8**) +

HO–C₆H₄–C(=O)O(CH₂)₅OC(=O)–C₆H₄–OH (**9**) —NaOH, BTBAB→ **M-C_5C_{10}**

M-C_5C_{10} + CH_2=C(CH_3)–C(=O)–O(CH_2)$_6$O–C₆H₄–COO–C₆H₄–O(CH_2)$_m$$CH_3$ (**10-12**) —70°C→ **Block copolymers 1-3**

m	0	3	5
methacrylate	**10**	**11**	**12**

SCHEME 1. Synthesis of block copolymers **1-3**.

of 2,6-di-*tert*-butyl-4-methylphenol. The resulting polyester **C_5C_{10}** was also studied as a model of the main-chain block in the copolymers. The molecular characteristics of the polymers and copolymers were studied by SEC. M_n = 11500 and M_w/M_n = 2.3 for **M-C_5C_{10}**, while M_n = 4800 and M_w/M_n = 1.9 for **C_5C_{10}** were found using the universal calibration method. Within each series, M_n increases as the concentration of

Table I. Composition and molar mass data of macroinitiator M-C_5C_{10}, polyester C_5C_{10}, and block copolymers 1-3

Sample	C_5C_{10}[a] (wt.-%)	M_n[b]	M_w/M_n[b]
M-C_5C_{10}	100	11,500	2.3
C_5C_{10}	100	4,800	1.9
1a	72	19,000	3.5
1b	49	33,000	2.2
1c	36	38,500	2.5
1d	24	50,000	3.0
2a	65	30,000	2.4
2b	45	32,000	2.1
2c	32	43,000	2.3
2d	29	37,000	2.5
3a	72	28,000	2.4
3b	47	36,000	2.2
3c	28	34,500	2.7
3d	19	50,000	3.0

[a] Polyester block **C_5C_{10}**, by ^1H NMR. [b]By SEC, in chloroform at 25 °C.

Table II. Liquid-crystalline properties[a] of polyester C_5C_{10}, block copolymers 1a-d and polymethacrylate 4

Sample	$T_{I\text{-}N}^{A,b}$ (K)	$\Delta H_{I\text{-}N}^{A,b}$ (Jg^{-1})	$T_{N\text{-}Sc}^{B,b}$ (K)	$T_{I\text{-}N}^{B,b}$ (K)	$\Delta H_{N\text{-}Sc}^{B,b}$ (Jg^{-1})	$\Delta H_{I\text{-}N}^{B,b}$ (Jg^{-1})
C_5C_{10}	-	-	411	430	4.7	2.9
1a	377	1.3	408	424	3.1	1.9
1b	378	1.9	408	426	2.7	1.5
1c	378	1.5	408	426	3.3	1.6
1d	377	1.6	408	427	4.3	1.2
4	378	2.1	-	-	-	-

[a]By DSC, at -10 $Kmin^{-1}$ scanning rate. [b]A: polymethacrylate block; B: polyester block.

from 30,000 to 43,000 for series **2** and from 28,000 to 50,000 for series **3**. M_w/M_n values comprised between 2.2 and 3.5 were usually observed. It is well known (*11,12*) that the free-radical polymerization of methacrylate monomers terminates by a disproportionation mechanism. Accordingly, in the present case starting from macroinitiator chains containing one reactive azo group, AB diblock copolymers were formed, but additionally ABA triblock copolymers could also derive from macroinitiator chains containing more than one reactive azo group.

Thermal Behavior. The LC behavior of block copolymer series **1-3** was studied by DSC, polarizing microscopy and X-ray diffraction. The relevant phase transition parameters were taken from the DSC cooling curves, on account of the better resolution of the transition peaks (Tables II-IV). Polymethacrylates **4**, **5** and **6** were amorphous and formed a nematic mesophase, a nematic and a smectic A mesophase and a smectic A mesophase, respectively. Polyester **C_5C_{10}** was also amorphous and exhibited smectic C (Sc) and nematic (N) mesophases. The DSC traces of all block copolymers showed two enthalpic peaks associated to the isotropic-nematic and nematic-smectic C transitions of the main-chain block and one or two enthalpic peaks relevant to the LC transition of the side-chain block. Two glass transitions, sometimes not well resolved, were also observed in the 20-30°C and 20-60°C regions attributed to the main-chain and side-chain block respectively. The mesophase nature attribution is confirmed by X-ray diffraction analysis. As a typical example, Figure 1 illustrates the DSC cooling curve of block copolymer **2c** with two insets. The low temperature inset shows the small angle region of the X-ray diffraction spectrum at temperatures in which the smectic mesophase of the main-chain block (interlayer spacing d =20 Å) coexists with the smectic mesophase of the side-chain block (interlayer spacing d = 29 Å). The high temperature inset represents the X-ray pattern of the smectic mesophase of the polyester block, which coexists with the isotropic phase of the side-chain block. The observation of two distinct signals relevant to both smectic mesophases of the main-chain and side-chain blocks clearly indicates that the chemically different blocks are phase-separated.

Figure 2 represents collectively the phase transition temperatures of both the main-chain and the side-chain blocks in the three series as functions of the main-chain block content. The phase transition temperatures of the main-chain block are quite constant, whereas the phase transition temperatures of the side-chain block slightly decrease as the main-chain block content increases. This decrease is probably connected to the parallel decrease of the side-chain block length. The molar mass of the polyester block is constant throughout the series, whereas the molar mass of the polymethacrylate block increases from samples **a** to samples **d** in each of the three series. Accordingly, the evolution of the transition parameters of the main-chain block can be taken as a better marker of the level of interaction between the chemically different blocks.

Considering the phase transition temperatures of both the chemically different blocks, it should be noted that in specific temperature ranges the smectic and the nematic mesophases of the main-chain block coexist with a nematic and an isotropic phase in series **1**, with a smectic, a nematic and an isotropic phase in series **2** and with a smectic and an isotropic phase in series **3** thus producing a variety of different interphases or interfaces. In particular, it appears that the smectic or, at higher temperature, the nematic domains of the main-chain blocks are surrounded by the isotropic domains of the side-chain blocks.

Table III. Liquid-crystalline properties[a] of polyester C_5C_{10}, block copolymers 2a-d and polymethacrylate 5

Sample	$T_{I-N}^{A,b}$ (K)	$\Delta H_{I-N}^{A,b}$ (Jg^{-1})	$T_{N-Sa}^{A,b}$ (K)	$\Delta H_{N-Sa}^{A,b}$ (Jg^{-1})	$T_{N-Sc}^{B,b}$ (K)	$T_{I-N}^{B,b}$ (K)	$\Delta H_{N-Sc}^{B,b}$ (Jg^{-1})	$\Delta H_{I-N}^{B,b}$ (Jg^{-1})
C_5C_{10}	-	-	-	-	411	430	4.7	2.9
2a	383	2.7	368	1.8	408	427	3.1	2.3
2b	385	3.2	373	2.7	410	429	3.2	1.9
2c	385	3.0	376	3.0	412	428	3.3	1.5
2d	387	3.2	370	3.3	411	428	3.2	1.4
5	382	2.8	372	3.0	-	-	-	-

[a]By DSC, at -10 $Kmin^{-1}$ scanning rate. [b]A: polymethacrylate block; B: polyester block.

Table IV. Liquid-crystalline properties[a] of polyester C_5C_{10}, block copolymers 3a-d and polymethacrylate 6

Sample	$T_{I-Sa}^{A,b}$ (K)	$\Delta H_{I-Sa}^{A,b}$ (Jg^{-1})	$T_{N-Sc}^{B,b}$ (K)	$T_{I-N}^{B,b}$ (K)	$\Delta H_{N-Sc}^{B,b}$ (Jg^{-1})	$\Delta H_{I-N}^{B,b}$ (Jg^{-1})
C_5C_{10}	-	-	411	430	4.7	2.9
3a	380	9.8	408	426	3.2	1.8
3b	385	15.3	409	428	3.2	1.7
3c	385	13.6	410	427	3.3	1.2
3d	386	15.3	410	426	2.7	0.9
6	388	15.5	-	-	-	-

[a]By DSC, at -10 $Kmin^{-1}$ scanning rate. [b]A: polymethacrylate block; B: polyester block.

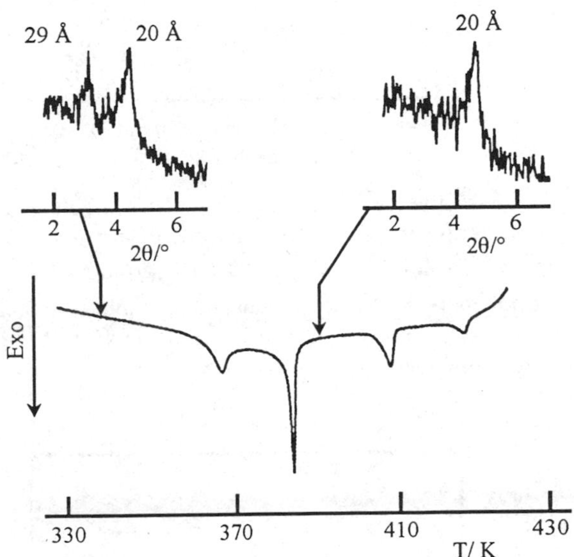

FIGURE 1. DSC second cooling curve (10 Kmin^{-1}) for block copolymer **2c** and the small angle region of the relevant X-ray diffraction spectra at 340 K (low temperature inset) and at 390 K (high temperature inset).

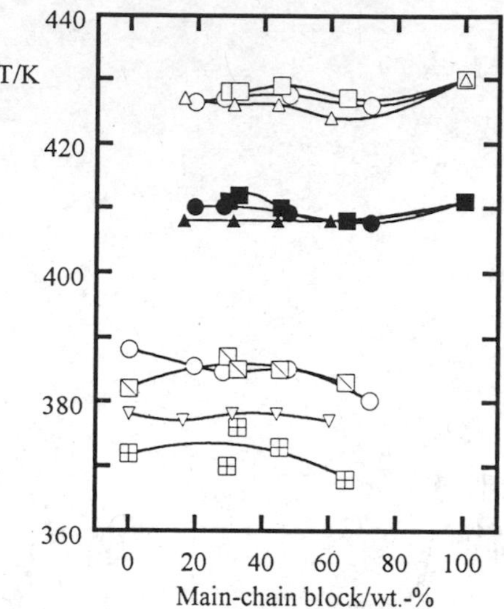

FIGURE 2. Trends of the isotropic-nematic (open symbols) and nematic-smectic C (full symbols) transition temperatures of the polyester block in block copolymers **1** (Δ,▲), **2** (○,●), and **3** (□ ,■) and of the isotropic-nematic (∇, copolymers **1**; ⧅ , copolymers **2**), nematic-smectic A (⊞, copolymers **2**) and isotropic-smectic A (○, copolymers **3**) of the side-chain block as functions of the main-chain block content.

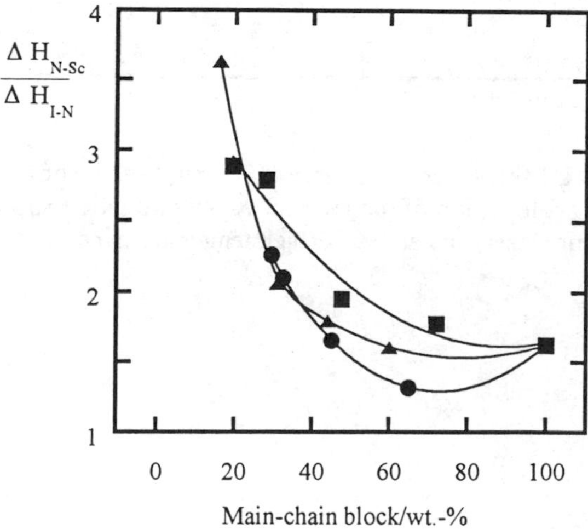

FIGURE 3. Trend of the ratio between the normalized nematic-smectic C and isotropic-nematic transition enthalpies of the polyester block in block copolymers **1** (▲), **2** (●), and **3** (■) as a function of the main-chain block content.

The normalized phase transition enthalpies of the side-chain blocks increase with increasing amount of the side-chain block (see Tables II-IV). The normalized enthalpy changes associated to the nematic-smectic C and to the isotropic-nematic transition ($\Delta H_{N\text{-}Sc}$ and $\Delta H_{I\text{-}N}$) of the main-chain block as a function of the main-chain block content displays a dual behavior. The former is quite constant, whereas the latter increases regularly as the main-chain block content increases. A better visualization of the distinct behavior of the main-chain block at the smectic C-nematic and nematic-isotropic transitions is obtained by plotting the ratio between the nematic-smectic C and isotropic-nematic normalized phase transition enthalpies of the main-chain block as a function of the main-chain block content for the three block copolymer series as illustrated in Figure 3. In each copolymer series, the ratio $\Delta H_{N\text{-}Sc}/\Delta H_{I\text{-}N}$ decreases as the main-chain block content increases. This behavior strongly points toward a differential effect of the side-chain isotropic melt on the LC mesophase structures, namely the nematic and the smectic C ones, of the main-chain block. As this effect may arise either from the existence of a disordered interface (partial miscibility) between the side-chain and main-chain blocks, or from unfavorable boundary conditions, we conclude that the smectic mesophase generated by the main-chain block results less miscible or less perturbable than the nematic mesophase. In addition, the size of the main-chain domains should depend on the overall block copolymer composition and decrease as the side-chain block content increases. A tentative schematic representation of the interphase boundary situation in the various block copolymer samples is reported in Figure 4.

Conclusion

We have synthesized and studied three series of block copolymers comprising both main-chain and side-chain LC blocks within the same polymer structure. Although thermal and X-ray diffraction results show that the two chemically different blocks are at least partly phase-separated, a differential effect of the side-chain isotropic melt on the LC mesophases, namely the nematic and the smectic C ones, of the main-chain block is observed. Analysis of the enthalpic data suggests that the extension of the interdomain boundary region resulting from the coexistence of the isotropic phase of the side-chain block and the nematic phase of the main-chain block is larger than the one relevant to the boundary region resulting from the coexistence of the isotropic phase of the side-chain block and the smectic phase of the main-chain block. In addition, the size of the main-chain domain should depend on the overall block copolymer composition and decrease as the side-chain block content increases.

Acknowledgment. This work was supported by the National Research Council of Italy (Progetto Chimica Fine 2 - Sottoprogetto Materiali Polimerici).

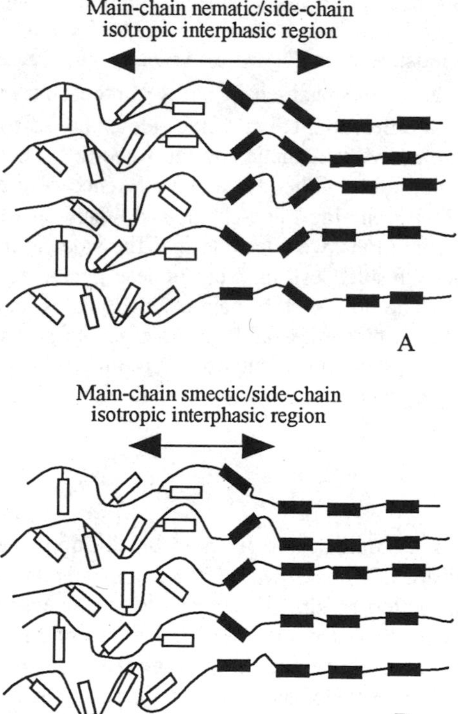

FIGURE 4. Schematic representation of the main-chain nematic side-chain isotropic interphasic region (A) and of the main-chain smectic side-chain isotropic interphasic region (B).

Literature Cited

1. Chiellini, E.; Galli, G.; Angeloni, A. S.; Laus, M. *Trends Polym. Sci.* **1994**, *2*, 244 and refs. therein.
2. Chiellini, E.; Galli, G.; Angeloni, A. S.; Laus, M.; Bignozzi, M. C.; Yagci, Y.; Serhatli, E. I. *Makromol. Chem., Makromol. Symp.*, **1994**, *77*, 349.
3. Galli, G.; Chiellini, E.; Yagci, Y.; Serhatli, E. I.; Laus, M.; Angeloni, A. S.; Bignozzi, M. C. *Makromol. Chem., Rapid Commun.*, **1993**, *14*, 185.
4. Angeloni, A. S.; Bignozzi, M. C.; Laus, M.; Chiellini, E.; Galli, G. *Polym. Bull.(Berlin)*, **1993**, *31*, 387.
5. Bignozzi, M. C.; Angeloni, A. S.; Greco, M.; Laus, M.; Chiellini, E.; Galli, G. In *Liquid Crystalline Polymers*, Carfagna, C., Ed.; Pergamon Press, NY, 1994 pp 61-68.
6. Galli, G.; Chiellini, E.; Laus, M.; Angeloni, A. S.; Bignozzi, M. C.; Francescangeli, O. *Mol. Cryst. Liq. Cryst.*, **1994**, *254*, 429.
7. Galli, G.; Chiellini, E.; Laus, M.; Bignozzi, M. C.; Angeloni, A. S.; Francescangeli, O. *Macromol. Chem. Phys.*, **1994**, *195*, 2247.
8. Chiellini, E.; Galli, G.; Laus, M.; Angeloni, A. S.; Francescangeli, O.; Yang, B. *J. Mater. Chem.*, **1992**, *2*, 449.
9. Francescangeli, O.; Yang, B.; Albertini, G.; Angeloni, A. S.; Laus, M.; Chiellini, E.; Galli, G. *Liq. Cryst.*, **1993**, *3*, 353.
10. Finkelmann, H.; Ringsdorf, H.; Wendorff, J. H. *Makromol. Chem.*, **1978**, *179*, 273.
11. Heitz, W. *Makromol.Chem., Macromol. Symp.*, **1987**, *10/11*, 297
12. Eastmond, G. C. *Makromol. Chem., Macromol. Symp.*, **1987**, *10/11*, 71.

Chapter 20

Synthesis and Chain Rigidity of Mesogen-Jacketed Liquid-Crystalline Polymers Based on Poly(2,5-disubstituted-styrene)

Qi-Feng Zhou, Xin-Hua Wan, Dong Zhang, and Xin-De Feng

Institute of Polymer Science, College of Chemistry, Peking University, Beijing 100871, China

The evolution of the concept "Mesogen-Jacketed Liquid Crystal Polymer (MJLCP)" is described. When mesogenic units attached through no or only very short spacer to the main chain with a side-on attachment, there should form a "jacket" of mesogenic units around each chain backbone. The polymer chains are forced to take extended conformations and show properties of rigid or semi-rigid macromolecules. MJLCPs thus may be termed also "Rigid SC-LCPs" because they are not only liquid crystal polymers of the side chain type when viewed chemically, but also of significant rigidity when viewed physically as are many of the main chain type liquid crystal polymers. The synthesis and properties (with emphasis on chain rigidity) of representative MJLCPs are discussed. This concept would have bridged the gap present in between the conventional side-chain LCPs and the main-chain LCPs.

It has been practice to classify liquid crystal polymers into two basic types, namely the side chain type and the main chain type. A side chain liquid crystal polymer (SC-LCP) has mesogenic units as side groups, while a main chain liquid crystal polymer (MC-LCP) has mesogenic units as main chain segments. Because of this kind of molecular constitution and the inherent rigidity of rod-like mesogenic units, MC-LCPs are usually rigid or semi-rigid. The liquid crystalline aromatic polyesters and polyamides are examples of this category. For SC-LCPs, on the other hand, flexible spacers are used to decouple the movements of the main chain and of the mesogenic side groups in order to realize the liquid crystallinity (*1*), such that the mesogenic side groups, however rigid, will have minor influence on the flexibility of the polymer backbones. The reverse is true so that the statistical movements and conformational changes of the main chain backbones will have only minor effects on the mesomorphic organizations of the mesogenic units. As a result, SC-LCPs are mostly flexible polymers and are thus expected to show rapid response to foreign electro-magnetic fields(*2*).On the other hand

0097–6156/96/0632–0344$15.00/0

MC-LCPs are applicable as high strength and high modulus materials with high heat deflection temperatures (3). Our efforts described in this article have been to contribute a third class of liquid crystal polymers that have chemical structures similar to conventional SC-LCPs, but have chain properties, chain rigidity in particular, similar to that of MC-LCPs. They are "Mesogen-Jacketed Liquid Crystal Polymers" or "MJLCPs" for short.

Chain rigidity may be evaluated by a variety of methods such as viscometry and light scattering. These methods will result in values for Mark-Houwink-Sakurada equation exponent α and for persistence length q of polymers. The parameter α can vary between 0.5 for non-draining random coils and 2 for stiff chains (*4*). On the other hand, a rigid polymer must have a large persistence length greater than 10 nm, a semi-rigid polymer may either have the value of q smaller than 10 nm or have a larger value coupled with very high molecular weight, while for a Gaussian chain the value is as small as about 1 nm (*5*). Yet for liquid crystal polymers the formation of lyotropic phases (*6*) and the formation of banded textures (*7*) are also indicative of significant chain stiffness. The chain rigidity of the MJLCPs will be discussed from these aspects.

In the following sections, the concept of MJLCPs is first described, followed by discussions on the synthesis, the liquid crystallinity, and chain properties of representative MJLCPs. Among the polymers are the newly synthesized poly-2,5-bis(4-methoxybenzamido)styrene (PA-1) and its homologues. The lyotropic liquid crystalline behavior of PA-1 and other MJLCPs is also reported for the first time in this paper.

The Concept

The concept "mesogen-jacketed liquid crystal polymer" (MJLCP) has been proposed to describe the side chain liquid crystal polymers that are of the side chain type but with mesogenic groups attached laterally to the main chain without or with only short spacers (*8-10*). With such a molecular constitution the mesogenic units would form a "jacket" around each chain backbone because of their high population in around the backbone, and because they are both bulky and rigid. The jacket would in turn force the main chain to extend and to show much higher rigidity than otherwise. Thus, another concept, "rigid side chain liquid crystal polymers", was further proposed for these polymers, with emphasis on the side chain type and on the stiffened chain conformation (*10-11*). A success of this effort would lead to polymers that may be synthesized by, say, radical chain polymerizations as for many conventional side chain type liquid crystal polymers, but also show the properties of rigid or semi-rigid polymers as represented by main chain type liquid crystal polymers. MJLCPs would distinguish themselves from the conventional main chain and side chain type liquid crystal polymers and should be justified as a unique class. So far, the mostly studied MJLCP in our lab is poly-2,5-bis[(4-methoxybenzoyl)oxy]styrene (*10*). For convenience, this polymer is denoted in this article by PE-1. The homologues of this polymer are denoted by PE-n with different n representing different R groups in the general formula:

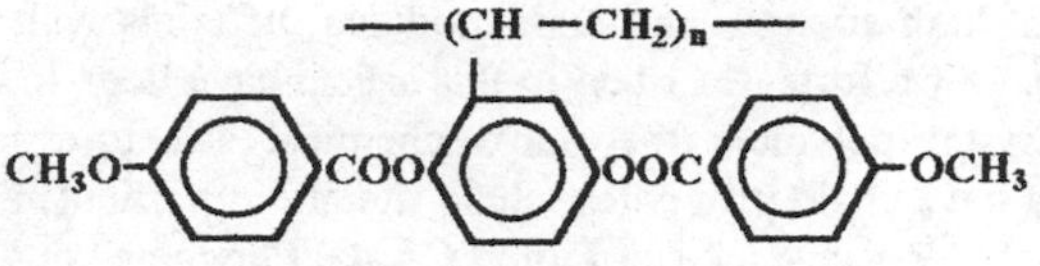

$—(CH—CH_2)_n—$

CH_3O–⟨⟩–COO–⟨⟩–OOC–⟨⟩–OCH_3

<PE-1>

$—(CH—CH_2)_n—$

R–⟨⟩–COO–⟨⟩–OOC–⟨⟩–R

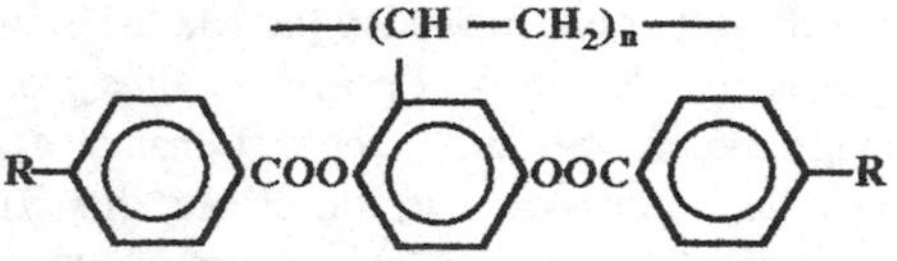

<PE-n>

The PE-n polymers were so designed that it has a kind of commonly used mesogenic units, but the connections between the main chain and the mesogenic units are at the waist of the rod-like mesogens and nothing more than a covalent bond is used for each connection. This closest relationship between the main chain and the rod-like units is distinctive from conventional SC-LCPs and is responsible for unique properties of MJLCPs. Note that the utmost length of the repeating unit in PE-n is about 2.5 A, but the length of the mesogenic unit bonded on each repeating unit is as large as 20 A or so. One can imagine how dense the mesogenic units are packed around the main chain, specifically when the waist (gravity center) is the location of attachment and no spacer is used to connect the two parts in a repeating unit. The second series of MJLCPs are the newly synthesized PA-n polymers that have the general formula shown below and are reported for the first time in this paper:

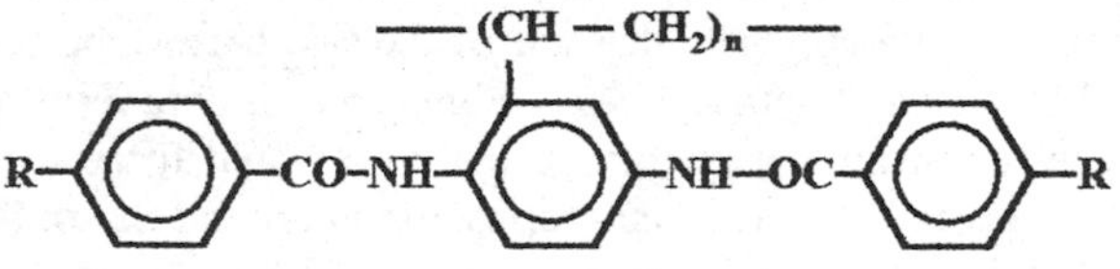

$—(CH—CH_2)_n—$

R–⟨⟩–CO-NH–⟨⟩–NH—OC–⟨⟩–R

<PA-n>

Poly-2,5-bis(4-methoxybenzamido)styrene, PA-1, with R the methoxy, is the most studied in PA-n series. PA-n polymers have same structural characters as PE-n but distinguish from the latter by amide (rather than ester) linkages in the rod-like units. Incorporation of amide rather than ester linkages in the mesogenic units is expected to increase stiffness of the units, which would in turn further stiffen the molecular chain.

Synthesis and Liquid Crystallinity of PE-n Polymers

The synthesis of poly-2,5-bis[(4-methoxybenzoyl)oxy]styrene and its homologues has been described (*9-10*). Table I shows some of the characterization results for these polymers.

All the polymers PE-n are non-crystalline as revealed by X-ray analysis. There are only a very diffuse halo at 20 degrees and a diffuse peak at 5.5 degrees for 2θ,

Table I. Molecular Weights and Tg of PE-n

PE-n	R	Mn ($\times 10^4$)	Mw ($\times 10^4$)	Mw/Mn	Tg (^{o}C)
PE-0	H	5.4	10	1.9	170
PE-1	OCH_3	13	30	2.3	160
PE-2	OC_2H_5	4.5	9.5	2.1	130
PE-3	OC_4H_9	24	88	3.7	80
PE-4	CN	a			200
PE-5	CH_3	7.7	22	2.9	172
PE-6	C_2H_5	6.8	17	2.5	160
PE-7	C_4H_9	2.8	8.6	3.0	130

a. It is not soluble in THF, intrinsic viscosity was given as 1.1 g/dl in DMF

corresponding respectively to the lateral packing of the chains and to the length of the side groups (16 A as calculated from $d = \lambda / 2\sin\theta$). The glass transition temperatures are shown in Table I. At temperatures above Tg the polymers all have a nematic phase (*10*). These mesophases are very stable and no isotropization temperature has ever been observed before decomposition starts to occur at about 300 ^{o}C or higher. Two remarkable differences between MJLCPs and conventional side chain liquid crystal polymers are noticed. Firstly, MJLCPs favor the formation of nematic phase rather than smectic phases. This is not only true for systems with no spacers as PE-n described here, it is also true even if flexible spacers are used to connect the main chain and the mesogens so long as the attachment is at the waist or nearby positions of the mesogens (*12-14*). For conventional SC-LCPs, It is well known that they often form smectic phases, in addition to nematics, especially when no spacers and when relatively long spacers are used (*2*). Secondly, the nematic phases of PE-n are so stable that no clearing point or isotropization temperature (Ti) can be observed before thermal decomposition of molecular structure occurs (*15*). On the other hand, Nematic phases of conventional SC-LCPs scarcely have Ti higher than 150 ^{o}C. By contrast, the thermal stability of MJLCPs is similar to many of the main chain type liquid crystal polymers, especially many of the semi-rigid or rigid aromatic polyesters. It is also interesting to mention the fact that semi-rigid and rigid aromatic polyesters form essentially only nematic liquid crystal phases. Therefore, we are able to notice here the first similarity between the properties of MJLCPs and the main chain type liquid crystal polymers.

At this point, we would like to mention again that the polymer PE-0, poly-2,5-di(benzoyloxy)styrene, was first synthesized some 40 years ago as a precursor of an electron exchange polymer (*16*). In this study we were lucky to have found that this polymer is liquid crystalline at temperatures above its glass transition (Table I). To our knowledge PE-0 is probably the eldest thermotropic liquid crystalline polymer ever synthesized by chemists.

It is also interesting to mention that in PE-0 the side group moiety, hydroquinone dibenzoate, is on its own not a liquid crystal. A question thus arises if one recalls that for liquid crystal formation of conventional side-chain type polymers it is the ordering of mesogenic side groups and not the ordering of the main chains that is principally concerned, even though the conformations of the main chains do change in accordance with the ordering of the side groups. What are the principal ordering elements for the formation of liquid crystalline phase of MJLCPs? Are they the side groups as are in the cases of conventional side chain type lcps? Or are they the rigid molecular segments as are in the cases of main chain type lcps describable by the worm-like or Kuhn chains? We are not able to answer this question for certain at this moment, but we would rather believe that it is the ordering of the rigid molecular segments which have the principal responsibility for formation of the liquid crystalline phase of our MJLCPs. In other words, we intend to believe that MJLCPs are worm-like and can be described by Kuhn chains (*5*).

Chain Rigidity of MJLCPs by Morphological Studies

As mentioned earlier, our work has been mostly on the MJLCPs having very short linkage (spacer) in between the main chain and the mesogenic rigid-rod groups. As represented by PE-n polymers, only a single covalent bond is used to bind the two parts of the molecules. These polymers should experience the strongest "jacket effect" created by the mesogenic jacket on chain conformation and properties . As a result, these polymer chains should become highly extended, rather rigid, and describable by Kratky-Porod worm-like chains (*5*). It is well known that rigid and semi-rigid main-chain type liquid crystal polymers, both lyotropic and thermotropic, can form banded textures by shear or elongational flow in their liquid crystal state(*7,17*). In the oriented liquid crystal specimens, the polymer molecules are packed in a parallel alignment to form fibrils. The fibrils travel parallely in a regular zigzag manner, resulting in the unique alternating light and dark pattern when being observed on a polarizing microscope. Thus the name "banded texture". No such textures have ever been observed for flexible chain polymer molecules including that of conventional side chain type liquid crystal polymers having mesogenic groups attached longitudinally (end-on) to a flexible backbone. Thus, the rigidity of mesogen-jacketed liquid crystal polymers may also be demonstrated by morphological study of oriented liquid crystal films (*15*).

Studies on the PE-n series of MJLCPs have revealed that all the polymers form banded textures when they are in nematic phases (*10*). Besides, their banded textures have been found to be thermally very stable, and remain almost unchanged in a wide temperature range (*15*).

In studies on banded textures of main-chain type liquid crystal polymers, a contraction effect has been proposed to explain the mechanism of band formation (*15*). Thus, the rigid or semi-rigid molecules of liquid crystalline polymers are packed in parallel in the form of fibrils under shear along the shearing direction. Formation of the banded textures after shear cessation is not the result of free thermal relaxation of individual molecules, but the result of contraction and zigzig rearrangement of the fibrils as a whole along the orientation direction. The elastic energy stored during shearing may have provided with driving force for contraction, but the high stability of the parallel alignment of neighbouring molecules due to very limited molecular motion of these polymers is a necessary condition for formation of the bands. By contrast, in the case of flexible chain polymers, molecules may also assume parallel orientation under shearing, however, the orientational relaxation takes place easily as the result of fast thermal motions of individual molecules. The external elastic force may even accelerate this process. As a result, no regular banded textures should be expected for flexible polymers.

The morphological studies have given clear evidence for a rather high degree of chain rigidity of MJLCPs. The very high capability of forming banded textures represents another similarity of properties of MJLCPs and the main chain type liquid crystal polymers. It is also interesting to mention the fact that this study has offered the first examples of side chain type liquid crystal polymers (though not of the conventional type) that form banded textures.

Chain Rigidity of MJLCPs by Study of Solution Properties

Chain rigidity of polymer molecules may be evaluated by studies of solution properties. According to Kratky and Porod (*5*), the persistence length of a polymer molecule in good solvent characterizes chain stiffness of the polymer. Theoretical studies and experimental evaluations of persistence lengths of variety of polymer molecules have given values in the order of 1 nm for the flexible or Gaussian chains, 10 nm for semi-rigid chains, and higher than 10 nm for rigid chains. Among liquid crystal polymers, lyotropic ones usually have higher rigidity than thermotropic aromatic polyesters which in turn are much stiffer than SC-LCPs. There have been reports on persistence lengths of LCPs (*18*). For example, a typical value of 30 nm has been found for poly-(1,4-phenylene teraphthamide), PPTA, in sulfuric acid. For thermotropic aromatic polyesters the persistence length data are also available. The value for a fully aromatic copolyester composed of 4-hydroxybenzoic acid, terephthalic acid, 2,6-naphthalene dicarboxylic acid, and 1,4- hydroquinone with the molar ratio of 6:2:2:4 has been reported to be 12 nm (*19*); that for an aromatic terephthalic acid copolyester in tetrachloroethane is 9 nm (*20*). Thus, in order to evaluate persistence length and chain stiffness of MJLCPs, a typical polymer sample, poly-2,5-bis[(4-methoxybenzoyl)oxy]styrene, PE-1, was studied (Wan, X.; Zhang, F.; Wu, P.; Zhang, D.; Feng, X.; Zhou, Q.F., Macromol. Symp. in press). Thus, by repeated fractionation precipitation with DMF as the solvent and methanol as the precipitant, nine fractions were obtained. The fractions were studied by GPC, DSC, viscometry and static light scattering (with a Photal DLS-700 Dynamic Light Scattering Spectro-photometer) techniques.

The experimental temperature was 25 °C and THF was used as the solvent. The results were collected in Table II. From the results and by a procedure of Bohdanecky (*21*), the persistence length q of this polymer was evaluated to be

$$11.5 \text{ nm} \leqslant q \leqslant 13.5 \text{ nm}$$

that is much higher than what a flexible polymer would have, and is in the range of values for rigid or semi-rigid polymers such as liquid crystalline aromatic copolyesters. Besides, the exponential parameter α of the Mark-Houwink-Sakurada equation for PE-1 was found to be 0.82, significantly higher than that for flexible polymers.

Table II. Results From Light Scattering and Viscometry for PE-1

Fractions	Mw g/mol	$A_2 \times 10^5$ mol·cm^3·g^{-2}	$(R_g^2)^{1/2}$ nm	[η] ml/g
1	194000	4.40	48.5	114.12
2	119000	1.73	37.4	76.81
3	103000	4.80	23.4	64.26
4	83000	5.02	22.2	52.28
5	63000	8.86	21.7	47.36
6	56000	6.36	17.9	41.79
7	48000	2.31	15.3	34.67
8	22000	5.70	11.2	23.34
9	11000	1.71	10.4	16.78

Both the persistence length and the α parameter have indicated that the molecules of this MJLCP are rather rigid and can be described as worm-like chains. The chain rigidity of this MJLCP is similar to that of the main-chain type liquid crystal aromatic copolyesters, but quite different from that of conventional side-chain type lcps.

Synthesis of Poly-2,5-bis(4-methoxybenzamido)styrene

We have shown that MJLCPs of the PE-n series, such as poly-2,5-bis-[(4-methoxybenzoyl)oxy]styrene and poly-2,5-di(benzoyloxy)styrene, are rather rigid. As a result, the melts of the polymers are very viscous even if at temperatures (near 300 °C) much higher than Tg (~ 166 °C). The high chain rigidity and melt viscosity have brought in difficulties for process and study. Nevertheless, we have been interested in design and synthesizing other types of MJLCPs so that the relationship between properties and molecular structure be better understood. One of the ideas has been to synthesize MJLCPs which are rigid enough and soluble enough so that a distinct lyotropic phase can be formed.

Accordingly, monomers and polymers with the following general formulae were designed and synthesized:

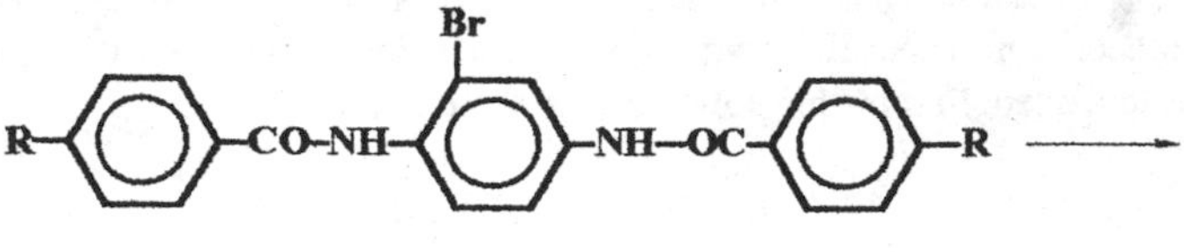

Br

R–⟨◯⟩–CO–NH–⟨◯⟩–NH–OC–⟨◯⟩–R ⟶

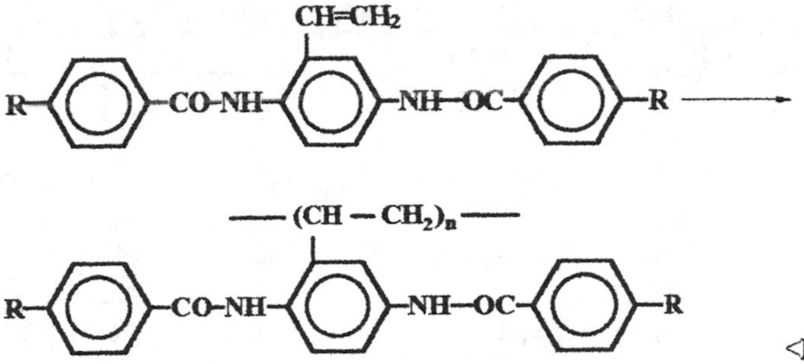

CH=CH$_2$

R–⟨benzene⟩–CO-NH–⟨benzene⟩–NH–OC–⟨benzene⟩–R ⟶

—(CH — CH$_2$)$_n$—

R–⟨benzene⟩–CO-NH–⟨benzene⟩–NH–OC–⟨benzene⟩–R

<PA-n>

Where R may be alkoxy, alkyl, nitro, cyano or other groups. When R is methoxy, the polymer is Poly-2,5-bis(4-methoxybenzamido)styrene and is denoted as PA-1.

The only difference in molecular structure between PA-n and PE-n is in the mesogenic rod-like units. For PA-n they are aromatic amides, while for PE-n they are aromatic esters. The idea of this change is to reinforce further the stiffness of the mesogenic unit and of the molecular chain as a whole.

The monomers were synthesized by reactions of the corresponding 2-bromo-N,N'-bis(4-R-substituted benzoyl)phenylene-1,4-diamines and vinyltributyltin in the presence of 1 mol % tetrakis(triphenylphosphine)palladium(0), as described below:

Bromo-N,N'-bis(p-alkoxybenzoyl)phenylene-1,4-diamine, a slightly excess amount of vinyltributyltin, the catalyst and DMF were added in a reaction tube, which was then sealed and put in a bath of 96 ^{o}C for 24 hrs. Filtered, the filtrate was poured into ice-water for precipitation of the product. The product was collected, washed in turn with deionized water and ethanol. Recrystallization was performed from DMF/Water with active carbon or aluminum oxide as decolorant to give a white product. The yield was usually 50-80 %. In table III are the characterization results for the monomers. NMR spectra of the monomers also confirmed their structures.

Some of the monomers are liquid crystal forming materials as suggested by results of polarizing microscopic studies. However, because of the high melting point the monomers undergo extensive polymerization and decomposition reactions at temperatures above melting, we have not been able to identify these liquid crystal phases. By contrast, 2-bromo-N,N'-bis(4-alkoxybenzoyl)phenylene-1,4-diamines, the precursors of the monomers, are thermotropic liquid crystalline compounds. These compounds represent first examples of aromatic amide molecules which form nematic phases after melting. Details for the chemistry and liquid crystallinity of these low mass compounds will be discussed else where.

The polymers were synthesized by solution polymerizations using AIBN as initiator and DMF as solvent. The polymerization was carried out at 60 ^{o}C. Methanol was used to precipitate the polymer. Poly-2,5-bis(4-methoxybenzamido)styrene, PA-1, the sample used for the following study on lyotropic liquid crystallinity of MJLCPs has an intrinsic viscosity of 0.58 g/dl. Characterizations of properties of the polymers are still on the way. Preliminary studies have indicated that the samples are non-crystalline as are the polymers PE-n. For example, X-ray studies of powder samples of PA-1 have shown at wide angles only a very diffuse halo from about 8 to 36 degrees for 2 θ with

Table III. Data for the Monomers of PA-n

monomers	R	formula	Mass	C	H	N
1	Methoxy	$C_{24}H_{22}O_4N_2$	402.45	71.87	5.62	7.01
	(m.p. 220)			(71.63	5.51	6.96)
2	Ethoxy	$C_{26}H_{26}O_4N_2$	430.50	72.70	5.99	6.47
	(m.p. 230)			(72.54	6.09	6.51)
3	n-Butoxy	$C_{30}H_{34}O_4N_2$	486.61	74.41	7.17	5.77
	(m.p. 206)			(74.05	7.04	5.76)
4	n-Octoxy	$C_{38}H_{50}O_4N_2$	598.83	76.09	8.55	4.70
	(m.p. 182)			(76.22	8.42	4.68)
5	R1: Ethoxy	$C_{25}H_{21}O_3N_3$	411.44	73.04	5.13	10.26
	R2: Cyano			(72.98	5.14	10.21)
	(m.p. 221)					
6	Ethyl	$C_{26}H_{26}O_2N_2$	398.50	78.07	6.59	7.09
	(m.p. 223)			(78.36	6.58	7.03)
7	n-Butyl	$C_{30}H_{34}O_2N_2$	454.61	79.38	7.58	6.12
	(m.p. 192.5)			(79.26	7.54	6.16)
8	Nitro	$C_{22}H_{16}O_6N_4$	432.35			
	(m.p. 316)			(61.11	3.73	12.96)

Notes: a. Masses of monomers were confirmed by a VG-ZAB-HS spectrometer;

b. Elemental analysis data are in per cent with the calculated data in parentheses for reference.

c. 5 is the homologous monomer unsymmetrically substituted by ethoxy and cyano: Vinyl-N-(p-ethoxybenzoyl)-N'-(p-cyanobenzoyl)phenylene diamine

d. Melting points(m.p.) are in ^{o}C.

the peak value at 20.2 degrees. In the lower angles region, there is only a diffuse peak centered at 5.2 degrees corresponding to a distance of 16.7 A. The diffractions at lower angles are believed to be from the mesogenic side groups:

CH_3O–⟨⟩–CO-NH–⟨⟩–NH–OC–⟨⟩–OCH_3

Lyotopic Liquid Crystallinity of MJLCPs

Liquid crystal polymers with sufficient high persistence length and sufficient solubility may form lyotropic phases (*6*). Examples of the lyotropic liquid crystal polymers include PPTA, polybenzothiazole, polybenzoxazole, and poly(benzyl glutamate). MJLCPs reported here have good solubility in normal solvents such as DMF and DMSO. Those in the PE-n series have been shown to have high values of persistence length and chain rigidity similar to that of liquid crystalline aromatic polyesters. On the other hand, the PA-n polymers have been expected to have even higher stiffness than the PE-n polymers (evaluation of persistence length of PA-n is on the way). Therefore it would be interesting to study the possibility of formation of lyotropic phases of both PE-n and PA-n types of MJLCPs.

We have now found that in dry DMF, the PA-n polymers indeed form lyotropic solutions with strong light scattering effect. The preliminary studies for PA-1 have shown that the scattering is observable at concentrations of about 0.2 wt %. It becomes stronger and stronger when the concentration increases to about 4 wt % when a gel is formed. The scattering of the gels is significantly weaker than the lyotropic solutions at concentrations of about 0.5 wt % and higher. Upon centrifugation, the solutions separate into two phases. The top phase is isotropic with no observable light scattering while the bottom phase scatters light strongly. The lyotropic phase has also thermotropic property. The temperature at which the liquid crystal to the isotropic liquid phase transition occurs for the solution of 0.20 wt % is 75 ^{o}C, that for the soution of 0.51 wt % is 90 ^{o}C, showing a very strong dependence on concentration of the solutions. The transitions are reversible.

PE-n solutions were also made to study the lyotropic phase. No solutions with concentrations up to 50 wt % have been observed to scatter lights. However, both the powder samples (obtained from solutions by addition of precipitant methanol and dried in vacuum) and the film samples (casted from solutions by evaporation of the solvent at ambient temperatures) are strongly birefringent when observed on a polarizing microscope. This is true not only for PE-n but also true for PA-n polymers. As discussed earlier, both PE-n and PA-n polymers are non-crystalline. Therefore, the birefringence must be a result of mesophase ordering formed during the precipitation or solvent evaporation. In other words, a mesophase is formed in the course of the preparations of the samples. Careful studies have found that with the process of concentration of the solutions of MJLCPs, a nematic phase with typical threaded and schlierene defects is developed. Figure 1 for PE-1 and Figure 2 for PA-1 are examples of the textures of these lyotropic liquid crystal phases obversable on a polarizing microscope.

In this section we have shown that PE-n and PA-n polymers are capable of forming lyotropic liquid crystals. The similarity in property of MJLCPs with rigid and semi-rigid main chain type liquid crystal polymers is further demonstrated by the fact.

Support for the Concept From Other Schools

Experimental supports of this concept were also given by other research groups. Side chain type liquid crystal polymers with mesogenic units laterally substituted on the main chains have been studied by Finkelmann (*12*), Keller (*13*), Gray (*14*), Pugh (*22*) and their coworkers. Mesogenic units used by Finkelmann and by Keller were the same as or very similar to what we have used in polymers PE-n. However, flexible spacers with different length and structure were used in their polymers. Finkelmann's polymers are polymethacrylates, while Keller's are polysiloxanes. Gray also used polysiloxanes as polymer backbones and polymethylene segments as flexible spacers, but the attachment

Figure 1. Photomicrography of the Texture of PE-1 formed in the process of solvent evaporation.

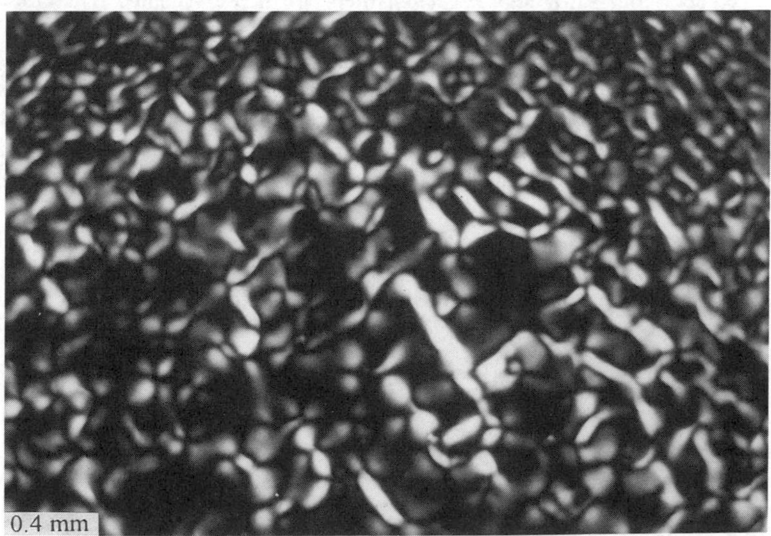

Figure 2. Photomicrography of the Texture of PA-1 formed in the process of solvent evaporation.

of the mesogens to the backbone is not at the waist but at the shoulder position of the mesogens. In spite of these differences in molecular structure, the polymers all form nematic liquid crystal phases, same as our PE-n and PA-n polymers.

With small angle neutron scattering, Hardouin and co-workers (*23*) have been able to study the global shape of the polymer molecules in nematic phase and the "Jacket Effect" imposed by mesogenic units on the molecular shape and property of the polymers. The polymers include those with mesogenic units laterally attached through flexible spacers of different length, those with different degrees of substitution of such mesogens in copolymers, and those having different degrees of polymerization (*13*, *23-25*). Because of the mesogenic jacket, according to these authors, these polymers all have strong tendency of chain extension. The highly prolated chain conformations are adopted in the nematic phase. Typical values of 4 or larger for the ratio of the apparent quadratic sizes parallel and perpendicular to the magnetic field were found for the polymers. The rusults contrast significantly with those for conventional side chain type polymers. Conventional side chain type lcps often take the oblate rather than the prolate chain conformations with quadratic size ratios smaller than 1. However, the jacket effect does vary for different molecules. In general, jacket effect is stronger for polymers that have higher degrees of mesogenic substitution, shorter spacers and lower degrees of polymerization.

Polymers studied by these groups all have flexible spacers, meaning the mesogenic jacket effect is smaller than it is in our PE-n and PA-n polymers in which no spacers are incorporated. As a matter of fact, all the polymers with flexible spacers have lower glass transition (40 ^{o}C or lower) and lower isotropization temperatures (below 120 ^{o}C). By contrast, PE-n polymers have Tg from ~80 to ~200 °C depending on the type of R groups at the two ends of the mesogenic units. As mentioned earlier, the PE-n polymers have very stable nematic phases. On the other hand, the PA-n polymers have even higher glass transitions so high that no definit value has been obtained before slow thermal decomposition starts.

Conclusions

Polymers with rigid mesogenic units as side groups attached through the waist or nearby positions to the main chain form a new class of liquid crystal polymers. The chain conformation and property of this class of polymers depend strongly on the presence of flexible spacer between the main chain and mesogenic units. In the polymers where no spacer or only very short spacers are used as are in the cases of PE-n and PA-n polymers, the rigid and bulky mesogenic units would form a dense "jacket" around the main chain, thus the term "Mesogen-Jacketed Liquid Crystal Polymers". The mesogenic jacket of a MJLCP will impose on the main chain a strong jacket effect and force the main chain to take extended conformations. The chain is thus stiffened and may be described by Kratky-Porod worm-like chains. If longer flexible spacers are used, there will be much more space for the mesogenic units to pack in, the density of the mesogenic units in the space around the chain will be smaller, the jacket will be looser. As the result, a weaker jacket effect should be expected. MJLCPs have shown chain conformation and properties of rigid or semi-rigid polymers, including high tendency (if not exclusive) of nematic phase formation, formation of nematic banded textures, high values of chain persistence length, and formation of lyotropic liquid crystalline phases. While in chemistry the MJLCPs are similar to conventional side chain type liquid crystal polymers by using mesogenic units as side groups, the similarity in physical aspects of MJLCPs and main chain type lcps is remarkable.

Acknowledgment:

This work was supported by NNSFC.

References

1. Finkelmann, H.; Ringsdorf, H.; Wendorff, H. J., *Makromol.Chem.*, **1978**, 179, 273
2. *Side Chain Liquid Crystal Polymers*, McArdle, C.B., Eds.; Blackie and Son Ltd, Glasgow and London, **1989**
3. Donald, A.M.; Windle, A.H., *Liquid Crystalline Polymers*; Cambridge Univ. Press: Cambridge, **1992**
4. M. Bohdanecky and J. Kovar: In *Viscosity of Polymer Solutions*, Jenkins, A.D., Ed., Elsdevier Amsterdam / New York, **1982,** Chapter 2.3
5. Kratky, O.; Porod, G., *Red. Trav. Chim. Paye-Bas*, **1949,** 68, 1106
6. Flory, P.J. , *Adv.Polym.Sci.*, **1984**, 59, 1
7. Chen, S.; Jin, Y.; Hu, S.; Xu, M., *Polym. Commun.*, **1987**, 28, 208.
8. Zhou, Q.F.; Li, H.M.; Feng, X.D., *Macromolecules*, **1987**, 20, 233
9. Zhou, Q.F.; Zhu, X.L.; Wen, Z.Q., *Macromolecules*, **1989**, 22, 491
10. Zhou, Q.F.; Wan, X.; Zhu, X.; Zhang, F.; Feng, X.D., *Mol. Cryst. Liq. Cryst.*, **1993**, 231, 107
11. Zhou, Q.F., *Polymer Bulletin (Chinese)*, **1991**, 3, 160
12. Hessel, F.; Finkelmann, H., *Makromol. Chem.*, **1987**, 188, 1579
13. Keller, P.; Hardouin, R.; Mauzac, M.; Achard, M., *Mol.Cryst.Liq.Cryst.*, **1988**, 155, 71
14. Gray, G.W.; Hill, J.S.; Lacey, D. , *Mol.Cryst.Liq.Cryst.*, **1991**, 197, 43
15. Xu, G.; Wu, W.; Shen, D.; Hou, J.; Zhang, S.; Xu, M.; Zhou, Q. F., *Polymer*, **1993**, 34(9), 1818
16. Ezrin, M.; Updegraff, I.H.; Cassidy, H.G., *J.Am.Chem.Soc.*, **1953**, 75, 1610
17. Dobb, M.G.; Johnson, D.J.; Saville, D.P., *J.Polym.Sci., Polym.Phys.Ed.*, **1977**, 15, 2201
18. Brelsford, G.L.; Krigbaum, W.R. , In *Liquid Crystallinity in Polymers*; Ciferri, A. , Ed.; VCH Publishers, Inc.: New York, **1991, pp 61-94**
19. Jackson, W.J. , *Chin.J.Polym.Sci.*, **1992**, 3, 195

20. Tsvetkov, V.N.; Andreeva, L.N.; Bushin, S.V.; Mashoshin, A.I.; Cherkasov, V.A.; Yedlinski, Z.; Sek, D., *Polym.Sci.USSR*, **1984**, 26, 2569
21. Bohdanecky, M., *Macromolecules*, **1983**, 16, 1483
22. Pugh, C.; Schrock, R.R., *Polymer Preprints*, **1993**, 34(1), 180
23. Hardouin, F.; Mery, S.; Achard, M.F.; Noirez, L.; Keller, P., *J.Phys.II*, **1991**, 1, 511
24. Leroux, N.; Achard, M.F.; Keller, P.; Hardouin, F., *Liquid Crystals*, **1994**, 16(6), 1073
25. Leroux, N.; Mauzac, M.; Noirez, L.; Hardouin, F. *Liquid Crystals*, **1994**, 16(3), 421

Chapter 21

Identification of Highly Ordered Smectic Phases in a Series of Main-Chain Liquid-Crystalline Polyethers

Yeocheol Yoon[1], Rong-Ming Ho[1], Edward P. Savitski[1], F. Li[1], Stephen Z. D. Cheng[1,3], Virgil Percec[2], and Peihwei Chu[2]

[1]Maurice Morton Institute of Polymer Science, University of Akron, Akron, OH 44325–3909
[2]Department of Macromolecular Science, Case Western Reserve University, Cleveland, OH 44106–2699

A series of polyethers have been synthesized from 1-(4-hydroxy-4'-biphenyl)-2-(4-hydroxyphenyl)propane and α,ω-dibromoalkanes having different numbers of methylene units [TPPs]. Both odd- and even-numbered TPPs [TPP(n=odd)s and TPP(n=even)s] exhibit multiple transitions during cooling and heating and they show little supercooling dependence. The phase diagrams for both TPP(n=odd)s and TPP(n=even)s have been constructed and the phase identification is the central part of this publication. The transition thermodynamic properties (enthalpy and entropy changes during the transitions) are studied based on differential scanning calorimetry experiments. The contributions of the mesogenic groups and methylene units to each ordering process can be separated and they indicate the characteristics of these processes thereby providing estimations of the transition types. Structural information can be obtained *via* wide angle X-ray diffraction powder and fiber patterns at different temperatures. In addition to the nematic liquid crystalline phase, highly ordered smectic F, G and H phases have been identified. These phase assignments are also supported by the morphological observations in the liquid crystals under polarized light and transmission electron microscopy. The concepts of the liquid crystalline phases defined for small-molecule liquid crystals are also applicable to main-chain liquid crystalline polymers.

Since the first discovery of the liquid crystalline phase over one hundred years ago, the classification of the distinct liquid crystalline phases in small-molecule liquid crystals has been well established (*1,2*). As shown in Figure 1, the least ordered liquid crystalline phase is the nematic phase that only possesses molecular orientational order due to the anisotropy of the molecular geometric shape. The next ordering level introduced is the layer structure in addition to the molecular orientation to form a smectic A (S_A) or a smectic C (S_C) phase. Following the S_A phase the hexatic B (H_B), smectic crystal B (S_B) and smectic crystal E (S_E) phases are observed. In this series the long axis of the molecules is oriented perpendicular to the layer surface while order is increasingly developed from positional order normal to the layer in S_A, bond

[3]Corresponding author

0097–6156/96/0632–0358$15.00/0

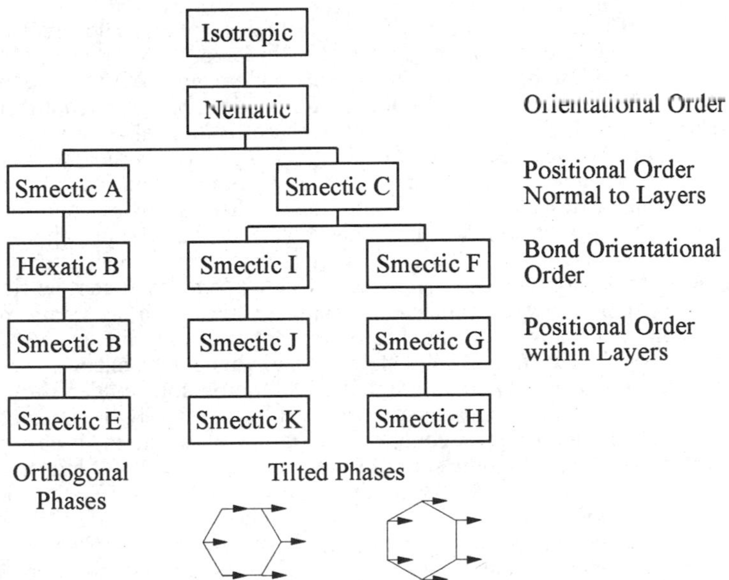

Figure 1. Classification of the liquid crystalline phases

orientational order in H_B, positional order within the layers in S_B and, finally, asymmetric axial site symmetry in S_E (*1,2*). Two separate highly ordered smectic phases are recognized after the S_C phase. Both of these phases possess long molecular axes that are tilted with respect to the layer surface normal. The developments of order in both cases are correspondingly identical as in the first series initiated from S_A to S_E. The difference between these two series lies in the tilt directions: the smectic F (S_F), smectic crystal G (S_G), and smectic crystal H (S_H) possess the long axis tilted towards one side while the long axis directions in the smectic I (S_I), smectic crystal J (S_J) and smectic crystal K (S_K) are tilted towards one apex (Figure 1). Furthermore, the S_F, S_G, S_I and S_J phases exhibit hexagonal (or pseudo-hexagonal) packing when viewed parallel to the long axis. The packing of the tilted long axis gives rise to a monoclinic lattice (for S_F and S_G phases $a>b$ while in S_I and S_J phases $a<b$). On the other hand, the lateral packing perpendicular to the molecular long axis becomes an orthorhombic, herringbone type in the S_H and S_K phases (*1,2*). The main experimental methods used to identify these highly ordered smectic phases in small-molecule liquid crystals are wide angle X-ray diffraction (WAXD), calorimetry, polarized light microscopy (PLM) and phase mixing experiments (*1*). The structural identification in small-molecule liquid crystals is convincing since, in many cases, monodomains of the small-molecule liquid crystals can be obtained *via* external force fields. Therefore, sharp WAXD reflections can be observed to determine the structural order and symmetry.

An enormous amount of effort has been made during the last twenty years on understanding the transition behavior of the main-chain liquid crystal polymers (*3*). It has been known that a liquid crystal transition from the isotropic melt usually exhibits a thermodynamic first-order transition (*4*) and occurs close to equilibrium. This behavior is reflected by the cooling-rate independence of the transition. Such behavior can be easily determined through differential scanning calorimetry (DSC). On the other hand, structural changes during the transition can be characterized *via* wide angle X-ray diffraction (WAXD) powder experiments. In nematic (N) liquid crystal transitions, only a shift of the *d*-spacing in the broad halo towards a higher reflection angle can be seen. This shift represents a decrease of the average lateral distance between chain molecules. In addition, for S_A and S_C transitions directly from the isotropic melt, an additional sharp diffraction peak in the low-angle region appears and represents the layer spacing in the smectic phase. For highly ordered smectic phases WAXD diffraction peak(s) can also be observed in the high-angle region. WAXD fiber patterns obtained from oriented samples are essential for distinguishing three-dimensional crystal packing from highly ordered smectic phases. They also provide further information about the detailed lateral chain packing, layer structure and order correlation. PLM is a unique experimental method to study liquid crystal order and morphology. In some cases, quantitative descriptions of defects may be obtained. Small angle light scattering is also used to study liquid crystalline morphology. For semicrystalline polymers having a liquid crystal state, the "lamellar decoration" method using transmission electron microscopy (TEM) (*5,6*) has been developed to establish relationships between molecular characteristics, such as chain rigidity, molecular weight, and Frank constants. Molecular motions in the liquid crystal states have also been investigated *via* ^{13}C-solid state nuclear magnetic resonance. During the last three decades, the thermodynamic properties and morphological identifications of main-chain liquid crystalline polymers have been progressively understood. The nematic phase has been the most commonly reported phase in main-chain liquid crystalline polymers. Recently, some studies have shown that S_A, S_C or higher ordered smectic phases may also be possible in these polymers. The difference between polymers and small-molecule liquid crystals is the connectivity existing in the polymer case. Despite this difference we can show that highly ordered smectic phases may also be found in main-chain liquid crystalline polymers based on our careful thermodynamic, structural and morphological analyses.

Experimental Section

Materials. The polyethers were synthesized from 1-(4-hydroxy-4'-biphenyl)-2-(4-hydroxyphenyl)propane (TPP) and α,ω-dibromoalkanes. The detailed synthetic procedure has been reported in an earlier publication (*7*). The chemical structure of the TPPs is shown below:

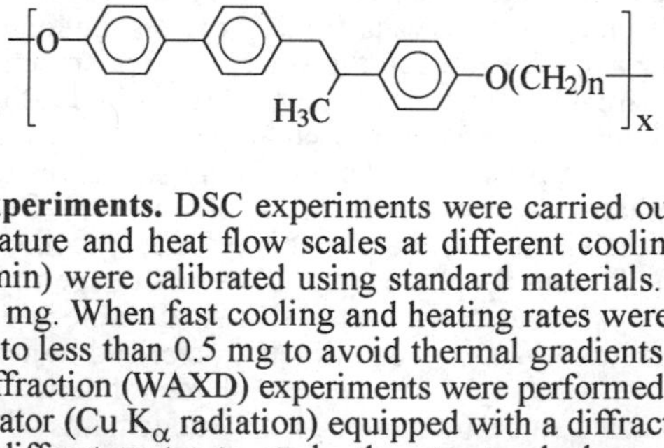

O
H3C
O(CH2)n
x

Instrument and Experiments. DSC experiments were carried out in a Perkin-Elmer DSC-7. The temperature and heat flow scales at different cooling and heating rates (2.5°C/min - 40°C/min) were calibrated using standard materials. Typically, the DSC sample size was 2-3 mg. When fast cooling and heating rates were applied the sample weight was reduced to less than 0.5 mg to avoid thermal gradients within the samples. Wide angle X-ray diffraction (WAXD) experiments were performed on a Rigaku 12 kW rotating anode generator (Cu K_α radiation) equipped with a diffractometer. A hot stage was set up on the diffractometer to study the structural changes during the phase transition at constant cooling and heating rates. The temperature can be controlled to better than ±1°C. The WAXD fiber patterns were also taken at different temperatures, particularly at those temperatures where the phase transitions occur, *via* a Siemens two-dimensional area detector with built in hot stage. Liquid crystalline morphology was examined *via* a PLM (Olympus BH-2) with a Mettler hot stage (FP-82) and a JEOL JEM-1200 EXII TEM at an accelerating voltage of 120 kV.

Results and Discussion

Phase diagrams in TPPs. Figures 2 and 3 show phase diagrams for TPP(n=odd)s and TPP(n=even)s, respectively (*9,10*). Due to the well-known odd-even effect, the

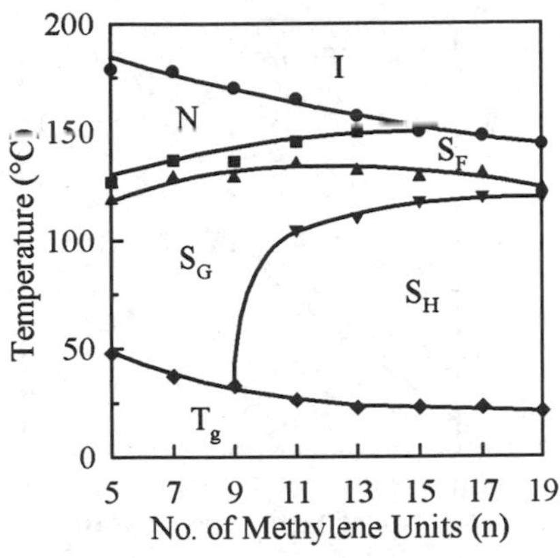

Figure 2. Phase diagram of TPP(n=odd)s

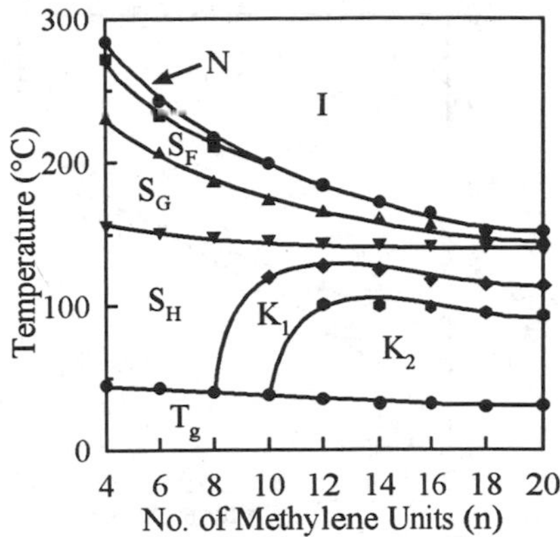

Figure 3. Phase diagram of TPP(n=even)s

transition temperature and phase boundary of the diagrams are significantly different. It is important that the transition temperatures in these phase diagrams should be in thermodynamic equilibrium. This can be achieved in this series of polymers since these phase transitions show little supercooling and superheating dependence (*9,10*). This is a clear indication that these transitions are close to equilibrium. Additionally, the enthalpy change for each transition is also supercooling independent (*8-10*), which further reveals the equilibrium status of the phase transitions. Since the DSC results provide information about the heat release or absorption during the transitions, sudden change in enthalpy, entropy and volume at the transition manifests the nature of the first-order transitions based on the thermodynamic definition (*4*). However, DSC experiments do not give rise to information of the structural changes during these transitions. As a result, the assignment of each phase in Figures 2 and 3 has to be identified *via* structural characterizations such as WAXD and ED experiments. Since ED experiments require a sufficiently large sized liquid crystalline monodomains which are difficult to obtain in polymers, the commonly used technique is WAXD of unoriented and oriented samples. The phase identification should also closely combine with the liquid crystal defect observations since different category of liquid crystal phases is associated with specific types of defects.

Identification of the nematic phase. The nematic phase is the most commonly observed phase in main-chain liquid crystalline polymers. This phase possesses one-dimensional orientational order along the long axis of the molecules while the lateral packing is still liquid-like. The isotropic to nematic (I→N) phase transition in the liquid crystalline polymers can be observed clearly *via* DSC and shows relatively small entropy and enthalpy changes during the transition (*12*). From Figures 2 and 3, the phase diagrams indicate that the I→N transition exists in odd-TPP(n≤13)s and even-TPP(n≤8)s. Figure 4 shows the relationships between the transition enthalpy change and the number of methylene unites for both TPP(n=odd)s and TPP(n=even)s. This kind of plot was first reported by Blumstein et al. for another main-chain liquid crystalline polymer (*13,14*). It is clear that linear relationships of the transition properties with respect to the number of methylene units are found and the enthalpy changes for the TPP(n=odd)s are relatively small compared with those in TPP(n=even)s. This is due to the odd-even effect which influences the mesogenic group packing and methylene unit conformation. Since the equilibrium transition temperatures are known from Figures 2 and 3, the entropy changes at the transitions can be calculated *via* $\Delta S=\Delta H/T$ and shown in Figure 5. Both figures indicate that the

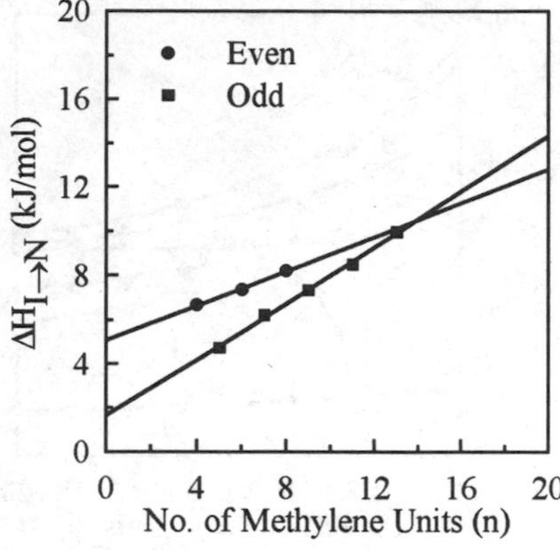

Figure 4. Relationships between the enthalpy changes at the I→N transitions and the number of methylene units for both TPP(n=odd)s and TPP(n=even)s

Figure 5. Relationships between the entropy changes at the I→N transitions and the number of methylene units for both TPP(n=odd)s and TPP(n=even)s

nematic structure in TPP(n=even)s is more ordered than that in TPP(n=odd)s. Furthermore, the slopes in Figures 4 and 5 roughly represent the contribution to the thermodynamic transition properties per mole of methylene units while the intercepts, which are extrapolated to zero methylene unit, should be the thermodynamic properties solely attributed to the mesogenic groups. The results are listed in Table I. The enthalpy and entropy changes of the mesogenic groups in TPP(n=even)s are much greater than those in TPP(n=odd)s [5.01 kJ/mol versus 1.64 kJ/mol for the enthalpy changes and 7.65 J/(K·mol) versus 2.80 J/(K·mol) for the entropy changes, respectively]. On the other hand, those changes of the methylene units in TPP(n=even)s are slightly smaller than those in TPP(n=odd)s [0.40 kJ/mol versus 0.63 kJ/mol for the enthalpy changes and 1.12 J/(K·mol) versus 1.54 J/(K·mol) for the entropy changes, respectively]. This reveals that the structural order in the nematic phase for TPP(n=even)s is mainly attributed to the mesogenic groups.

However, the structural change at the transition detected by WAXD is rather trivial. Since the N phase does not possess two- or three-dimensional order in the structure, no

Table I. The transition enthalpy and entropy changes for methylene units and mesogenic groups for the liquid crystalline transitions

Sample	Transition	ΔH (kJ/mol)		ΔS (J/K·mol)	
		Mesogene	Methylene	Mesogene	Methylene
Odd	I→N	1.64	0.63	2.80	1.54
	$N \to S_F$	0.50	0.43	1.78	0.96
	$I \to S_F$	2.14	1.06	4.58	2.50
	$S_F \to S_G$	0.00	0.42	0.00	1.04
	$S_G \to S_H$	0.00	0.92	0.00	2.35
Even	I→N	5.01	0.40	7.65	1.12
	$N \to S_F$	7.15	0.16	11.43	0.73
	$I \to S_F$	12.16	0.56	19.14	1.81
	$S_F \to S_G$	2.45	0.32	3.85	0.89
	$S_G \to S_H$	0.00	0.95	0.00	2.25

shape reflection is expected. A broad reflection halo at around 2θ=18° represents the average distance between the chains in the lateral direction. This indicates that the lateral packing in this phase is still liquid-like. The correlation length of this broad halo is less than 2 nm as calculated using the Scherrer equation. A careful study leads to the recognition that, at the I→N transition temperature observed in DSC, there is a sudden shift of the *d*-spacing of this broad halo towards a smaller value for each TPP in addition to the thermal effects of lattice expansion during heating and shrinkage during cooling. Figure 6 shows the *d*-spacing changes for TPPs with temperature during cooling. This phenomenon has also been reported for several main-chain liquid crystalline polymers exhibiting a N or S_A phase (*15,16*).

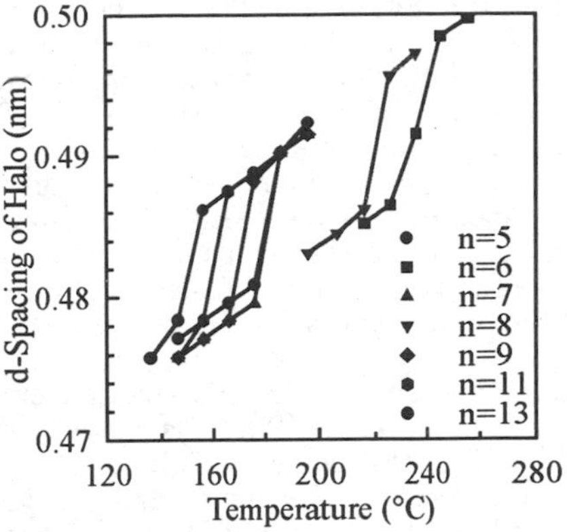

Figure 6. The *d*-spacing changes of the broad halo at the I→N transitions for TPPs detected *via* WAXD powder experiments

In principle, the nematic phase can also be identified from PLM observations. Since the defect morphology of high molecular weight main-chain liquid crystalline polymers is usually too small in size to be clearly recognized under PLM, the TEM "lamellar decoration" technique is commonly used (*5,6*). PLM observations of the samples under mechanical shear also help to identify low-ordered liquid crystalline phases such as N, S_A and S_C phases in which fluidity is still retained. Figure 7 shows sheared odd-TPP(n=7-13)s as examples. The banded texture in the sheared sample is obvious and it is perpendicular to the shear direction. As extensively studied a decade ago, this kind of banded texture is caused by the continuous oscillation of the

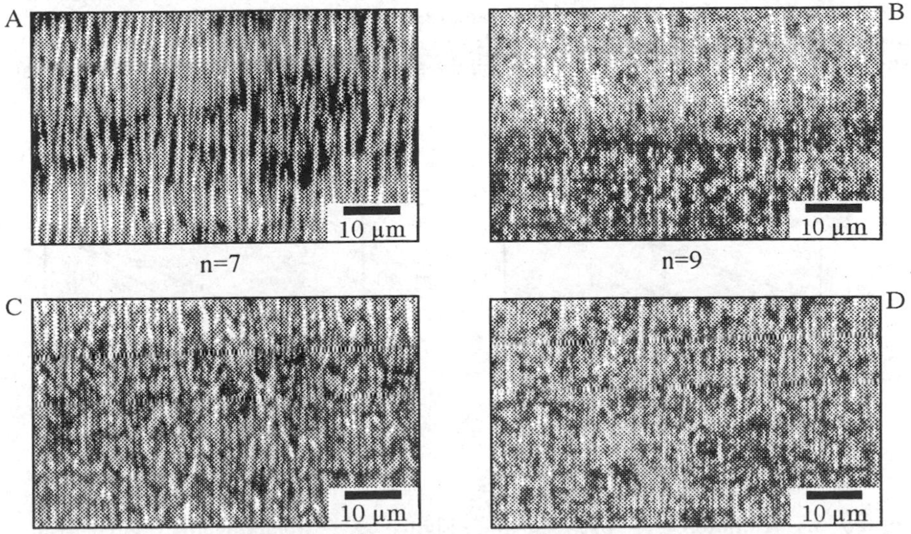

Figure 7. PLM morphological observations of sheared odd-TPP(n=7-13)s which show a banded texture at the I→N transition temperature (Shear direction is horizontal.)

molecules about the direction imposed by the previous flow after allowing slight relaxation. Even for TPP(n=6), the N phase is only stable within a temperature region of 10°C and the mechanical shear generates a clear banded texture. As a result, this banded texture becomes a specific characteristic of the low-ordered liquid crystalline polymers. In odd-TPP(n≥15)s and even-TPP(n≥10)s the sheared samples at the highest transition temperatures (Figures 2 and 3) do not show any banded texture. Note that this highest transition represents a higher ordered smectic transition from the isotropic melt (see below). The TEM observations for this I→N transition can also be found through the defect morphology. However, this morphology may be disturbed by successive high-ordered transitions since each TPP exhibits multi-transition behavior as shown in Figures 2 and 3 (see below).

Identification of the S_F phase. The thermodynamic transition properties of the N→S_F transition for odd-TPP(n≤13)s and even-TPP(n≤8)s as well as the I→S_F transition for odd-TPP(n≥15)s and even-TPP(n≥10)s can be studied using DSC experiments. Figures 8 and 9 represent the linear relationships between these properties and the number of methylene units (similar to Figures 4 and 5). Again, the slopes and intercepts possess the same physical meanings as described above. Specifically, it is important to note that in these two figures the transition enthalpy and entropy changes for odd-TPP(n≥15)s and even-TPP(n≥10)s are only attributed to the single transition process directly from the isotropic melt (the I→S_F transition), while for odd-TPP(n≤13)s and even-TPP(n≤8)s these changes are the summation of the contributions from the two transition processes, namely, the I→N and N→S_F transitions. Surprisingly enough, all the data fit well into the same linear relationships. It is thus expected that an additive scheme holds in these thermodynamic properties. This indicates that Ostwald's law of successive states can be used to describe this scheme which says that a phase will occur step-by-step through successively more stable polymorphs (*17*). Furthermore, the quantitative data show that the mesogenic group contributions to the enthalpy and entropy changes of this transition in TPP(n=even)s are almost five times greater than those for TPP(n=odd)s. On the other hand, the methylene unit contributions to these changes for TPP(n=odd)s are more than two times of those in TPP(n=even)s. The detailed results are listed in Table I for comparison.

A S_F phase has been assigned for TPPs in both phase diagrams (Figures 2 and 3). The identification of this phase is based on the structural order and symmetry. The first

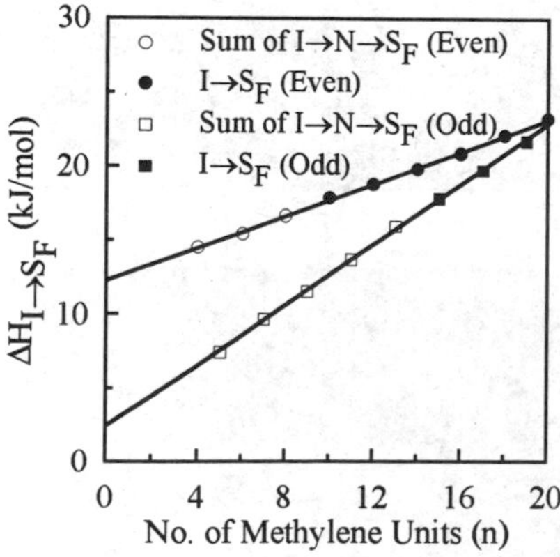

Figure 8. Relationships between the enthalpy changes at the I→S_F transitions and the number of methylene units for both TPP(n=odd)s and TPP(n=even)s

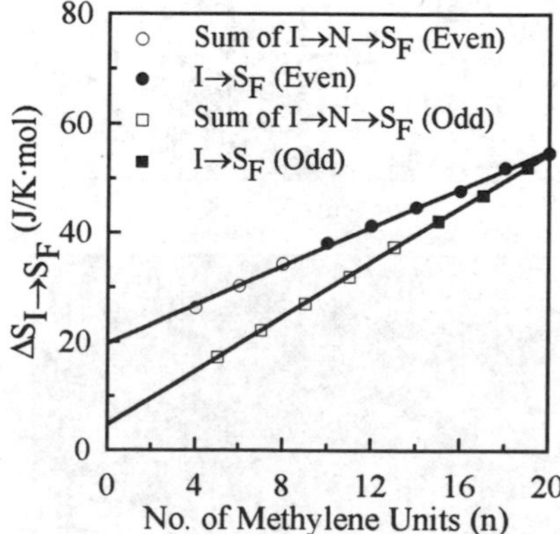

Figure 9. Relationships between the entropy changes at the I→S_F transitions and the number of methylene units for both TPP(n=odd)s and TPP(n=even)s

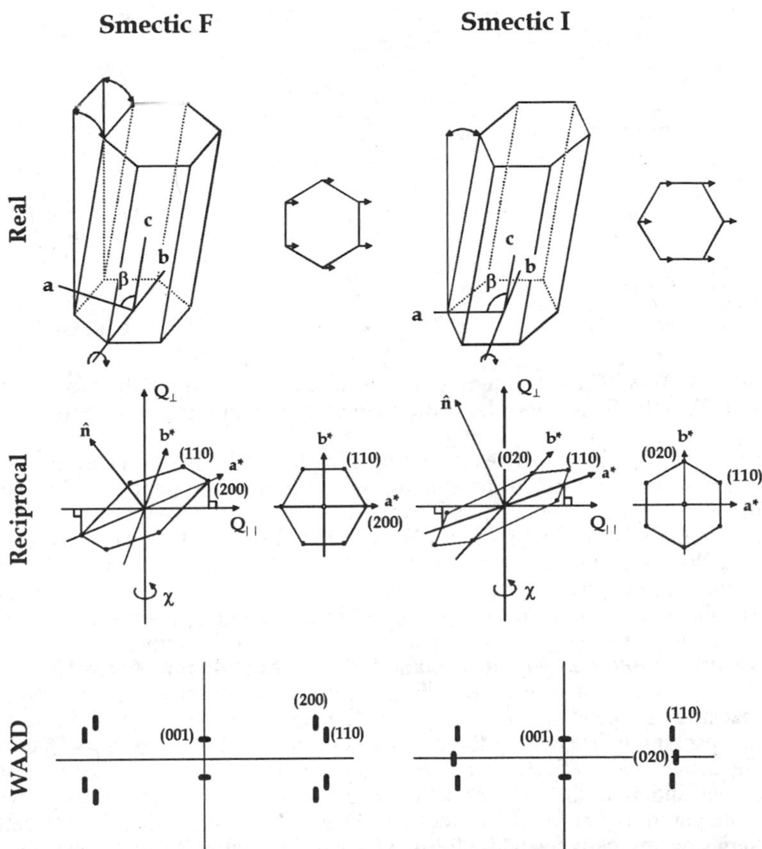

Figure 10. Schematic drawing of the S_F and S_I phases represented by unit cells in the real space, reciprocal lattices and WAXD patterns. (*a, b, c* and β are real lattice parameters. a^*, b^* and c^* are reciprocal lattice parameters. $Q_\perp$ and $Q_\parallel$ represent the components of the momentum transfer which are perpendicular and parallel respectively to the smectic layers. $\hat{n}$ is the director vector and χ is an average rotation angle.)

step is to distinguish the S_F from the S_I phases. As it is known, the difference between these two phases is in the tilting directions of the molecular long axis (*1*,*2*). Figure 10 shows schematic representations of these two phases for the real spaces, reciprocal spaces, and the predicted uniaxially oriented polydomain WAXD patterns. It is evident that in the S_I phase the (020) plane is on the equator in the fiber pattern since the normal vector of this plane is parallel to the incident X-ray beam. For the S_F phase both reflections of the (200) and (110) planes are in the quadrant. Neither of them possess a normal vector which is parallel to the X-ray beam. On the other hand, since the (00l) planes in both unit cells are parallel to the equator, the layer spacings thus appear on the meridian. In polymers, however, the commonly obtained fiber patterns often possess a chain axis that is parallel to the meridian direction. As a result, the (00l) layer planes are not parallel to the equator and the layer spacing reflections are thus no longer on the meridian but in the quadrant. In this orientational mode, the cross-section perpendicular

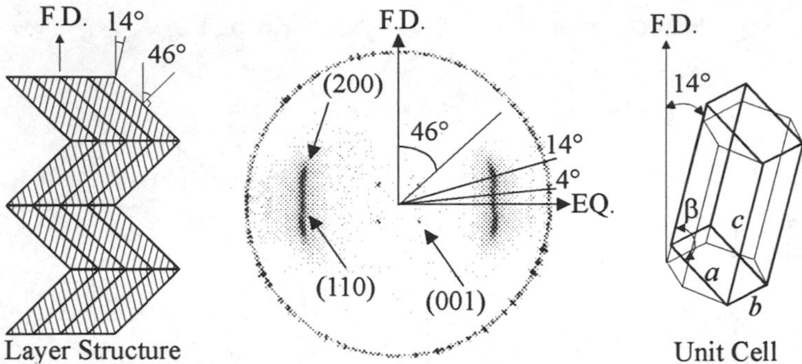

Figure 11. WAXD fiber pattern, unit cell structure and morphology for oriented TPP(n=7) in the S_F phase (F.D. is the fiber direction and EQ. is the equator.)

to the fiber direction indicates hexagonal packing for both phases. Therefore, both the 200 and 110 reflections are on the equator and they are superimposed. Consequently, one cannot distinguish the S_F phase from the S_I phase based only on these two reflections in the WAXD fiber patterns. However, in the odd-TPP(n≤13) cases, we have found that the chain molecules are not parallel to the fiber direction and an angle between them is observed to range between a few degrees to fifteen degrees. With increasing the number of methylene units, this angle gradually decreases towards zero, namely, the molecular axis becomes parallel to the fiber direction. For example, this angle is 14° for TPP(n=7) (Figure 11), and it decreases to 4° for TPP(n=13). Since the molecular axis is tilted in reference to the fiber direction, the cross-section of the chain lateral packing perpendicular to the fiber direction is not hexagonal (note that the hexagonal packing exists only in the cross-section perpendicular to the molecular long axis). The 200 and 110 reflections of the hexagonal lattice are not on the equator but in the quadrant and thus the 110 reflection possesses less tilted angle away from the equator compared to that for 200 reflection. We have also found that the 110 reflection in this series of unit cells for odd-TPP(n≤13)s have a smaller 2θ angle than that of the 200 reflection. Hence, the *a*-dimension of the unit cell is greater than the *b*-dimension under the condition that the tilting angle of the molecular long axis with respect to the fiber direction is smaller than 15°. This is a clear indication of a S_F phase. However, when the molecular and fiber axes are parallel to each other, such as in the cases of odd-TPP(n≥15)s and all TPP(n=even)s, it is important to look for other reflections in the quadrant. For example in TPP(n=12) as shown in Figure 12, the 200 and 110

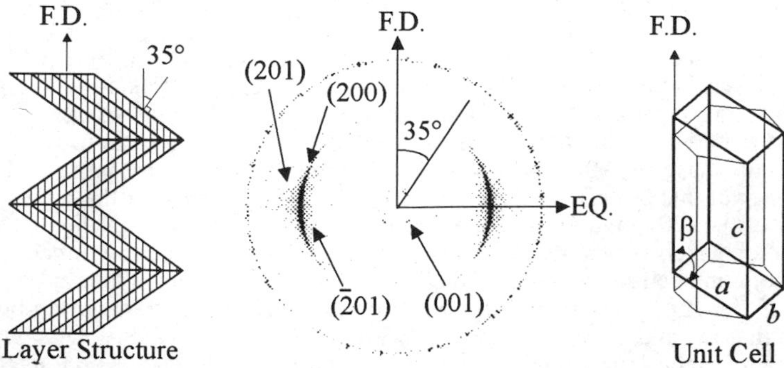

Figure 12. WAXD fiber pattern, unit cell structure and morphology for oriented TPP(n=12) in the S_F phase (F.D. is the fiber direction and EQ. is the equator.)

reflections are superimposed and located on the equator. Additional reflections of $\bar{2}01$ and 201 at 2θ=17.2° and 21.4°, respectively, can be assigned if one uses a S_F assignment with a monoclinic unit cell of a=1.14 nm, b=0.54 nm, c=2.95 nm and β=125°. When one uses a S_I unit cell to fit the reflections, a unit cell of a=0.68 nm, b=0.93 nm, c=2.95 nm and β=125° can be found, and the $\bar{2}01$ and 201 reflections in the S_F phase now become the $\bar{1}11$ and 111 reflections. Nevertheless, the standard deviation after refinement for the S_I phase is more than two orders of magnitude greater than that of the S_F phase (0.548×10^{-1} *versus* 0.261×10^{-3}), indicating that for TPP(n=12) this liquid crystalline state should be a S_F phase. Similar determinations can also be found in other TPPs (*10,11*).

One important parameter in the identification of highly ordered smectic phases is the correlation length of the ordered structure. Traditionally, this can be obtained by measuring the width of half-height on the WAXD reflection peaks using the Scherrer equation as a first approximation. In liquid crystalline polymers this parameter is usually smaller than the corresponding ones in small-molecule liquid crystals. We have defined the following criterion to identify the structural correlational order. Namely, short range order is represented by a correlation length of less than 2 nm while a long range order shows a correlation length of greater than 10 nm. Values between 2 nm and 10 nm define a quasi-long range order region (*9-11*). In the N phase, as indicated previously, the lateral packing of the chain molecules indicates short range order. However, in the S_F phase of TPP(n=odd)s, both the high-angle reflections possess correlation lengths ranging from 8 nm to 15 nm thus revealing that the correlation lengths are at the boundary between quasi-long range and long range order. On the other hand, for TPP(n=even)s the high-angle reflections are all in the long range order region. The low-angle reflection which represents the layer structure usually shows a correlation length in a range of 4 nm to 15 nm and it increases with the number of methylene units. This reveals that, in the S_F phase, the layer structures also possess limited correlation lengths in-between the quasi-long range and long range order regions.

PLM observations of the morphological changes during this transition are not drastic in both unsheared and sheared samples compared with those in Figure 7. The only noticeable difference is a change in the birefringence (*9-11*). However, under TEM observations, the liquid crystalline banded texture as well as defects can be seen. Figure 13 shows a mechanically sheared TPP(n=7) thin film sample. The shear direction is also indicated in the figure by the arrow. It is clear that the layer structure having a typical size of 0.5 to 1 μm corresponds to the banded texture observed in PLM (Figure 7). The layer structure is perpendicular to the shear direction (for a schematic drawing see Figure 11). The thin layers within the banded texture are generated by lamellar decoration and they possess an angle of ±46° from the shear direction. An ED pattern included in this figure corresponds well with the WAXD fiber pattern. The thickness of these lamellae is around 15 nm and may be representative of the correlation length along the layer normal (see below). On the other hand, if the S_F phase is directly entered from the isotropic melt, no large layer structure (banded texture) can be found as shown in Figure 14 for TPP(n=15). This provides additional proof that the large layer structure is attributed to the banded texture.

Identification of the S_G phase. Further cooling the TPP samples leads to the appearance of transition processes which indicates continuous ordering (see Figures 2 and 3). The $S_F\rightarrow S_G$ transition can be observed in all TPP(n=odd)s and TPP(n=even)s. Figures 15 and 16 show the linear relationships between the thermodynamic transition properties with the number of methylene units and again, the data are listed in Table I. An important issue here is that for TPP(n=odd)s the extrapolations of both the relationships to zero methylene unit are close to zero. This indicates that the ordering process of this $S_F\rightarrow S_G$ transition is basically attributed to the methylene units and the mesogenic groups do not contribute. However, in the cases of TPP(n=even)s, the

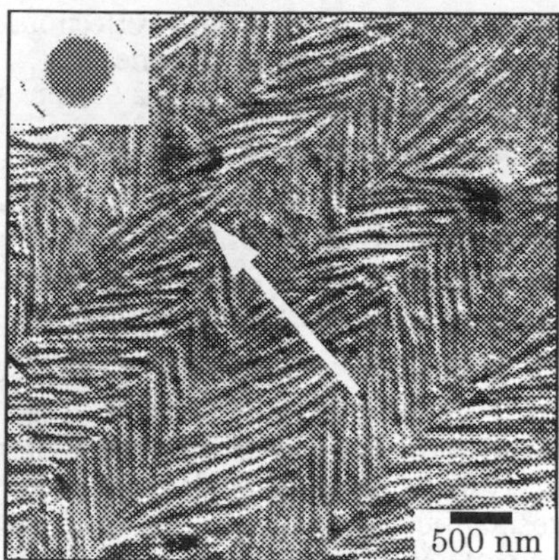

Figure 13. TEM micrograph of the mechanically sheared TPP(n=7) (The arrow indicates the shear direction.)

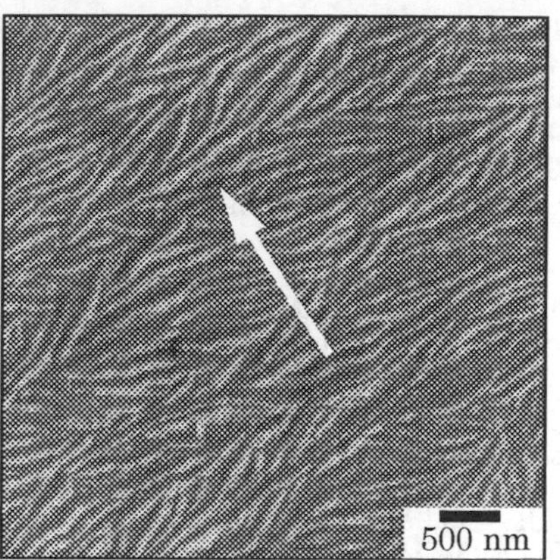

Figure 14. TEM micrograph of the mechanically sheared TPP(n=15) (The arrow indicates the shear direction.)

transition enthalpy and entropy changes for the mesogenic groups are 2.45 kJ/mol and 3.85 J/(K·mol), respectively. Also note the non-zero contribution from the methylene units. This reveals that the formation mechanism of this S_G phase in TPP(n=odd)s is intrinsically different from that in TPP(n=even)s. Note that the enthalpy changes in this transition are relatively small compared with other transitions and fit well with the observations of this transition in small-molecule liquid crystals (*1*).

Moreover, it is also interesting to find that the unit cell dimensions along the *a*- and *b*-axes show small, but sudden, changes at the transition temperature of the $S_F \rightarrow S_G$. Their first derivatives with respect to the temperature, the coefficients of thermal expansion of those axes, also exhibit discontinuous changes (*8,11*). This clearly indicates that this transition possesses the characteristics of a thermodynamic first-order transition (*4*).

To obtain the structural information about the S_G phase in TPPs, WAXD fiber patterns still serve as the central piece of evidence for the phase identification. The

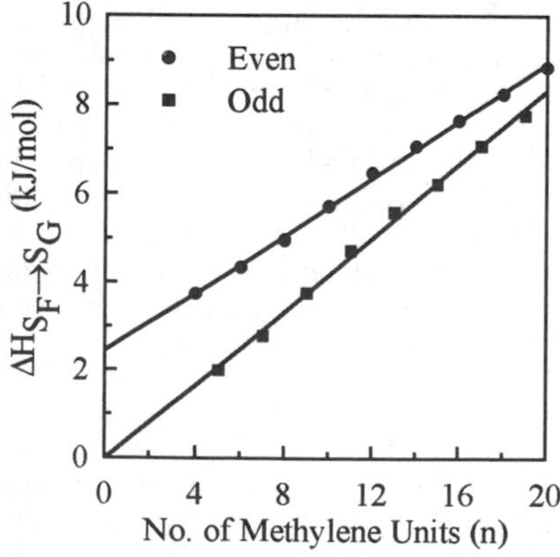

Figure 15. Relationships between the enthalpy changes at the $S_F \rightarrow S_G$ transitions and the number of methylene units for both TPP(n=odd)s and TPP(n=even)s

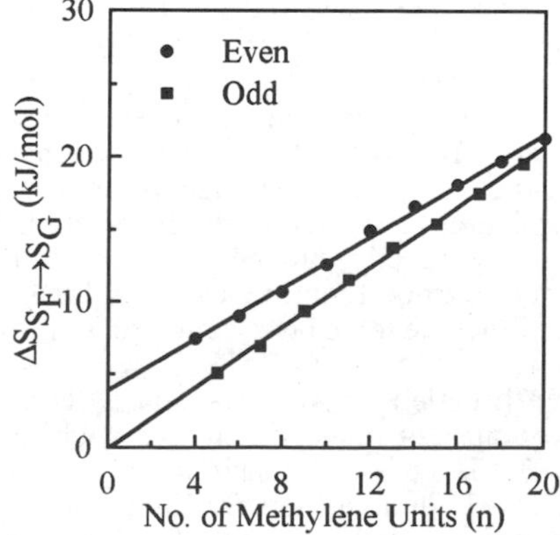

Figure 16. Relationships between the entropy changes at the $S_F \rightarrow S_G$ transitions and the number of methylene units for both TPP(n=odd)s and TPP(n=even)s

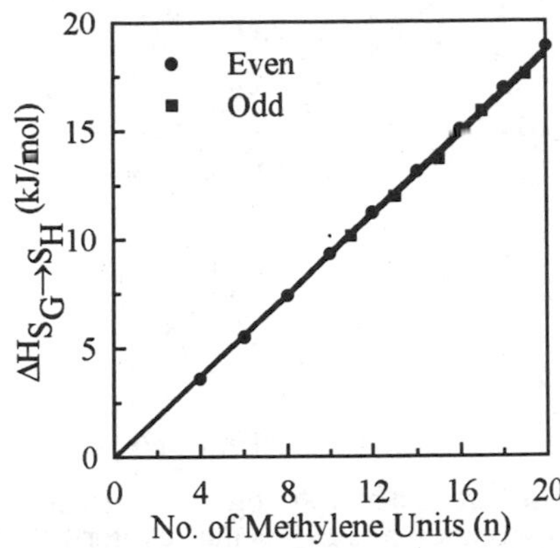

Figure 17. Relationships between the enthalpy changes at the $S_G \rightarrow S_H$ transitions and the number of methylene units for both TPP(n=odd)s and TPP(n=even)s

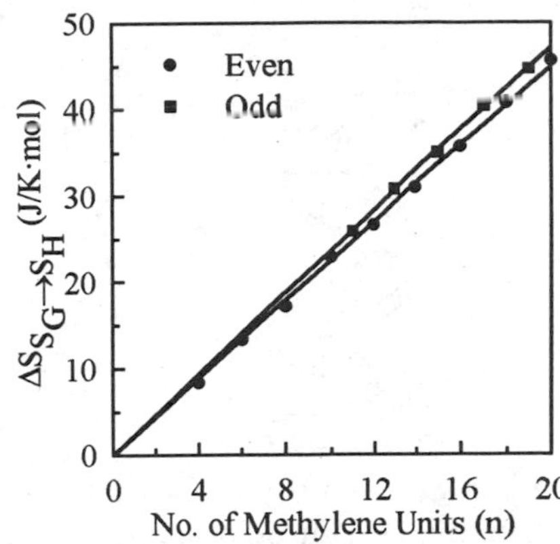

Figure 18. Relationships between the entropy changes at the $S_G \rightarrow S_H$ transitions and the number of methylene units for both TPP(n=odd)s and TPP(n=even)s

number of reflections in S_G is the same as those in S_F phase but at slightly different 2θ positions. This indicates that the basic unit cell lattice in S_G is retained and it is the monoclinic unit cell. Only the dimensions are slightly changed (*9-11*). The most important difference in the S_G from the S_F phases is the correlation lengths of the reflection planes. All of them show a substantial increase in the S_G phase, revealing that the order correlation is enhanced. The most significant increase in the correlation length is along the layer structures. Typically, it exhibits a 50% to over two times improvement and this change is more pronounced when the number of the methylene units is lower. For example, the S_G phase in TPP(n=7) is the most ordered phase down to the glass transition temperature and the improvement of the correlation length in the S_G from the S_F is more than 2.2 times (11 nm versus 5 nm). For TPP(n=15), the improvement is about 1.3 times (20 nm *versus* 15 nm). The improvement in TPP(n=even)s is not as drastic as that in TPP(n=odd)s but still is clearly recognizable.

The morphological observations of the S_G phase under PLM and TEM are apparently the same as those in the S_F phase. For example, TEM morphology of TPP(n=7) shown in Figure 13 is actually taken at room temperature and therefore, it is in the S_G phase. The correlation length of 11 nm corresponds well with the observation of the lamellar-like spacing in Figure 13 (~15 nm). Note that the correlation length of the layer structure in the S_F phase in TPP(n=7) is only 5 nm.

Identification of the S_H phase. For TPP(n=odd)s, the S_G phase is retained down to their glass transition temperatures for n≤9. Only odd-TPP(n≥11)s show the $S_G \rightarrow S_H$ transition (see Figures 2). On the other hand, all TPP(n=even)s possess this transition. Figures 17 and 18 show the linear relationships between thermodynamic transition properties and the number of methylene units. Again, it is important to find that the mesogenic groups in all the TPPs do not contribute to the enthalpy change during this transition. This indicates that the ordering process is solely attributed to the methylene units. Moreover, the quantitative values of the contributions for both series of TPP(n=odd)s and TPP(n=even)s are very close to each other (see Table I).

The structural analyses indicate a substantial change of the unit cell lattice symmetry during this transition. This is expected since based on the definition of the S_H phase, the lateral packing perpendicular to the chain direction is no longer hexagonal, but is orthorhombic with a herringbone type (*1,2*). It can be seen from the WAXD

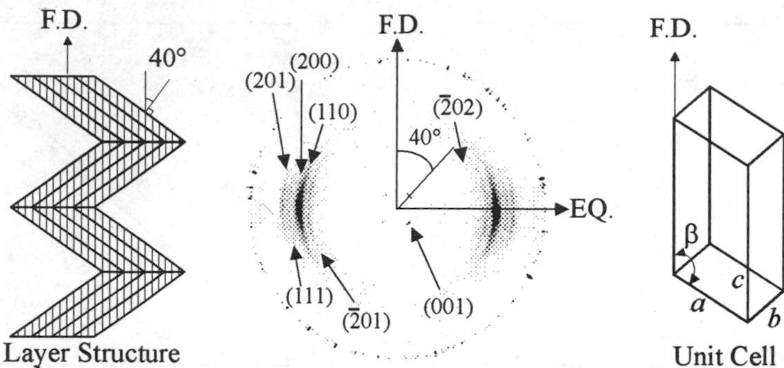

Figure 19. WAXD fiber pattern, unit cell structure and morphology for oriented TPP(n=11) in the S_H phase (F.D. is the fiber direction and EQ. is the equator.)

patterns that the layer structures still exist in the quadrant with a further enhanced correlation length [Figures 19 and 20 for TPP(n=11 and 12) for examples]. This implies that the layer normal is still not parallel to the molecular direction. On the other hand, the number of reflections in the high-angle region is substantially increased. In the quadrant, up to second layer reflections can be identified, indicating that true three-dimensional order thus exists within the layers. Monoclinic unit cells can be determined for all TPPs, and calculations back to the lateral packing perpendicular to the molecular chain axis based on these unit cell dimensions indeed lead to an orthorhombic packing having $a>b$. The identification of the S_H phase in odd-TPP(n≥15)s and even-TPP(n≤10)s is relatively strict forward since the 110 and 200 reflections are clearly separated on the equator, indicating a lateral orthorhombic packing. They can thus be used to calculate the unit cell lattice combined with other reflections. For even-TPP(n≥12)s, on the other hand, it is important to recognize the tendency for the separation of the 110 and 200 reflections on the equator in the S_H phase. This tendency is represented by a substantial decrease of the correlation length of this reflection during the transition from the S_G phase to the S_H phase. For example, the unit cell dimensions of S_H in TPP(n=12) are a=1.07 nm, b=0.53 nm, c=3.22 nm and β=121° and the cross-section perpendicular to the chain axis possesses an orthorhombic packing of a=0.92 nm and b=0.53 nm. The correlation length of the 200 reflection in the S_G phase is 26 nm while it decreases to 18 nm in the S_H phase. This is an indication that the initiation of the separation from one reflection to two reflections (the 200 and 110

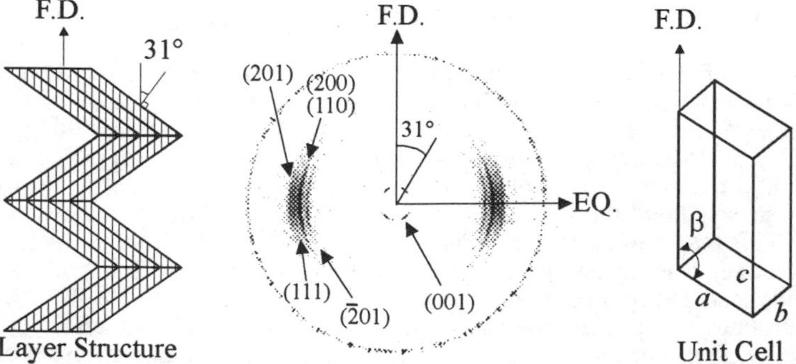

Figure 20. WAXD fiber pattern, unit cell structure and morphology for oriented TPP(n=12) in the S_H phase (F.D. is the fiber direction and EQ. is the equator.)

reflections) leads to a transformation of hexagonal into orthorhombic packing (*11*). Detailed molecular conformation analysis is necessary in order to illustrate the herringbone nature. It is generally observed that not only the number of reflections increases in the S_H phase, but their correlation lengths are also enhanced from those in the S_G phases [except for the 200 reflection in the S_G of even-TPP(n≥12)s] (*11*).

In the S_H phase, PLM and TEM morphological observations do not show substantial change from the S_G phase. For TPP(n=even)s with high numbers of methylene units, further cooling leads to two crystalline phases appearing at even lower temperatures. The structure determinations of these phases are beyond the scope of this publication and further discussion along this research will be carried out elsewhere.

Conclusion

In summary, TPPs show complicated phase transition behaviors. Their phase diagrams are established and various phases are identified *via* the thermodynamic transition properties obtained from DSC, the structural order and symmetry determined by WAXD, and morphology and defects observed under PLM and TEM. In particular, the WAXD fiber patterns in different phases play the most important role in determining the phase structures and symmetry. It is evident that the concepts of highly order smectic phases developed in small-molecule liquid crystals can also be utilized in the main-chain liquid crystalline polymers.

Acknowledgment. This work was supported by the SZDC's Presidential Young Investigator Award from the National Science Foundation (DMR-9175538). The generous donation of a DSC-7 from Perkin Elmer Inc. is also greatly appreciated.

Literature Cited

1. Gray, G. W.; Goodby, J. W. G. *Smectic Liquid Crystals;* Leonard Hill: London, 1984.
2. Pershan, P. S. *Structure of Liquid Crystal Phases;* World Scientific: NJ, 1988.
3. For a recent review, see for example, Percec, V.; Tomazos, D. In *Comprehensive Polymer Science;* First Supplement; Allen, G.; Aggarwal, S. L.; Russo, S. Eds.; Pergamon: Oxford, 1992, pp. 300-356. In this review, over four hundred references were collected.
4. Ehrenfest, P. *Proc. Acad. Sci. Amsterdam* **1933,** *36,* 153.
5. Thomas, E. L.; Wood, B. A. *Faraday Discuss. Soc.* **1985,** *79,* 229.
6. Wood, B. A.; Thomas, E. L. *Nature* **1986,** *324,* 655.
7. Percec, V.; Chu, P.; Ungar, G.; Cheng, S. Z. D.; Yoon, C.-Y. *J. Mater Chem.* 1994, *4*, 719.
8. Cheng, S. Z. D.; Yoon, Y.; Zhang, A.-Q.; Savitski, E. P.; Park, J.-Y.; Percec, V.; Chu, P. *Macromol. Rapid Commun.* in press.
9. Yoon, Y.; Zhang, A.; Ho, R.-M. and Cheng, S. Z. D.; Percec, V.; Chu, P. *Macromolecules* in press.
10. Yoon, Y.; Moon, B.; Kim, D.; McCreight, K. W.; Harris, F. W.; Cheng, S. Z. D.; Percec, V.; Chu, P. *Polymer* in press.
11. Yoon, Y. Ph.D. Dissertation, Department of Polymer Science, The University of Akron, Akron, Ohio, 44325-3909, 1995.
12. Wunderlich, B.; Grebowicz, J. *Adv. Polym. Sci.* **1984,** *60/61,* 1.
13. Blumstein, A.; Thomas, O. *Macromolecules* **1982,** *15,* 1264.
14. Blumstein, R. B.; Blumstein, A. *Mol. Cryst. Liq. Cryst.* **1988,** *165,* 361.
15. Ungar, G.; Feijoo, J. L.; Keller, A.; Yourd, R.; Percec, V. *Macromolecules* **1990,** *23,* 244.
16. Yandrasits, M. A.; Cheng, S. Z. D.; Zhang, A.-Q.; Cheng, J.-L.; Wunderlich, B.; Percec, V. *Macromolecules* **1992,** *25,* 2112.
17. Ostwald, W. *Z. Physik. Chem.* **1897,** *22,* 286.

Chapter 22

Ordered Liquid-Crystalline Thermosets

B. A. Rozenberg and L. L. Gur'eva

Institute of Chemical Physics in Chernogolovka, Russian Academy of Sciences, Chernogolovka, Moscow Region 142432, Russia

The paper is an overview of the authors' works on the regularities of formation of ordered networks via the epoxy liquid crystalline thermoset (ELCT) curing in the magnetic field. The changes of order parameter during ELCT cure in the magnetic field are the focus of this work. The result of the preliminary study of the influence of a magnetic field on ELCT cure kinetics is reported. The effect of the magnetic field on the properties of cured ELCT is discussed.

Liquid crystalline thermosets (LCT) with a rigid rod-like mesogenic group capped from both ends by reactive functional groups can be used as matrices for production of advanced composites, coatings or adhesives *(1, 2)*. In this case ordering of the chains of the final network polymer is the only way to improve its physical and mechanical properties.

Using LCT as a starting material gives an advantage in comparison to the high molar mass linear LC polymers because high degree of orientation can be easily achieved even in comparatively weak magnetic fields *(3)*.

One could expect that the process of LCT cure in the LC temperature range under the action of a magnetic field can proceed both with preservation of high initial degree of orientation of LCT and with disordering of the interknot chains of forming network polymer. The understanding of the character of changes of the order parameter for a curing LCT is the key problem of the study.

One of the major questions arising in this case was how the evolution of the degree of orientation during cure depended on the method of the network formation (polymerization or polycondensation) and structure of polymer formed.

The second problem was that our interests in this study consisted of clearing up the influence of macroscopic ordering under the action of magnetic field of the thermoset molecules on cure kinetics. Only few publications were known, in which the influence of magnetic field on the rate of the LC molecules' reactions were described *(4-6)*. Contradictory results on the influence of magnetic field on the rate of

0097–6156/96/0632–0372$15.00/0

radical polymerization of monofunctional LC methacrylates were obtained in *(4, 5)*. An accelerating effect of magnetic field was found in work *(4)* while no effect was reported in *(5)*. At the same time practically similar polymerization rates of bifunctional diacrylate in LC and isotropic states were observed *(6)*.

The LC diepoxy monomer with a thermostable mesogenic group, diglycidyl ester of terephthaloyl-bis-(1,4-hydroxy benzoic acid) (DGET), has been synthesized by us earlier (7):

$$\underset{\backslash O /}{CH_2-CH}-CH_2-O\underset{\|}{\overset{}{C}}PhO\underset{\|}{C}Ph\underset{\|}{C}OPh\underset{\|}{C}O-CH_2-\underset{\backslash O /}{CH-CH_2}$$

The temperature range of existence of DGET in the LC nematic state is 158-250°C *(7)*. As reported in *(8, 9)*, this monomer can be cured by heating in the LC or isotropic state without any curing agent by the so called molecular "insertion" mechanism or by anionic polymerization under the action of tertiary amines or quaternary ammonium salts. It was found *(1, 8)* that there is no difference in the reactivity of DGET in LC and isotropic state (the activation energy of the cure reaction in the LC and isotropic state are the same). This result makes us to expect a weak influence of magnetic field on the reactivity of the epoxy groups in the considered case.

Structural, thermal and mechanical characteristics of network polymers formed by polymerization of DGET in the LC nematic phase within and without magnetic field are also the subject of the work.

Experimental Section

DGET was synthesized and purified according to reported technique *(7)*.

Dimethyldibenzyl ammonium chloride used as catalyst for anionic polymerization of DGET was obtained by reaction of benzyl chloride with equimolar quantity of dimethylbenzyl amine and used without further treatment *(8)*. Mixture of DGET with 0.09 wt.% of the catalyst was prepared from the solution of methylene dichloride.

Investigation of the DGET cure kinetics without a magnetic field effect was performed with an isothermal calorimeter. Kinetic curves of heat release were processed using the Thian equation *(10)*.

The phase transition temperatures during DGET polymerization were determined using the Boetius polarized light microscope.

Evolution of alignment during the DGET catalyzed and self-cure was studied by ^1H NMR method using a PI-2303 multipulse spectrometer with resonance frequency of protons equal to 60 MHz. The samples of DGET were placed into the NMR spectrometer cell and heated quickly to the necessary temperature at the LC state. Then the samples were exposed at that temperature and the NMR spectra were recorded during cure. After the exposition the samples were cooled at a rate of 4°C/min to room temperature. The order parameter of the obtained polymer samples was determined by the NMR spectra and by X-ray analysis using mono- and polychromatic radiation with ionization registration and photography of the diffraction patterns.

Glass transition temperature (T_g), coefficients of thermal expansion (CTE) at temperatures below and higher T_g, and the modulus of elasticity in the rubbery state (E_∞) were investigated. In the case of aligned polymers the thermal and mechanical properties were measured in the direction perpendicular and parallel to the magnetic field applied to the samples.

The Mechanism of DGET Cure

We used two ways of DGET cure in this work. One of them is anionic polymerization of DGET under the action of quaternary ammonium salt. Such a cure reaction easily proceeds in the LC temperature range even without specially added catalyst due to the presence of traces of the tertiary amine and its quaternary ammonium salt formed during DGET synthesis. The mechanism of this reaction is well known *(11,12)*.

Nevertheless, DGET, precisely purified from the traces of tertiary amine and its quaternary salt, self-cures at much lower rate in the same temperature range without addition of any catalyst. From the kinetic point of view this reaction obeys the regularities which are characteristic of the polycondensation type of polymer formation *(13)*. As found in *(9)*, DGET self-cure proceeds by the molecular mechanism of insertion of the epoxy group into the ester group. Investigation of the mechanism of this reaction using low molecular model compounds showed that the reaction proceeded through a cyclic orthoester intermediate that was directly identified *(9)*.

$$\text{—C(=O)—OR} + \text{Ar—CH}_2\text{—}\underset{\text{O}}{\text{CH—CH}_2} \longrightarrow \text{—C(OR)}\genfrac{}{}{0pt}{}{\text{—O—CH(CH}_2\text{—OAr)}}{\text{—O—CH}_2} \qquad (1)$$

$$\longrightarrow \text{—C(=O)—O—CH}_2\text{—CH(CH}_2\text{—OAr)—OR}$$

Self-cure was the second method that was used in this study. This reaction gave us some new opportunities at experimental study of the problems mentioned above. Low cure reaction rate allowed us to measure precisely all phase transformations of the monomer. Large difference in the cure rate between the catalytic and noncatalyzed reactions and small difference in the structure of network polymer formed from the same monomer (see below) allowed us to obtain interesting information on the behavior of DGET in the magnetic field during cure.

Formation of Aligned Structures during DGET Cure in the Magnetic Field

There is no model which could describe the changes in NMR spectra in the course of LCT cure and predict the degree of macroscopic ordering of the polymer formed in the magnetic field. One could use the Stroganoff's model *(14)* for measurement of macroscopic alignment in LC systems placed into a magnetic field. However, this model is suitable for chemical systems which do not undergo any changes. It does not take into consideration drastic chemical and conformational changes taking place during formation of network polymer. These changes can lead to the disappearance of the orientation order. However, from the other hand, formation of network can lead to the topological fixation of macroscopic ordering.

In this work the changes in the NMR spectra during DGET cure are described as superposition of the spectra, corresponding to the original ones of aligned monomer and to the final polymer. Such a simple approach allowed us to describe the NMR spectra of the curing system with a good accuracy.

Self-cure of DGET. When a DGET sample is placed into a magnetic field at a temperature of the LC range (169°C in our particular case), its quick melting occurs and a transition from the wide line corresponding to the solid phase into the narrow one, which is characteristic of liquids, is observed in the NMR spectrum (15). Due to orientation of the DGET molecules in the magnetic field the spectrum line is splitted and progressive narrowing of the doublet line occurs within one minute (Figure 1a). The value of the ^{1}H NMR spectra splitting, $\Delta\gamma$, is equal to 14.5 kHz and points to a dipole-dipole interaction between the neighboring protons of the aromatic ring. It can be represented by the following expression:

$$\Delta\gamma = \frac{3\mu\gamma}{2\pi r^3}(3\cos^2\theta - 1) \quad (2)$$

where μ, γ, and r are the dipole moment, gyromagnetic proton ratio and the distance between interacting protons respectively; θ is the angle between the vector joining the interacting protons and the direction of an external magnetic field. It is evident that $\Delta\gamma$ in (2) linearly depends on the order parameter *(16)*:

$$S = \langle(3\cos^2\theta - 1)\rangle/2 \quad (3)$$

The evolution of NMR spectra during DGET cure in the magnetic field is presented in Figure 1. One can observe the formation of a less ordered phase due to the consecutive transformation of a monomer to an oligomer and then to an amorphous network polymer. The ratio between the ordered and disordered phases is changing during DGET cure. The observed NMR spectra of the curing system (Figure 1b) at any moment of time are successfully described by superposition of the spectrum of the initial ordered monomer phase (Figure 1a) and low mobile phase of the final amorphous polymer (Figure 1c).

A standard spectrum of the initial monomer in the LC state drawn by a dashed line (Figure 1a) represents a doublet consisting of Gaussian lines of the same width and a single wide Gaussian line. The doublet corresponds to the NMR line of the proton pairs oriented along the magnetic field *(17)*. The wide singlet corresponds to the dipole-dipole interaction between the spacer's protons having random distribution of angle θ. The ratio of the integral intensities of the doublet and singlet corresponds to the ratio of the protons number of the benzene ring and the spacer. The width of the line in the

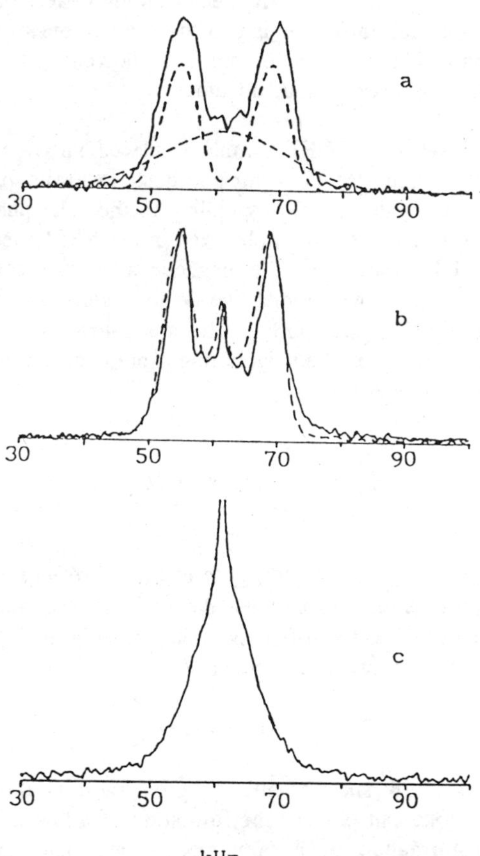

Figure 1. NMR spectra recorded in the process of DGET cure in the magnetic field at 169°C. Sample is exposed for (hours): 0.1 (a); 1 (b), and 12 (c). (Adapted from ref. 19).

doublet and half-width of the singlet are determined from the initial experimental spectrum (Figure 1a). They were used as adjustment parameters in the construction of the calculated spectra. The splitting value, $\Delta\gamma$, was defined as the distance between the doublet lines for each spectrum.

The second standard spectrum is related to the experimental spectrum after 12 hours which does not change during further exposition of the sample in the magnetic field (Figure 1c). Its view corresponds to the spectrum of an amorphous polymer.

Thus, the calculated spectrum is the superposition of an aligned part of the system, expressed as a doublet, and an unaligned part, shown by a central peak.

The evolution of the fraction of the unaligned phase, which is determined as the ratio of the intensity of the unaligned phase to the sum of intensities of unaligned and aligned parts of the observed spectrum, is presented in Figure 2. As expected, the evolution of ordering does not coincide with the cure kinetic curve. The changes in alignment of the polymer formed stop at conversion 0.85 while the cure reaction develops further. It is interesting to note that the rate of disorientation drastically increases at the region of gelation point (α_g =0.7).

Cooling of the cured sample to room temperature results in widening of the spectrum (Figure 3). Figure 3 shows also that definite anisotropy of the lines' form is preserved that indicates to an alignment of the final polymer.

For evaluation of the order parameter of the curing system from the NMR data an X-ray analysis was used for the calibration of the NMR method by determining the constant factor equal to $3\mu\gamma/2\pi r^3$ in (2). The disorientation function, *F*, *(18)* can be used as an analogue of the order parameter *S* in expression (3):

$$F=(3\cos^2 \langle\varphi\rangle -1)/2 \qquad (4)$$

where φ is the angle of azimuth distribution of the wide angle X-ray scattering intensity on the texture patterns (Figure 4).

The validity of such approach is supported by the assessment of the *r* value in (2) equal to 1,6Å what is close to the interatomic distance between the protons of the benzene ring. The direct evaluation of the macroscopic ordering of the curing thermosets by an NMR technique gives an essential advantage in comparison with an X-ray method.

It is seen from Table I that parameters F and S are essentially decreased during the DGET cure to the limit value equal to 0.5 for ultimate conversion *(19)*. The close value of the order parameter for network polymers based on diglycidyl ethers of 4,4'-dihydroxy-α-methylstilbene were obtained by Ober and coworkers *(2, 20)*.

As expected, the higher the cure temperature the lower the order parameter of polymer formed.

Anionic Polymerization of DGET. The NMR spectra obtained by direct monitoring during anionic polymerization of DGET at 169°C (Figure 5) are similar to those for self-cure of DGET (Figure 1). However, the rate of evolution of the NMR spectra as well as the cure rate in the case under consideration is much higher than that for the DGET self-cure. Even at the initial moment of polymerization the central peak of the less aligned phase already exists. It slightly increases with simultaneous narrowing of the doublet line up to conversion α = 0.45 (Figures 5a and 5b). Then fast growth of the

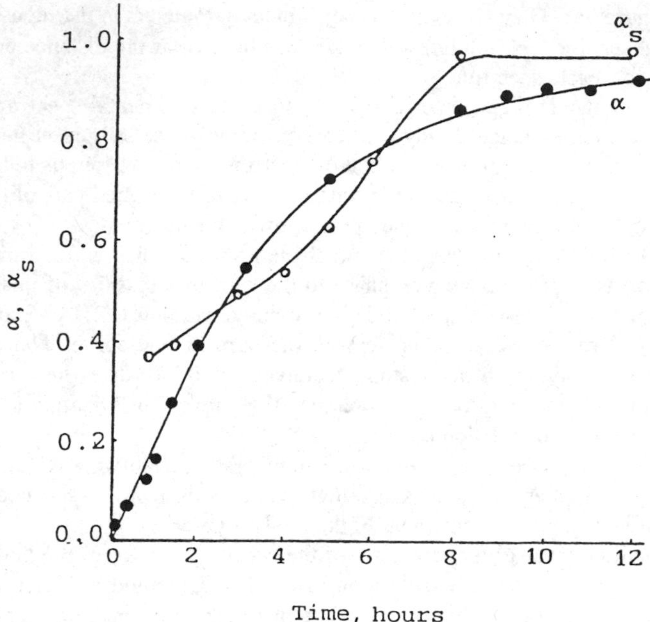

Figure 2. Chemical conversion (α) and value of the unaligned fraction (α_s) versus time on DGET self-cure at 169°C.

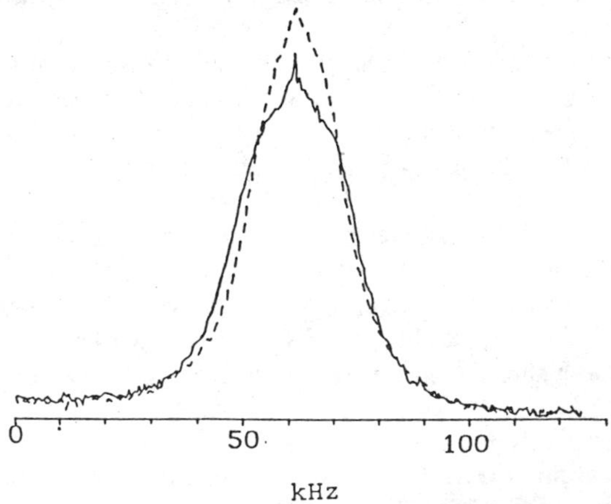

Figure 3. NMR spectra at room temperature of the sample (see legend for Figure 1c) parallel (solid line) and perpendicular (dashed line) to the direction of the applied magnetic field.

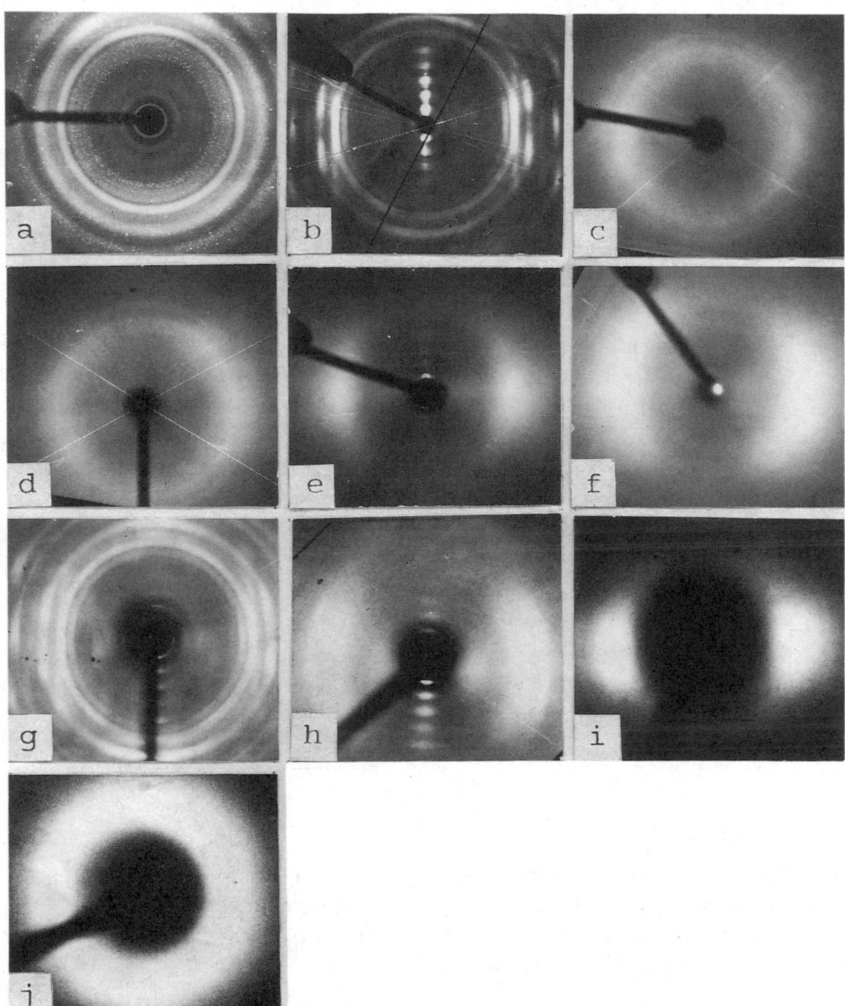

Figure 4. Texture-patterns recorded at room temperature versus duration of exposition in the magnetic field during DGET self-cure (a-f) and during DGET anionic polymerization (g-i) at 169°C (b, c, d, g, h, i, j) and 180°C (e, f) for (min): 0 (a); 10 (b), 180 (c, f), 720 (d), 60 (e, i), 15 (g), 35 (h) and 190 (j).

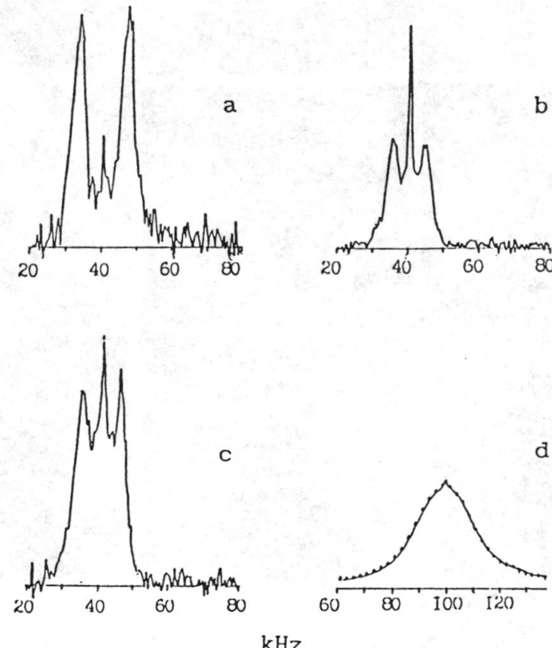

Figure 5. NMR spectra recorded during DGET anionic polymerization in the magnetic field at 169°C for (min): 5 (a), 55 (b), 68 (c), 300 (d). The last NMR spectrum is recorded at room temperature parallel (solid line) and perpendicular (dots) to the direction of the applied magnetic field.

central peak and simultaneous sharp decrease of the doublet take place. This indicates to the fast disappearance of the LC phase during cure. The texture patterns of the corresponding samples are shown in Figures 4g-4i. Parameters *S* and *F*, presented in Table I, are gradually decreased to the values of 0.34 at α=0.45 and then they quickly vanish.

The curing system can be returned to the LC state again by cooling an absolutely disordered sample (see Figure 5c and Table I). Further exposition of this sample in the magnetic field up to the ultimate conversion at this temperature leads to further decrease of the orientation parameter (Table I).

Table I. Order Parameter (*S*) and Function of Disorientation (*F*) in the course of the DGET Cure in the Magnetic Field

Type of Cure Reaction	*Cure Temperature, °C*	*Curing Time, min*	*Conversion*	*F*	*S*
Self-cure					
	169	5	0.025	0.80	0.80
		60	0.15	-	0.75
		90	0.29	-	0.74
		180	0.52	0.60	0.65
		300	0.71	-	0.58
		360	0.78	-	0.47
		720	0.92	0.50	-
	180	180	0.58	0.45	-
		540	0.90	0.45	-
Anionic polymerization					
	169	5	0.05	0.50	0.50
		15	0.09	0.50	0.50
		30	0.21	-	0.51
		55	0.42	-	0.37
		60	0.47	-	0*
	160	65	0.49	-	0.32
		68	0.51	-	0.39
		120	0.68	-	0.11
		150	0.75	-	0.14
		180	0.80	-	0.14
		540	0.87	-	0.14
	169	300	0.93	0	0

* Temperature jump from 169°C to 160°C.
SOURCE: Adapted from ref. 19.

The presented data indicate that the character of formation of oriented structure is quite different for anionic polymerization and for self-cure of DGET. Structural and kinetic factors can influence the orientation of interknot chains of a polymer formed in the magnetic field. The network polymer structure generated in anionic polymerization is shown in Scheme 5.

The network polymer structure generated by the insertion mechanism of DGET cure has the same fragments as shown in the previous scheme as well as the fragments with structures shown in Scheme 6a and 6b. The new types (6a and 6b) of knots are formed due to the possibility of opening of both orthoesters bonds of the intermediate cyclic structure (see Scheme 1), which is equivalent to the abnormal addition of the epoxy groups to the ester bonds *(9, 12)*.

Two different mechanisms of formation of network polymers ordered in the magnetic field can be realized:

1. Ordered structure of network polymer formed by cure of a thermoset oriented in the magnetic field can be generated by the original orientation of a thermoset if the cure reaction does not disturb it. It can be realized if the end reactive groups are attached to the mesogen by a flexible spacer of a sufficient length. This mechanism can be called a "hereditary" one. The order parameter during cure in this case must be constant and determined by the order parameter of an initial thermoset without any dependence on the cure kinetics. This case is not present in our study.

If the spacer is too long, the ability to orientation of the initial thermoset and, as a result, of the final polymer is decreased up to a complete loss of the ability to local and macroscopic ordering *(20)*.

2. If the spacer length is too small, the chemical reaction of the functional end groups can proceed only after deformation of their spatial arrangement and will undoubtedly lead to disorientation of the thermoset molecules during cure. Further ordering of the polymer generated in the magnetic field depends on its structure and cure kinetics. Exactly this case is present in our study.

It is probable that the presence of the neighboring rigid mesogenic units providing a strong intermolecular interaction is favorable to the orientation of the polymer chains during cure according to the insertion reaction mechanism. It can be supposed also that a longer and more flexible spacer (Scheme 6a) in comparison to the structure shown in Scheme 5 is favorable to orientation of mesogens of the first mentioned structure.

The second important factor that determines the capability of a curing system to keep orientation can be connected with the cure kinetics. High rate of cure and, as a consequence, quick changes of rheological properties of the curing system prevents the orientation of mesogenic units of forming polymer in the magnetic field. The rate of the DGET anionic polymerization is several times higher in comparison to that of the insertion reaction *(8)*. High rate of the DGET anionic polymerization fixes the nonequilibrium structure of polymer chains disoriented by the reaction

The order parameter can be increased (Table I) if the DGET cure is carried out by the programmed reduction of temperature in the process of thermoset cure.

Thus, the process of network formation in DGET cure reduces the degree of orientation of the polymer formed. The final degree of orientation dramatically depends on the structure of the polymer formed and cure kinetics.

The promising way to increase the order parameter of network polymers during thermosets cure in the magnetic field is to find the optimal balance between the rigidity of the thermoset mesogen and the flexibility of the spacer in order to realize the "hereditary" mechanism.

```
|
O         O     O
|         ||    ||           |
CH—CH2 —O—C—M—C—O—CH2—CH—
|
CH2
|
O                                          (5)
|
CH—CH2 —O—C—M—C—O—CH2 —CH—
|         ||    ||           |
CH2       O     O
|
```

```
|
O
|
CH2       O     O
|         ||    ||           |
CH—CH2 —O—C—M—C—O—CH2—CH—
|
O                                          (6a)
|
CH—CH2 —O—C—M—C—O—CH2 —CH—
|         ||    ||           |
CH2       O     O
|
```

```
|
O         O     O
|         ||    ||           |
CH—CH2—O—C—M—C—O—CH2—CH—
|
CH2
|
O                                          (6b)
|
CH2
|
CH—CH2—O—C—M—C—O—CH2 —CH—
|         ||    ||           |
          O     O
```

```
                    O        O
                    ||       ||
Here M is —Ph—O—C—Ph—C—O—Ph— .
```

Effect of the Magnetic Field on the DGET Cure Kinetics

The influence of the magnetic field on the DGET cure rate was determined by comparison of cure kinetics curves in the absence and presence of a magnetic field (Figure 6). The kinetic curve of the DGET anionic polymerization in the presence of magnetic field was built by exposition of the sample in the magnetic field and a subsequent measurement of the concentration of unreacted epoxy groups using isothermal calorimetry in the absence of a magnetic field. The whole curve could not be found because of a quick disappearance of the LC phase at $\alpha = 0.45$ as it was shown above. As expected, the magnetic field slightly slows down the process of the DGET anionic polymerization.

A weak effect of the magnetic field on the cure rate observed in this work is likely to be connected with the low value of the magnetic field strength available in the particular study. Perhaps, with higher values of the magnetic field strength higher kinetic effects could be expected.

Properties of Network Polymers Prepared by DGET Cure in the Magnetic Field

The goal of this study was to compare the properties of network polymer formed by DGET self-cure at a temperature of the isotropic state (250°C); in the liquid crystalline state (169°C) without a magnetic field, which provides anisotropy on domain level; and in the presence of the magnetic field, that provides anisotropy on the macroscopic level. A typical view of thermomechanical curves of the polymers obtained by TMA is presented in Figure 7. A dashed line corresponds to the thermomechanical curve without any loading (dilatometric condition).

As one can see thermomechanical curve has a bimodal view at the first scan that disappears at the second scan of the sample. Glass transition temperature at the second scan is slightly increased. These results point to incomplete conversion of the polymers cured at 169°C due to diffusion control of the cure reaction at the latest stages (at $\alpha > 0.85$). It is interesting to note that such view of the thermomechanical curve is characteristic even of the polymer obtained at the DGET isotropization temperature (250°C).

The observed bimodality of the thermomechanical curves of cured DGET testifies for the fact that the forming polymers are structurally essentially inhomogeneous. The inhomogeneity could be on a molecular or/and supramolecular level. On a molecular level the observed bimodal thermomechanical curve could be explained by bimodality of the molar mass distribution of the polymer tested. There is, however, no chemical reason to expect the molar mass distribution of the polymer formed to be bimodal in the case under consideration.

Structural inhomogeneity of the polymers obtained at LC state temperatures in the presence of a magnetic field consists in the presence of LC and amorphous phases. The observed structural inhomogeneity of the polymer obtained at the DGET isotropization temperature can be explained by partial preservation of the LC phase due to the growth of isotropization temperature of the thermoset during cure. This fact was confirmed by direct measurements of the phase diagram during DGET self-cure *(8)*.

Thus, the observed bimodality of thermomechanical curves can be attributed to the essentially different deformation behavior of the amorphous and LC phases of the polymer.

The values of glass transition temperatures (T_g), modulus of elasticity in the rubbery state (E_∞), and coefficients of thermal expansion of the polymers in the glassy (CTE_1) and rubbery (CTE_2) state are presented in Table II.

One can see from Table II that the conditions of polymers preparation essentially influence their thermal and mechanical properties. The data of Table II show an evident anisotropy of the properties of polymer aligned in the magnetic field. First of all, noteworthy is the high value of elasticity modulus for the aligned polymer in the direction perpendicular to the direction of the magnetic field. It is about four times

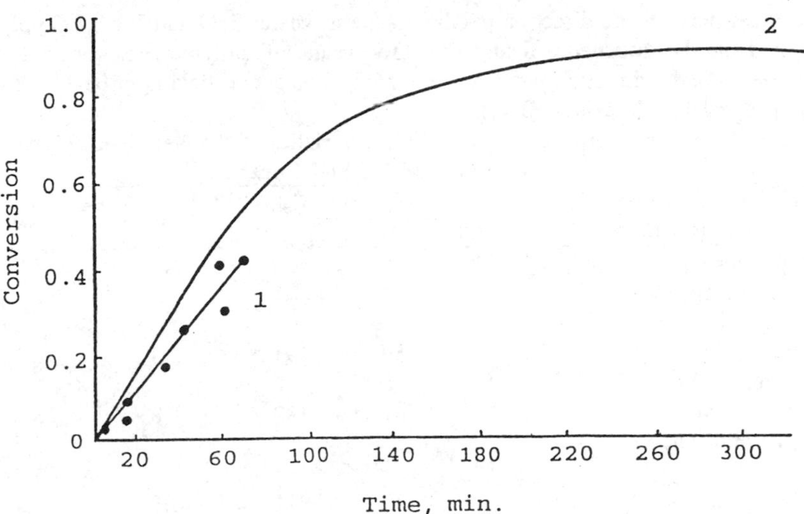

Figure 6. Kinetic curves of the DGET anionic polymerization at 169°C in the presence (1) and absence (2) of the magnetic field.

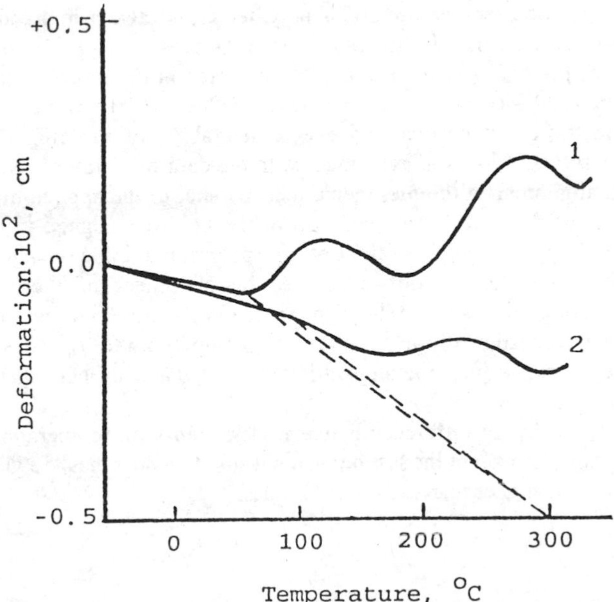

Figure 7. Thermomechanical curve of the polymer in the direction perpendicular to the direction of the applied magnetic field. First scan (1); second scan (2). Dashed line is a dilatometric curve.

higher than those in the direction parallel to the magnetic field and for a LC polymer unaligned in the magnetic field. The low value of polymer resistance to the compression in the direction parallel to the applied magnetic field is, probably, due to the loss of stability (buckling effect).

Table II. Thermal and Mechanical Characteristics of the Network Polymers Formed by DGET Self-Cure

Conditions of Polymer Samples Preparation	*Direction of Measument*	T_g, °C	E_∞, kg/cm²	$CTE_1 \cdot 10^5$, °C^{-1}	$CTE_2 \cdot 10^5$, °C^{-1}
Isotropic phase in the absence of magnetic field		65	450	5.8	35.7
LC phase in the absence of magnetic field		59	191	2.3	18.2
LC phase in the presence of magnetic field	‖	61	182	1.8	15.2
LC phase at the presence of magnetic field	⊥	75	911	6.1	16.4

A surprisingly large value of elastic modulus is observed for the polymer obtained at DGET isotropization temperature. This polymer shows also a high value of CTE_2, which is two times higher than for the other tested samples.

As expected, the CTE_1 value of the polymer tested in the direction parallel to the applied magnetic field is rather low in comparison with the CTE_1 value in the direction perpendicular to the magnetic field and even remarkably low in comparison with the unaligned LC network. This is in agreement with the data of X-ray analysis (Figure 4) that reveals the alignment of the mesogenic units parallel to the applied magnetic field. The temperature of thermal destruction (T_d) of the polymer aligned in the magnetic field is about 50°C higher than T_d of the isotropic polymer and the unaligned one.

Among the interesting features provided by alignment of thermosets in the magnetic field during cure is the stability of macroscopic ordering and corresponding anisotropy of the properties of polymers for temperatures up to T_d. These facts mean that the polymer network fixes the aligned interknot chains and keeps their orientation on heating of the polymer.

There is practically no difference between glass transition temperatures of all the tested polymer samples except the last one from Table II, which has the glass transition temperature 10-15°C higher than the other samples.

Conclusions

It was found that the thermosets decrease their order parameter during cure in the magnetic field. The extent of diminution depends on the structure of the polymer formed and cure kinetics.

The order parameter on the anionic polymerization of DGET decreases up to zero. However, in the course of the reaction proceeding by insertion it falls down to

0.5 and then remains constant. The explanation of the influence of the formed polymer structure on the order parameter is proposed. The method of increasing the order parameter of the network by a nonisothermal proceeding of thermoset cure is found.

It was shown that the magnetic field slightly retards the anionic polymerization of DGET. Further study of the DGET cure kinetics in stronger magnetic fields is needed in order to determine the effect of magnetic field on cure kinetics of thermosets.

As expected, the aligned network polymer shows a substantial anisotropy of its properties. High value of elastic modulus in the rubbery state at compression of the aligned polymer in the direction perpendicular to the applied magnetic field and low value of CTE_1 in the direction parallel to the magnetic field are observed. The reduction of CTE_1 parallel to the direction of the applied magnetic field in comparison with the polymer networks with randomly oriented LC domains is found. No effect of the magnetic field on the of CTE_2 value is observed.

High stability of the macroscopic ordering of the network polymers obtained by LCT cure in the magnetic field at high temperatures (above T_d) is revealed.

Acknowledgments

We greatly acknowledge the support of this work to the International Science Foundation, Projects REA000 and REA300, and the Russian Fund for Fundamental Research, Project 93-03-4705, and to Drs. Ju.A.Olkhov and E.A.Dzhavadyan for the measurements of thermal and mechanical characteristics of the polymers.

Literature Cited

1. Rozenberg, B.A.; Gur'eva, L.L.; In "*Synthesis, Characterization, and Theory of Polymeric Networks and Gels*"; Aharony, S.M., Ed.; Proceedings of an American Chemical Society Division of Polymeric Materials Science and Engineering symposium on Synthesis, Characterization, and Theory of Polymeric Networks and Gels, held April 5-10, 1992, in San Francisco, California; USA; Plenum Press: New York and London, **1992**, 147-164.

2. Barclay, G.G.; Ober, C.K.;.Papathomas, K.; Wang, D. *J. Polym. Sci., PartA: Polym. Chem.*, **1992,** *30,* 1831-1841.

3. Tsukruk, V.V.; Gur'eva, L.L.; Tarasov, V.P.; Shilov, V.V.; Erofeev, L.N.; Rozenberg, B.A. *Vysokomolek. Soed.*, **1991,** *33B,* 168-172.

4. Perplies, E.; Ringsdorf H.; Wendorff, J.H. *J. Polym. Sci. Polym. Lett. Ed.*, **1975,** *13,* 243-246.

5. Paleos, C.M; Labes, M.N; *Mol. Cryst. Liq. Cryst.*, **(1970)**. *11,* 385-393.

6. Doorkamp, A.T; Alberda van Ekenstein, G.O.R.; Tan, Y.Y. *Polymer,* **1992,** *33,* 2863-2867.

7. Gur'eva, L.L. ;Belov, G.P.;.Boiko, G.N. ; Kusch, P.P.; Rozenberg, B.A. *Polym. Sci.,* **1992,** *34A,* 593-597.

8. Gur'eva, L.L.; Dzhavadyan, E.A.; Rozenberg, B.A. *Polym. Sci.,* **1993,** *35A,* 1341-1344.

9. Kosikhina, S.A.; Gur'eva, L.L.; Kusch, P.P.; Rozenberg, B.A. *Polym. Sci.*, **1993,** *35B,* 14-17.

10. Kalve, E.; Prat, A. *Microcalorimetry*; Foreign Literature: Moscow, Russia, **1963**; p. 477.
11. Kusch, P.P.; Lagodzinskaya, G.V.; Komarov, B.A.; Rozenberg, B.A. Vysokomolek. Soed., **1979,** *21B,* 708-713.
12. Rozenberg, B.A. *Adv. Polym. Sci.,* **1985,** *75,* 113-165.
13. Rozenberg, B.A., Irzhak, V.I.; In "*Polymer Networks '91*"; Dusek, K; Kuchanov, S.I., Eds.; Proceedings of the International Conference, 21-26 April, Moscow, Russia ; VSP: Utrecht, The Netherlands and Tokyo, Japan, **1992,** 7-24.
14. Stroganov, L.V.;. Prokhorov, A.N.; Galliullin, R.A.;.Kireev, E.V.; Shibaev, V.P.; Plate, N.A. *Polymer Science USSR,* **1992,** *34A,* 89-95.
15. Gur'eva, L.L; Ermolaev, K.V.; Tarasov, V.P.; Ponomarev, V.I.; Erofeev, L.N.; Rozenberg, B.A. *Polym. Sci.,* in press.
16. Blumstein, R.B.;.Stickles, E.M.; Gauthier, M.M.; Blumstein, A.; Valino, F. *Macromolecules,* **1984,** *17,* 177-183.
17. Doscosilova, D.; Schneider, B.; Trekoval, J. J. Collect. Chech. Chem. Commun., **1974,** *39,* 2943-2948.
18. Tsukruk, V.V; Shilov V.V. *Strukture of Polymeric Liquid Crystals*; Naukova Dumka: Kiev, USSR, **1990**, p.57 (in Russian).
19. Rozenberg, B.A.; Gur'eva, L.L. *Polymeric Materials Sciemce and Engineering*; Proceedings of the American Chemical Society Division of Polymeric Materials: Science and Engineering; ACS Spring Meeting 1995, Anaheim, California, *72, 243-244.*
20. Barclay, G.G.; Ober C.K. Prog.Polym.Sci. **1993**, *18*, 899-945.

Chapter 23

Role of Curing Agent on the Nature of the Mesophase and the Properties of Mesogenic Epoxy Resins

E. Amendola[1], C. Carfagna[2], and M. Giamberini[2]

[1]Institute of Composite Materials Technology, National Council of Research, Piazzale Tecchio 80, 80125 Naples, Italy
[2]Department of Materials and Production Engineering, University of Naples "Federico II", Piazzale Tecchio 80, 80125 Naples, Italy

Reaction of poly-functional epoxy with suitable curing agent is a well established practice to produce thermosetting materials with chemical and physical properties strongly depending upon the chemical nature of the reacting molecules as well as on the processing conditions. Among the curing agent most extensively used and investigated are the primary and secondary amines, both aromatic and aliphatic, carboxylic acids and anhydrides.
The suitability of epoxy group to react via ring opening mechanism catalyzed by a wide variety of experimental conditions is the primary reason for the wide applications based on epoxy resins. On the other hand the draw back of the ready reactivity is the difficulty to control the crosslinking density and other structural parameters affecting the properties of the materials.
This statement holds true especially in the case of liquid crystalline epoxy resins. In fact, the structural features and distribution of crosslinks along the molecular backbone strongly influences the liquid crystalline phase developed during the curing process and the performances and the applicability of the resulting material.

In recent years liquid crystalline thermosetting polymers have been the object of extensive research and patent activity *(1-8)*. The interest in these materials lies in their challenging theoretical implications and in the noticeable potential applications.
Highly crosslinked liquid crystalline thermosets turned out to be promising candidates for thin film applications in electronic packaging and as matrix materials for advanced composites due to their superior mechanical properties and heat resistance *(4-6)*; on the other hand, lightly crosslinked networks can be easily oriented by mechanical stress and this results in a change of their optical properties *(9)*. Applications as

0097–6156/96/0632–0389$15.00/0

optical switches, nonlinear optics (NLO), waveguides, and matrices in polymer dispersed liquid crystals (PDLC) can be therefore foreseen. Many theoretical predictions have been proposed, since the formulation of the Landau-de Gennes theory for the behaviour of these materials under mechanical stress *(10-13)*. From experimental evidence it is now clear that actually LC elastomers behave in qualitatively new ways with respect to isotropic elastomers *(2,14,15)*. Good agreement has been found between theory and experimental results, but investigations are still in progress in order to understand these phenomena.

A transition from a LC phase to an isotropic phase can still be observed in the case of networks with poor crosslinking density. These systems can actually be considered as highly viscous liquids in the LC state, as proposed by Finkelmann *(16)*; they are therefore expected to exhibit such a transition. In the case of networks with higher crosslinking density, the molecular organization, typical of the LC phase is locked by the crosslinks. Therefore, no clearing temperature can be observed, and this ordered structure can be destroyed only upon thermal decomposition.

In previous works (*5,6,17,18*) the synthesis and the physical properties of highly crosslinked liquid crystalline networks obtained by reaction of rigid rod epoxy resins with aromatic diamines have been described by the authors. In this case, the mesophase develops during the initial curing step, when the epoxy monomers and the curing agent reacting together give rise to a a lightly branched linear prepolymer. The mesophase is then stabilized upon the gelation of the thermoset *(19)*.

The thermal characterization of the curing reaction of LC epoxy resins reveals an unusual behaviour *(20,21)*. In fact, the isothermal curing reaction, followed by differential scanning calorimetry (DSC) showed a double peak exotherm. Kinetic analysis performed by differential scanning calorimetry and infra-red spectroscopy showed a two steps curing reaction: in the first stage, the addition of the primary amine to the epoxy group is responsible for the increase in molecular weight; chain extension enlarges the range of thermal stability of the mesophase, also reducing the crystallization of the polymer. Finally, the further addition of the secondary amine proceeds quickly. When these systems are cured above their isotropization temperature, crosslinking locks the macromolecules in an isotropic structure and no double peak can be observed by DSC.

The materials with a LC structure showed better mechanical properties with respect to the isotropic ones. In particular, a considerably higher value of fracture toughness was found *(22)*. This was ascribed by the authors to the microstructure of LC epoxy resins, which consists of anisotropic domains with properties, such as strength, different along and across their molecular orientation. This inhomogeneity and the localized anisotropy of the LC structure can therefore cause the deviation of crack propagation from straight line, thus increasing fracture toughness.

The orientation of the nematic domains through the sample can be carefully controlled by surface phenomena in the case of thin film. In fact, the presence of an aligning layer deposited on the surface of the substrate supporting the resin during the curing reaction can produce the propagation of the orientation through the whole sample, thus contibuting to the formation of a monodomain. Aligning layers are available for uniform parallel or homeotropic orientation. Once the reaction is complete, the tridimesional structure is stable over a wide range of temperature up to the thermal decomposition of the epoxy resin. Another possibility of inducing uniform orientation in thin film applications is to exploit the interaction of liquid crystalline monomers with electro-magnetic fields during the curing reaction. In fact, the low viscosity of the resin during the first stage of the curing reaction allows the molecules to orient when exposed to an electro-magnetic field. The great unisotropy of dielectric constant or magnetic susceptibility is responsible for the interaction with the external field and the presence of the liquid crystalline phase stabilize the orientation achieved.

An interesting aspect concerns the properties of LC epoxy resins cured with aliphatic diacids, having a lower crosslinking extent. These thermosets can in fact be mechanically oriented at a temperature higher than their glass transition temperature (Tg), and then quickly cooled, thus resulting in polymers with promising optical and mechanical properties. Barclay et al. *(23)* reported the synthesis and thermal characterization of networks prepared from glycidyl endcapped oligoethers, based on 4,4'-dihydroxy-α-methylstilbene and α,ω-dibromoalkanes, cured with aromatic diamines. Their work consisted of the synthesis of oligoethers containing the mesogens and followed by subsequent glycidyl termination. Curing was finally accomplished with aromatic diamines. The same authors *(24)* discussed the effects of processing these materials under tensile stress and in applied magnetic fields. Better mechanical orientation resulted in the case of less crosslinked networks, while a good orientational stability above Tg was accomplished only for the networks cured in applied magnetic fields.

In this paper we describe the synthesis and the properties of LC epoxy thermosets from p-(2,3-epoxypropoxy)-α-methylstilbene (DOMS) cured with the aromatic diamine 2,4-diaminotoluene (DAT) or with decanedioic acid (SA). Polyfunctional amines are widely used in adhesives, castings and laminating applications, while polybasic acids and acid anhydrides are the second most commonly used curing agent after amines. They are preferred to polyamines in some applications because they give long useful pot lives, low peak exotherms, good electrical properties and are less irritating to the skin, being also less carcinogenic. Polybasic acids have found application in castings, laminates, adhesives, molding powders, filament windings and are widely used in coatings.

Experimental Part

The synthesis of p-2,3-(epoxypropoxy)-α-methylstilbene (DOMS) has been previously described by the authors *(20)*. The 2,4-diaminotoluene (DAT) and the decanedioic acid (SA) were purchased by Aldrich and used without any further purification.

The preparation of the highly croslinked samples was accomplished by mixing the glycidyl terminated compound (DOMS) and the tetrafunctional amine (DAT) in a molar stoichiometric ratio of 2/1. The mixture was prepared by dissolving the amine with acetone at room temperature. Being the epoxy compound not completely soluble, only a fine suspension can be obtained. The solvent was subsequentely evaporated and the mixture dessiccated under vacuum.

Lightly crosslinked networks were prepared by mechanically mixing finely grounded DOMS and SA in molar ratio 1curing/1 and subsequently curing them in a test tube or DSC pan at 180°C for the selected time. DOMS-SA films were prepared by curing the molten mixture between two glass slides, previously treated with a surfactant agent (SurfaSil, Pierce), in oven at 180°C for 90 minutes.

The concentration of residual carboxyl groups in the samples cured for different times before gel point was determined by dissolving a weighed amount of reaction products in a tetrahydrofuran/methanol mixture at room temperature and by subsequently titrating with a NaOH/methanol solution. Phenolphthalein was used as indicator.

The number average molecular weights of the oligomers, obtained by reacting DOMS-SA mixtures at different times, were calculated by means of vapor phase osmometry in chloroform at 37°C with a Knauer osmometer. Monodispersed polystyrene standards were used to generate the calibration curve.

Differential scanning calorimetry (DSC) data were recorded on a Du Pont DSC (mod. 2940) in the dynamic mode with a heating rate of 10°C/min or in isothermal mode pre-selecting the final curing temperature. Nitrogen was used as purge gas.

X-ray diffraction patterns were recorded by the photographic method with a Rigaku mod. III/D max generator, a Ni-filtered CuK_{α} radiation was used.

Optical investigations under crossed polarizers were performed on a Reichert-Jung, mod. Polyvar microscope equipped with a hot stage Linkam, mod. TH 600.

Results and Discussion

The crosslinking reaction of DOMS with the tetrafunctional amine DAT is assumed to be consistent with the reaction scheme reported for conventional epoxy compound *(21,25-28)*. The reaction proceeds through a nucleophilic attack of the primary amine onto the epoxy ring. The step determining the reaction kinetic is strongly catalyzed by presence of traces of water, alcohols or acid protons. After the formation of the secondary amine a competition is established between the primary amine still present in the reaction medium and the secondary one formed so far. The addition of the primary amine is reported to be favored with respect to the secondary one, essentially for the steric hindrance determined by the substituent on the nitrogen atom. Nevertheless, it can not be assumed that the addition of the secondary amine starts only after the complete exhaustion of the primary one. The reaction of the glycidyl terminated compound with the primary amine is responsible of the increase of molecular weight of essentially linear molecular fragments, associated with the increase in viscosity observed if the reaction is carried out in testing tube. The subsequent addition of the secondary amine to unreacted DOMS molecules is responsible of the formation of a tridimensional network, as can be deduced by the analysis of FT-IR spectra *(21)*.

The thermal characterization of DOMS-DAT mixture performed in the temperature range between 130 and 180°C reveals an unusual bimodal peak. The formation of the liquid crystalline phase during the curing reaction is assumed to increase of the reaction rate. This fact is sustained by the microscopic investigation performed on the epoxy mixture during the curing reaction. In fact it is possible to correlate the maxima of the second peaks in the DSC thermogram obtained during the curing reaction with the appearance of a birefringent phase as observed by means of optical microscopy.

The kinetic analysis performed in the temperature range between 150 and 170°C has led to the resolution of the two peaks and to the determination of the kinetic parameters of the phenomena related to the different peaks. The method is essentially based upon a trial and error procedure aimed to split the reaction enthalpy into the contributions attributed to the addition of the primary and secondary amine. The spectroscopic analysis performed by means of FT-IR sustained the hypothesis that during the first stage of curing reaction the addition of secondary amine to the epoxy groups occurred, thus leading to crosslinking, was negligible *(21)*.

Despite the complexity of the reaction mechanism involved in the cure of epoxy resins by polyfunctional amines, the overall reaction rate can be expressed by means of a relatively simple kinetic equation. Under the assumption that the heat flow relative to the instrumental baseline is proportional to te reaction rate, the fractional conversion i s expressed by equation 1:

$$\alpha_{(t)} = \frac{\int_0^t \Delta H dt}{\int_0^\infty \Delta H dt} \qquad (1)$$

where DH is the enthalpy of complete reaction.

Several kinetic models have been proposed *(29-31)* to account for the autocatalytic behavior according to the mechanism proposed in Figure 1. The following equations expressing a kinetic order of m+n has been adopted:

$$\frac{d\alpha}{dt} = k \cdot \alpha^m \cdot (1-\alpha)^n \qquad (2)$$

where k is the kinetic constant, m and n the empirical order of reaction

$$k = k_0 \cdot \exp(-\frac{E_a}{R \cdot T}) \qquad (3)$$

where k_0 is the preexponential factor and E_a is the overall activation energy.

The analysis has confirmed an higher activation energy for the addition of the secondary amine with respect to the addition of the primary one. Nevertheless, the ratio $\frac{k_2}{k_1}$ equals approximately the value of 4, while the data reported in literature for conventional epoxies range between 0.2 and 0.4.

In Figure 2 the thermal characterization of the reacting mixture performed in the temperature range between 130 and 190°C is represented. The details of the deconvolution of the two peak for the isothermal reaction at 150°C are reported in Figure 3. The peaks associated to the addition respectively of the primary and secondary amines are related to the overall heat flow. The agreement with the experimental data is very satisfactory, confirming the validity of the proposed model.

The X-ray pattern of samples cured (see Figure 4) and the microscopic observation lead to the characterization of the liquid crystalline phase as beeing

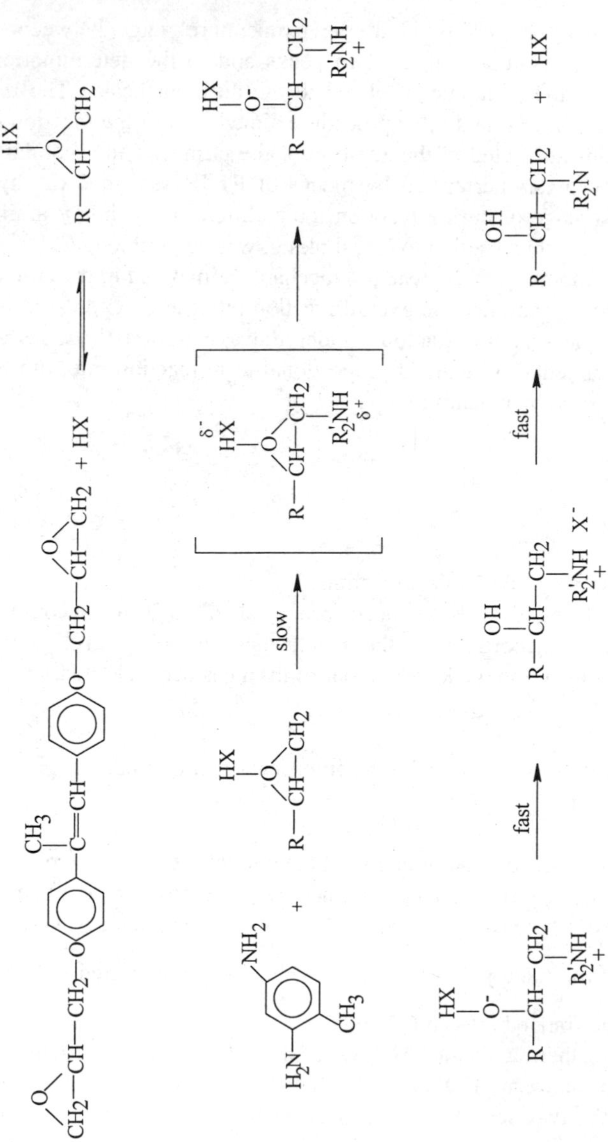

Figure 1. Reaction scheme of p-(2,3-epoxypropoxy)-α-methylstilbene (DOMS) and 2,4-diaminotoluene (DAT).

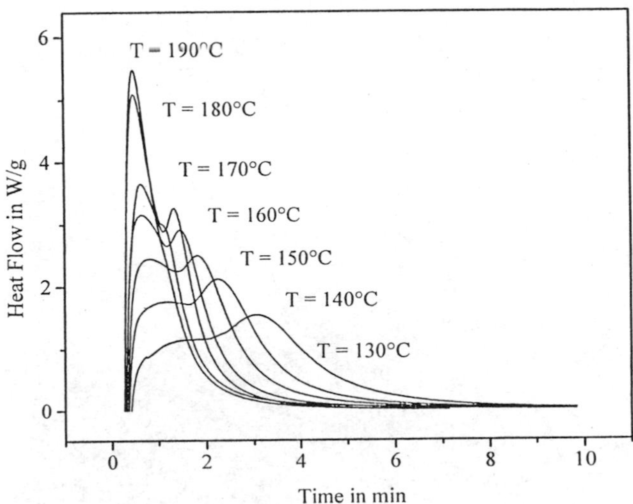

Figure 2. DSC of DOMS/DAT mixture recorded in isothermal mode at 130, 140, 150, 160, 170, 180 and 190 °C.

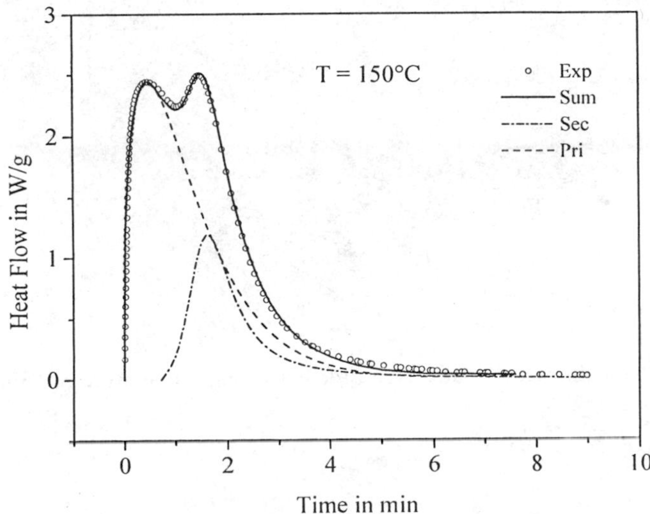

Figure 3. DSC spectra of DOMS/DAT mixture recorded at 150 °C (open circles) and comparison with calculated curves for primary (dashed line) and secondary (dash-dotted line) amine additions. The sum of primary and secondary curves is reported as solid line.

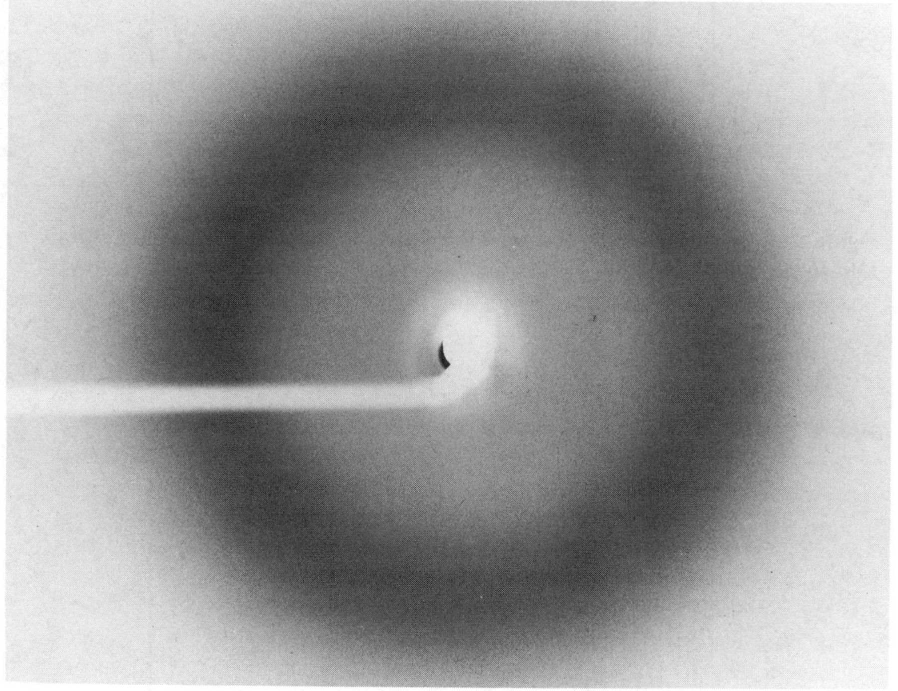

Figure 4. X-ray diffraction pattern recorded at room temperature of DOMS/DAT.

nematic. Moreover, the distribution of the nematic director is random through the sample specimen, i.e. no preferred macroorientaion is detectable, as can be inferred by the lack of polarization of the nematic halo in the x ray diffraction pattern.

The glass transition of the cured sample is in temperature range of 100-110°C. The uncertainty of the determination arise from the high crosslinking density that affects the broadness of the Tg transition, as observed by DSC studies.

In Figure 5 the reaction scheme between DOMS and SA is reported. The first reaction involves the opening of the epoxy ring by the carboxyl group. The resulting hydroxyl group is subsequently involved in the crosslinking reaction which can take place either through the formation of ester linkages by condensation with excess carboxylic acid molecules, or, in the presence of catalytic amounts of proton donors, through homopolymerization with not yet reacted epoxy molecules. The prevailing reaction path strongly depends on the reaction conditions and has not been clearly established *(32)*. In our case, FTIR studies confirm this reaction path *(33)*, though a complete identification of the chemical nature and of the extent of the crosslinks is still under investigation. Figure 6 reports the number average molecular weight and the concentration of residual carboxyl groups versus curing time before gel point. From this trend it can be inferred that crosslinking takes place when the molecular weight of the oligomers is relatively low, namely about 1500 atomic mass units. This value corresponds to an oligomer formed at least after the reaction of two rigid rod epoxy molecules with four molecules of dicarboxyl acid (oligomer A).

In Figure 7 the DSC scan of DOMS-SA mixture is reported. At 127°C an endothermic peak, which is due to the melting of the components, can be observed; the exotherm starting at 170°C and reaching a maximum at 213°C indicates the development of the curing reaction. In order to control the reaction kinetics, 180°C was selected as curing temperature.

Figure 8 reports the DSC heat flow graph recorded in isothermal mode at 180°C of DOMS-SA mixture. From the diagram it is clearly evident that at this temperature the curing reaction is complete after 90 minutes, as also confirmed by dielectric analysis *(33)*. In addition, no further heat release could be detected by subsequent DSC scan up to 200°C, thus excluding a postcure. This is due to the high curing temperature, well above the glass transition temperature (T_g) of the cured thermoset, which is 27°C, as can be assessed from the dynamic scan (Figure 9). The endotherm centered at 81°C with a ΔH of 17.8 J/g can be attributed to the transition from a liquid crystalline to an isotropic phase, as indicated by optical observation. Iannelli et al. *(34)* reported an isotropization enthalpy value of 16.2 J/g for semiflexible linear mesophasic polymers having α-methylstilbene as rigid core; in the case of side chain nematic elastomers, an isotropization enthalpy of 1.2-1.3 J/g was reported in the literature *(2)*.

The X-ray pattern of the cured DOMS-SA system is reported in Figure 10.The sharp reflection at low diffraction angle, corresponding to a spacing of 16,2 Å, together with the high value of liquid crystalline-isotropic transition enthalpy and the texture observed under crossed polarizers, indicate that the fully cured sample develops a smectic phase upon curing.

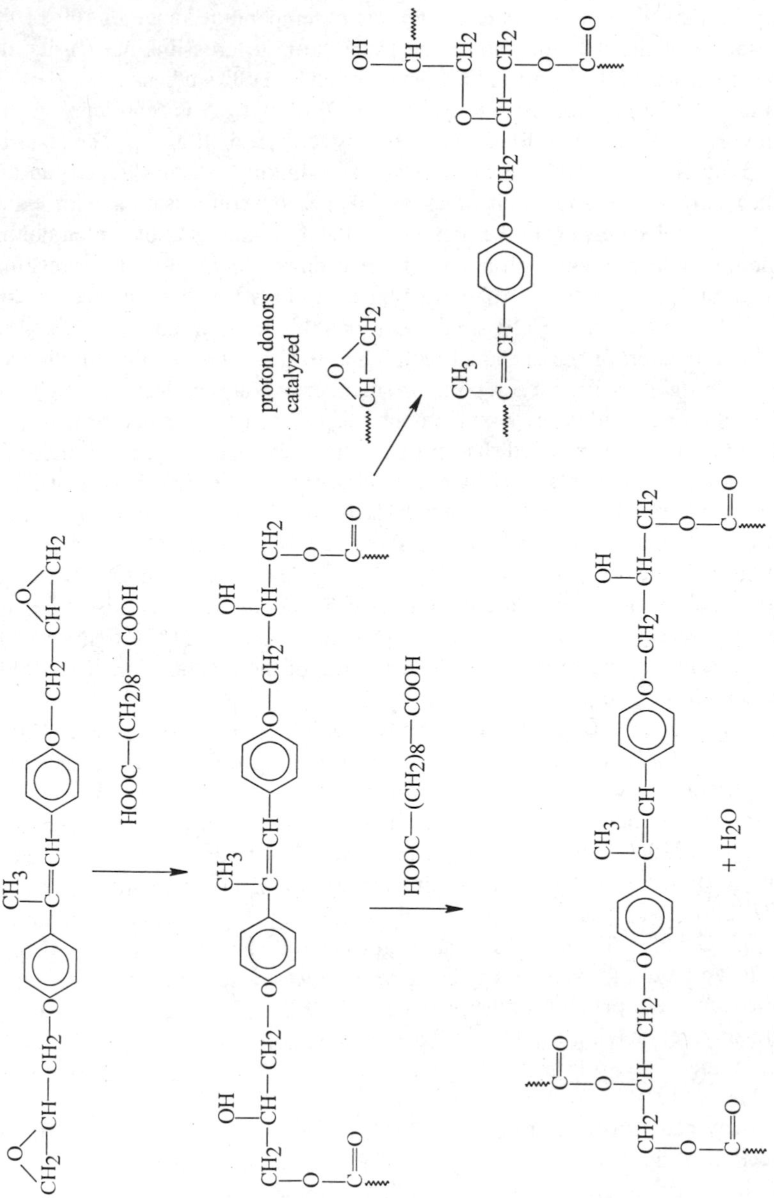

Figure 5. Reaction scheme of p-(2,3-epoxypropoxy)-α-methylstilbene (DOMS) and decandioic acid (SA).

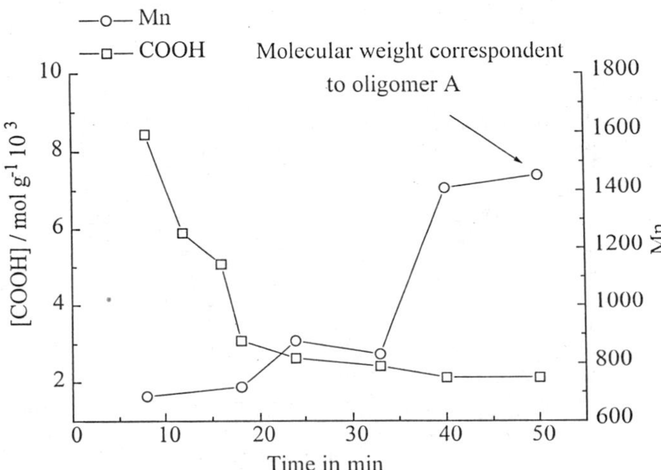

Figure 6. Number average molecular weight and concentration of residual carboxyl groups versus curing time before gel point for DOMS/SA system.

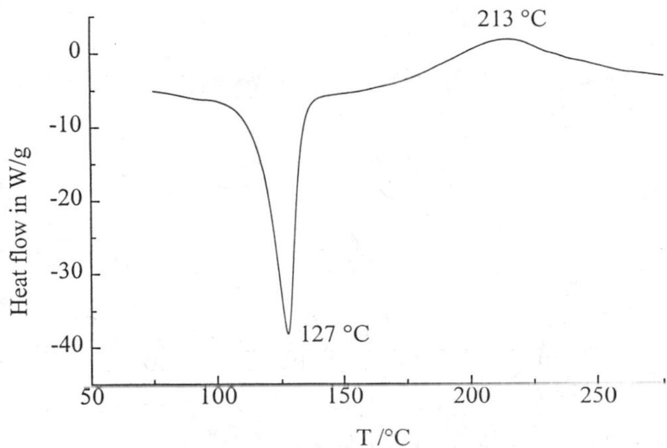

Figure 7. DSC of DOMS/SA mixture recorded at 10 °C/min.

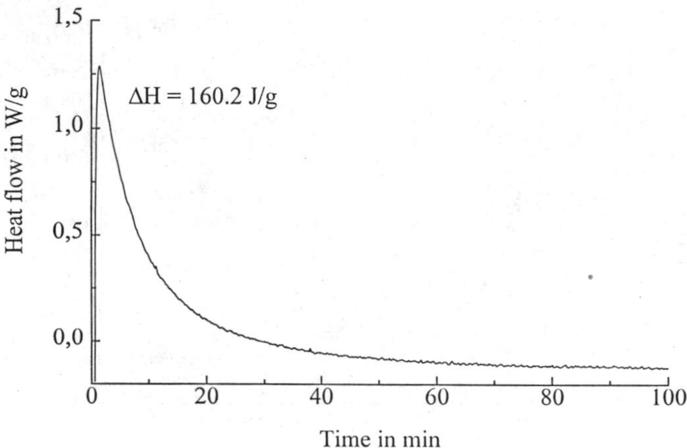

Figure 8. DSC of DOMS/SA mixture recorded in isothermal mode at 180 °C.

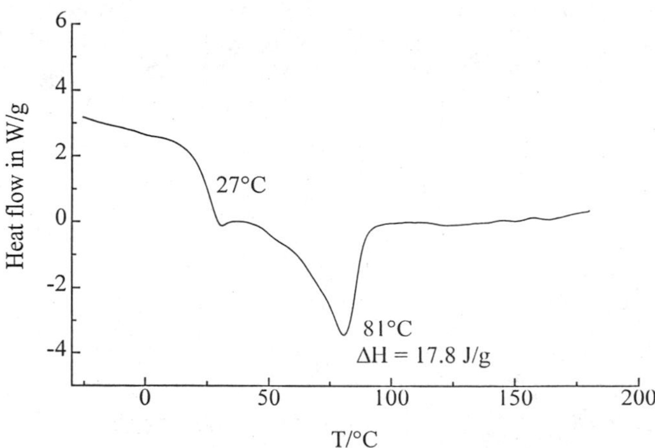

Figure 9. DSC of cured DOMS/SA recorded at 10 °C/min.

In the case of DOMS-SA, no double peak could be put in evidence from the heat flow graph related to the isothermal cure performed in the DSC pan (Figure 8). This can be easily explained if one considers that in this case the reaction occurs at a temperature higher than the isotropization temperature of the growing thermoset, and therefore no kinetic effect due to the liquid crystalline state of the polymer can be expected on cure.

DOMS-SA films were heated above Tg and stretched up to λ=1.8 (where λ is defined as $\frac{(l-l_0)}{l_0}$, with l_0 = initial sample length and l= sample length after stretching), then rapidly cooled down to the glassy state. Under a macroscopic observation, the sample turned from white and turbid to colourless and transparent. This indicated that light was no longer scattered, as it is expected when the size of the liquid crystalline domains in the polydomain sample has increased. Polarized light microscopy confirmed a uniform orientation of the stretched polymer in a monodomain *(33)*.

The X-ray pattern taken at room temperature from a stretched sample is reported in Figure 11. A broadened wide angle reflection at the equator is evident. The smectic layer reflections can be observed at the meridian, indicating a perpendicular orientation of the smectic layers to the stress direction. The orientation is therefore dominated by the polymer chains, as it was reported also in the case of highly stretched fibers of a smectic main chain elastomer *(35)*.

The stretched samples were allowed to relax at room temperature without any stress applied; then, their structure was again examined by X-ray. No differences in the shape nor in the intensity of the liquid crystalline reflections could be detected, thus indicating that the orientation of the polymer chains was fully retained.

Conclusions

Liquid Crystalline Epoxy Resins can be cured in different ordered states depending on the curing agent used. With aromatic amines the thermoset is crosslinked in a nematic structure. No isotropization temperature can be detected. The reaction can be monitored by means of DSC analysis. A double peak exotherm indicates that the crosslinking reaction forms a liquid crystalline resin. Kinetic models can be applied to separate the contributions of the different reactions.

If a diacid is mixed with the glycidyl terminated rigid rod monomer the reaction is much slower. The resulting cured material will exhibit a smectic phase with high values of isotropization enthalpy. If stressed above Tg, the smectic planes will orient, and macroscopically this implies a transition from an opaque state to a transparent state for the anisotropic film. The oriented polymer will relax back to the unoriented state if heated above Tg. No double peak exotherm can be observed by DSC analysis, since at the reaction temperature the forming thermoset is above its isotropization temperature.

As a final comment it has to be outlined that liquid crystalline epoxies represent a novel class of thermosetting polymers with unique properties. The relative easiness of synthesis, their production costs and eventually the possibility to taylor their properties make them potential candidates for specific applications.

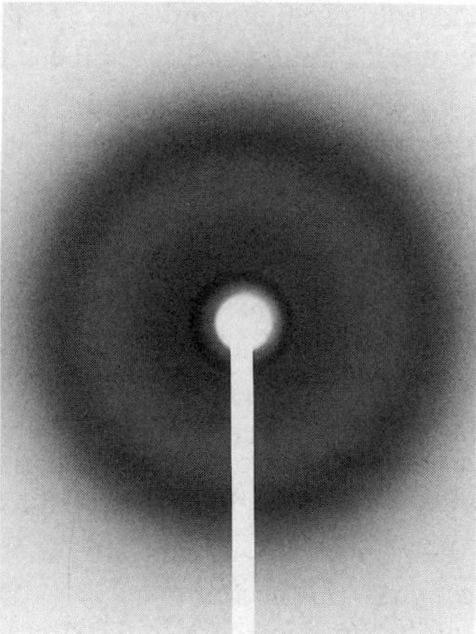

Figure 10. X-ray diffraction pattern recorded at room temperature of cured DOMS/SA.

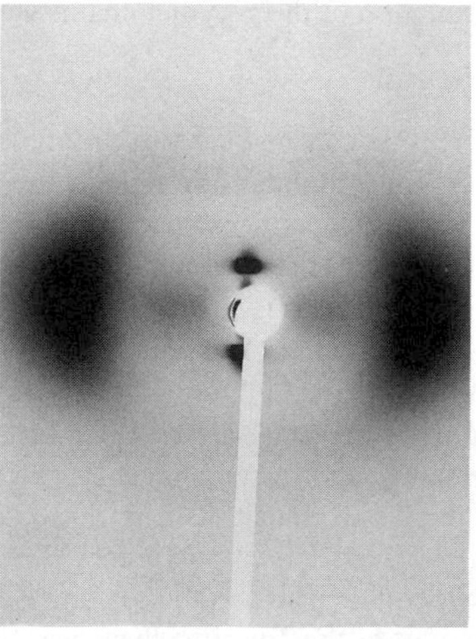

Figure 11. X-ray diffraction pattern recorded at room temperature of cured DOMS/SA film stretched at 35 °C up to $\lambda = 1.8$ and quenched.

Literature Cited

1 Finkelmann H.; Kock, H.J.; Rehage, G.; *Macromol. Chem. Rapid Commun.* **1981**, *2*, 317
2. Schätzle, J.; Kaufhold, W.; Finkelmann, H. *Macromol. Chem.* **1989,** *190*, 3269
3. Küpfer, J.; Finkelmann, H. *Macromol. Chem. Rapid Commun.* **1991,** *12*, 717
4. Hoyt, A.E.; Benicewicz, B.C.; Huang, S.J. *Polymer Preprint*, **1989,** *30*, 536;
5. Carfagna, C.; Amendola, E.; Giamberini, M. *J. Mat. Sci. Lett.* **1994,** *13*, 126
6. Carfagna, C.; Amendola, E.; Giamberini, M. *Composite Structures*, **1994,** *27*, 37
7. Ger. 36 22 613 (1988/86) Bayer AG, invs.: Dhein, R.; Mueller, H.P.; Gipp, R.; Meier, H.M.; *Chem. Abstr.* **1988,** *109*, 7136b
8. Ger. 36 22 610 (1986) Bayer AG, invs.: Dhein, R.; Mueller, H.P.; Gipp, R.; Heine, H.; *Chem. Abstr.* **1988,** *109*, 38835h.
9. (Finkelmann, H.; Schätzle, J.; Kaufhold, W.; Pohl, L. *Freiburger Arbeitstagung Flussigkristalle*, 15-17 Marz 1989, Abstracts of Communication)
10. (Ref. Landau, L. *Phys. Z. Sowjetunion* **1937,** *11*, 26;
11. de Gennes, P.G. *Mol. Cryst.* **1971,** *12*, 193;
12. de Gennes, P.G. *The Physics of Liquid Crystals*; Clarendon Press: Oxford , UK, 1974
13. de Gennes P.G., In *Polymer Liquid Crystals;* Ciferri, A.; Krigbaum, W.R.; Meyer, R.B. Eds.; Academic Press: New York, USA, 1982; pp. 115-131
14. Mitchell, G.R.; Davis, F.J.; Ashman, A. *Polymer* **1987,** *28*, 639
15. Kaufhold, W.; Finkelmann, H.; Brand, H.R. *Macromol. Chem.* **1991,** *192*, 2555)
16. Finkelmann, H. *Adv. Polym. Sci.* **1984,** *60/61*, 99
17. Carfagna, C.; Amendola, E.; Giamberini, M. *Macromol. Chem.*, **1994,** *195*, 279
18. Carfagna, C.; Amendola, E.; Giamberini, M., *Macromol. Chem.*, **1994,** *195*, 2307
19. (Kirchmeyer, S.; Karbach, A.; Muller, H.P.; Meier, H.M.; Dhein, R. *International Conference on Crosslinked Polymers*, Proceedings of the Conference, Conference Director Prof. A.V. Patris, **1990**, Luzern, CH, pp. 167-176)
20. Carfagna, C.; Amendola, E.; Giamberini, M. *Liq. Cryst.*, **1993,** *13(4)*, 571
21. Amendola, E.; Carfagna, C.; Giamberini, M.; Pisaniello, G. *Macromol. Chem.*, **in press**
22. Carfagna, C.; Amendola, E.; Giamberini, M., In *Liquid Crystalline Polymers*; Carfagna, C. Ed.; Pergamon Press: Oxford, UK, 1994; pp.69-85
23. Barclay, G.G.; Ober, K.I.; Papathomas, C.K.; Wang, D.W. *J. Polym. Sci., Part A: Polym. Chem.*, **1992,** *30*, 1831
24. Barclay, G.G.; McNamee, S.G.; Ober, C.K.; Papathomas, K.I.; Wang, D.W. *J. Polym. Sci., Part A: Polym. Chem.*, **1992,** *30*,1845
25. Smith, I.T.; *Polymer*, **1961** , *2*, 95
26. Barton, J.M., *Polymer*, **1980** , *21*, 604
27. Barton, J.M. *Advances in Polymer Science;* Springer-Verlag: Berlin, D, 1985; p.72
28. Rozemberg, B.A *Advances in Polymer Science;* Springer-Verlag: Berlin, D, 1986; p.75

29. Lee, W.I.; Loos, A.C.; Springer, G.S. *J. Comp. Materials,* **1982**, *16*, 510
30. Dusi, M.R.; Lee, W.I.; Criseidi, P.R.; Springer, G.S. *J. Comp. Materials,* **1987**, *21*, 243
31. Eichler, J.; Dobas, I.; *Collection Czechoslov. Chem. Commun.,* **1973,** *38*
32. Saunders, K.J. *Organic Polymer Chemistry*, 2nd edition, Chapman and Hall: London,. UK, 1988; 424-427
33. Giamberini, M.; Amendola, E.; Carfagna, C. *Macromol. Chem. Rapid Commun.* **1995**, *16(2)*, 97
34. Iannelli. P.; Roviello, A.; Sirigu, A. *Eur. Polym. J.* **1982**, *18*, 759
35. Canessa, G.; Reck, B.; Reckert, G.; Zentel, R. *Macromol. Chem., Macromol. Symp.* **1986**, *4*, 91

INDEXES

Author Index

Affiliation Index

Subject Index

M